D1061945

THE ELECTRIC WAR
The Fight Over Nuclear Power

The
ELECTRIC
WAR

The Fight
Over
Nuclear Power

by Sheldon Novick

Sierra Club Books
San Francisco
1976

Portions of this book appeared in somewhat different form in *Environment* magazine, published by The Scientists Institute for Public Information.

Acknowledgment is made for permission to reprint material from *Pyramids of Power: The Story of Roosevelt, Insull and the Utility War*, by M. L. Ramsay. Copyright © 1937 by M.L. Ramsay. Reprinted by permission of the Bobbs-Merrill Co.

Library of Congress Cataloging in Publication Data

Novick, Sheldon.
      The electric war.

      Includes index.
      1. 'Atomic power.   2.   Atomic power-plants.
3.   Atomic energy industries—United States.   I.   Title.
TK9153.N68   333.7   76-13440
ISBN 0-87156-148-4

Design by Richard Schuettge

Production by David Charlsen & Others

Printed in the United States of America

TK
9153
.N68

THE TIME IS NOW come, in which every Englishman expects to be informed of the national affairs; and in which he has a right to have that expectation gratified. For, whatever may be urged by Ministers, or those whom vanity or interest make the followers of ministers, concerning the necessity of confidence in our governours, and the presumption of prying with profane eyes into the recesses of policy, it is evident that this reverence can be claimed only by counsels yet unexecuted, and projects suspended in deliberation. But when a design has ended in miscarriage or success, when every eye and every ear is witness to general discontent, or general satisfaction, it is then a proper time to disentangle confusion and illustrate obscurity; to shew by what causes every event was produced, and in what effects it is likely to terminate; to lay down with distinct particularity what rumour always huddles in general exclamation, or perplexes by indigested narratives; to shew whence happiness or calamity is derived, and whence it may be expected; and honestly to lay before the people what inquiry can gather of the past, and conjecture can estimate of the future.

Samuel Johnson

"Observations On The Present State of Affairs"

THIS BOOK is dedicated to my parents, Ruth and Irving Novick, with love, gratitude and admiration.

Acknowledgments

L IKE MANY MORE important works, this book owes its existence to the late
Marie Rodell, a great woman now greatly missed.

Sherwood Novick assisted with invaluable advice, editing and patient attention; Bob Novick, from whom I learned a great deal over the years, made a complex enterprise possible with his timely help.

My coeditors for many years at *Environment* magazine, Julian McCaull and
Kevin Shea, have been of assistance in many ways. Much of what is presented
here evolved in discussions among us; I am particularly grateful to Julian McCaull
for help on the materials relating to the history and democratization of industry,
and to Kevin Shea for help with material relating to reactor safety; both reviewed
large portions of this book before publication. I have borrowed freely from their
work without acknowledgment elsewhere in this book.

Marvin Madeson and Jack J. Schramm have been my tutors in many of the
matters touched on here, and I am grateful for past lessons and for Jack's patient
review of the manuscript before publication. Professor Arnold W. Reitze, Jr. and
James Harding also reviewed the manuscript and made helpful suggestions. Professor David R. Newburger read and commented on certain chapters.

Alan McGowan gave me some important leads; Peter and Lucia Montague and
Sandra Simon were of great help to me in finding my way about the Southwest,
as were others in the uranium-mining country.

My thanks also to Behavioral Analysis, Inc. and their able staff for transcribing
miles of tape recordings; and to Gladys Boles, who typed the book quickly and
accurately and corrected many of my errors.

Finally, I am grateful to the dozens of people I interviewed, most especially
those whose names I cannot give here.

Table of Contents

Introduction

MUCH OF WHAT is said publicly about nuclear power is untrue; much else is secret. Accounts of the controversy concerning it seem confusing and are often contradictory.

The fight over nuclear power is not an orderly engagement between well-organized forces—it is a scattered, individualistic, very American struggle. The industry over which the contending forces fight is a curious hodgepodge of the modern and the backward, of advanced technology and pick-and-shovel labor. The electric-power companies that build and own nuclear powerplants are relics of an earlier time, and the opponents of nuclear power sometimes seem to be leading a religious revival.

This book is an attempt to display as much of the battle as can be comfortably viewed, enough for a reasonable person to form an opinion about nuclear power. The protagonists in the struggle are asked to speak for themselves, and the larger part of the following pages is given to these people and the things they have to say. They are men and women who mine uranium, machine plutonium and direct great companies. People who think nuclear power is a mistake, and people who are fighting to dismantle a multibillion-dollar industry.

We all work in the limits set by others and by what has already been done. The talk about nuclear power makes very little sense unless we know something about what is being discussed. The chapters which follow therefore alternate between the present and the past, between the actors and the settings in which they contend.

Three great forces, acting over decades, have shaped the nuclear-power industry, and the first three sections of this book are devoted to these. To put names to them is to make judgments instead of showing them at work: the Cold War; the struggle between governments and corporations; the creation of the private electric-power monopolies. In each section, modern protagonists alternate on the stage with those of the past.

In section four, the nuclear-power industry shaped by these great forces is described. In section five, the present struggle to control this industry is portrayed. And section six contains some speculations about the future effect of the forces acting on the industry.

The author has tried to be a passive eye, observing: The chapters have been arranged to give, as nearly as possible, a simultaneous view of the people who make up the great struggle over nuclear power and of the setting in which they contend.

I. MILITARY STRENGTH

Barry Commoner (Chapter 2)

Model of the bomb "Fat Man" that destroyed Nagasaki, at the
National Atomic Museum (Chapter 3)

Carl Walske, president, Atomic Industrial Forum (Chapter 4)

Display of tactical atomic weapons at the National
Atomic Museum (Chapter 4)

Uranium mill near Grants, New Mexico (Chapter 6)

CHAPTER ONE

Hiroshima

FROM 1945 TO 1949, Harry Truman bluffed. America challenged the Soviet Union in Greece and Turkey, rebuilt Western Europe and organized it into a NATO alliance to oppose the USSR; in 1948, when the Russians blockaded Berlin, we responded with an airlift and an apparent willingness to take further military action.

In fact, however, the United States had no militarily significant arsenal of nuclear weapons. The atom bomb which exploded over Nagasaki on August 9, 1945, was the last of three bombs built and detonated during the war. If the Japanese had failed to surrender, another atom bomb could not have been assembled for weeks. It was not until well into 1949 that the US possessed any significant number of nuclear weapons, and even then the small number of weapons we had would not have been decisive in a renewed war in Europe. In any case our nuclear monopoly was almost ended; in August, 1949, the Soviet Union exploded a nuclear weapon.

The wartime atom-bomb project, conducted by the Army under the code-name Manhattan Engineer District, had cost $2.2 billion, much of which was wasted. It produced only three nuclear explosives during the war, all of which were expended by August, 1945. No production capability had been built. There was no assembly line producing atom bombs; the temporary facilities which had been built were already disintegrating and would be largely abandoned within two years. The wartime military effort did produce those three climactic weapons, but it was otherwise a very mixed success, and very unlike the effort which its publicists portrayed. Some of the nation's best-known scientists cooperated, perhaps unknowingly, in the postwar propaganda myth of our nuclear strength. Our entire foreign policy seemed to be based on the reality of nuclear weapons; our domestic political struggles revolved for some time over the question of civilian as against military control of the supposed nuclear arsenal. Even now we congratulate ourselves on the United States' selflessness in offering to turn our supposed nuclear arsenal over to international control, an offer which the Soviet Union spurned. But until 1949, all this was fantasy.

The primary misconception, which still persists, concerns what was actually accomplished during the Manhattan District effort in World War II. The romantic side of this story has been told repeatedly: how European refugee scientists attempted to alert the US government to the threat of the bomb; how Albert Einstein was persuaded to write to Franklin Roosevelt; how the world's greatest physicists gathered in scattered locations across the country; how Enrico Fermi built the first nuclear reactor and, in Chicago, December, 1942, for the first time demonstrated the feasibility of a controlled nuclear chain reaction; how J. Robert Oppenheimer, another distinguished physicist, directed the effort to assemble the first nuclear weapons; the dramatic dawn test, at Alamagordo, New Mexico, of the first nuclear explosion (which the Army wisely had arranged for sympathetic newspaper reporters to observe); the devastation of Hiroshima and Nagasaki the following month; then peace.

All of this is usually told from the viewpoint of the distinguished scientists who were central to the effort. Shortly after the war's close, Henry DeWolf Smyth, a well-known physicist and later a commissioner of the Atomic Energy Commission, published the Army's authorized account of the Manhattan District, emphasizing the scientific successes.

Those scientific accomplishments cannot be denigrated, and the air of self-congratulation, or remorse, among the scientists is not unsuitable. After Enrico Fermi's group demonstrated a chain reaction in a rough pile of uranium and purified carbon at the University of Chicago in 1942, the theoretical speculations of the past three years were confirmed: the "pile" could be used to produce plutonium.

Accordingly, shortly after the Chicago experiment, plans were put underway to construct three very large versions of Fermi's pile in a remote area in the state of Washington. These reactors produced just enough plutonium for the weapon which destroyed Nagasaki, but the Du Pont Corporation—which constructed the reactors—although it allowed large margins for the unforeseen, could not know that graphite swells when bombarded with radiation. Shortly after the end of the Second World War, graphite swelling had so distorted the functioning of the three plutonium-production reactors that one was shut down entirely, to be kept in reserve, and the remaining two were operated at only half power.

Uranium 235, it was known, could also be used in a weapon. And since it was not known whether the techniques for plutonium production would be successful in the short time available, General Leslie Groves, head of the Manhattan District, also authorized several different approaches toward the production of uranium 235. Several such techniques were imaginable, and all were attempted, but none was completely successful. The two techniques which produced the uranium that exploded over Hiroshima—thermal diffusion and electromagnetic separation—were not very practical and were abandoned shortly after the war. No further uranium bombs were constructed for several years. (A third and more expensive technique of obtaining uranium 235—gaseous diffusion—while it did not contribute very directly to the war effort, is still in use, in a form much improved over that developed by the Manhattan District.)

When small quantities of plutonium and uranium became available, the design of the weapons began. J. Robert Oppenheimer organized the group of scientists and engineers who would actually assemble the nuclear weapons at a remote laboratory at Los Alamos, New Mexico.

By the spring of 1945, all of these efforts had converged. The reactors were producing a trickle of plutonium and could be counted on to produce the few pounds needed for a nuclear explosive by early summer. All of the uranium-production techniques taken together would also produce, by summer, a few pounds of uranium 235, enough for a single explosive. The design efforts for both types of weapons were progressing well. President Truman was notified that a single test weapon might be ready by July, 1945.

The war in Europe was then over; Germany had been occupied by the Allies, and Japan was asking for our terms of surrender. It would take heroic efforts to complete the bomb project while the war was still being fought, and the result would not be an arsenal but only two or three weapons, at the most, by midsummer of 1945; we would have only the most limited capability to produce more. Nevertheless, the project went forward. The pattern of events at that time may give some indication of the reason. President Truman deferred until July, when the atom-bomb test was scheduled, a meeting with Stalin and Churchill at Potsdam, to settle plans for the postwar world. During the Potsdam meeting, Truman received news of the first explosion of a nuclear weapon in New Mexico; according to all reports, he was greatly cheered by the news, which he shared with Churchill, but not with Stalin. Stalin was told, casually, only that the US had a new weapon.

A major item on the agenda at Potsdam was the question of Japanese surrender. Stalin relayed the Japanese peace overtures which had been made through Moscow; he and Truman discussed the terms under which Russia would enter the war in the East. Truman was unhappy about concessions which had been granted to Stalin in exchange for a promise to enter the war against Japan after the conclusion of hostilities in Europe: Stalin was to have great latitude in Europe and a share of the Japanese Empire.

Russia was to enter the war against Japan by August 10. On August 6, 1945, we dropped a single nuclear bomb on Hiroshima. The Japanese government did not become aware of the attack for some time. At first, the military command knew only that communications had been cut with Hiroshima; it took a day for the news to arrive and be believed: Hiroshima had been destroyed. On August 8, the Soviet Union entered the war against Japan and moved into Manchuria. Before the government of Japan could respond to either development, a second nuclear weapon exploded, destroying Nagasaki. Japan surrendered unconditionally.

The bomb which exploded over Hiroshima contained all of the uranium 235 which months of unstinting effort had produced; we had no more uranium with which to build weapons. Slightly more plutonium had been produced, and this supply was used to build two weapons. The first was exploded in the test in New Mexico on July 17, 1945. By August 9, working at highest intensity, we produced only one more such weapon, which was exploded over Nagasaki.

The destruction of two cities with two bombs in the space of three days implied a vast destructive power. Although the Japanese—and our allies—were unaware of the fact, we could not have produced more than about one atom bomb per month; the appalling implication of the attacks on Hiroshima and Nagasaki, that a new city could be destroyed every three days, was false. The use of the two bombs to destroy two cities seems to have been calculated to deceive the Japanese as to our real capability. (Some now think, however, that Japan would have surrendered unconditionally in the face of the Russian attack, and that the atomic bombing was unnecessary, and perhaps ineffective. A single day's fire-bombing of Tokyo had caused a hundred thousand casualties, more than were suffered at either Hiroshima or Nagasaki, but the Japanese had not yet found such damage unacceptable.)

The psychological effect of the two atomic attacks was presumably felt elsewhere. Whether Truman principally intended to impress the Russians, the Japanese, or simply the world at large with our new nuclear armaments will probably never be known. Truman reportedly gave the decision to use the bomb little thought, and he may not himself have understood how limited our nuclear capacity was at the time.

CHAPTER TWO

Scientists

BARRY COMMONER is one of the world's best-known scientists. When *Time* magazine put him on its cover, the accompanying story called him the "Paul Revere of ecology," meaning that Commoner had been among the first to alert the United States to the dangers of pollution. While Commoner objected to the cover story, he was later pleased by the Paul Revere designation. Commoner has come to prominence, not by virtue of his laboratory discoveries, but by his efforts to make political use of scientific knowledge and by his continued attacks on his own profession's failures to be of service to society. Trained as a plant physiologist, and for a time chairman of Washington University's (St. Louis) botany department, Commoner's best-known works are his writings for the layman on the environment. In articles and books, he has presented his theory of the political and economic causes of environmental problems; in the summer of 1975, he was at work on a book *(The Poverty of Power)* for the general public which would present a theory about the interlocking economic and energy problems of society.

Commoner's office on the Washington University campus, where he is now head of the Center for the Biology of Natural Systems, seems to symbolize his relationship with the established scientific disciplines. He works in a large room in the basement of what was once the botany building of the university, a department which has long since been absorbed into a modern biochemistry division dominated by the university's medical school. The former botany building, a quaint small-scale copy taken from the colleges of Oxford and Cambridge (there are slits for archers, presumably to assist in defending the campus against James II), is surrounded and overshadowed by the new laboratory buildings built in modern cracker-box style to house the expensive equipment of modern biochemical research. Commoner's own research makes use of modern equipment, but it is housed in cramped quarters in his basement; above ground, the greenhouse on which he relies for plant materials is constantly being shifted to make way for new laboratory construction.

Like the office of any important man, however, Commoner's is guarded by an anteroom in which an attractive dark-haired woman types, takes telephone calls and screens visitors. The office itself is large and pleasant, furnished with cast-off chairs and homemade bookshelves; an analyst's couch of 1920s vintage stands against one wall. Covering every chair, a worktable and much of the floor are heaps of books, papers and correspondence. Commoner, a man in his late fifties, with a suntanned seamed face and short gray hair, shifts a handful of papers from one armchair and invites his guest to be seated in a second, identical chair. Both chairs creak disconcertingly.

As a private citizen, Commoner says, he is opposed to nuclear power simply because there are cheaper and less hazardous ways of obtaining electricity. His principal concerns are related to the dispersal of radioactive materials into the environment during routine operations of the industry. He explains that his concern with civilian nuclear power is quite recent, although his contact with atomic energy extends back to 1945. In the early postwar years, Commoner observed the struggle between the Manhattan Project scientists and the military heads of the project, a struggle that merged into the larger fight for civilian control of atomic energy. The scientists knew quite well in those days that atomic energy had only limited and distant promise for peacetime use; the shroud of secrecy surrounding the atom seems to have kept this fact from Congress, however, just as the scientists themselves were deceived about the nation's military atomic arsenal.

In 1945, Commoner had already made a reputation by leading the Navy's development of techniques for aerial application of DDT, important when American troops occupied the malaria-ridden islands of the Pacific. He now ruefully recounts his responsibility for the world's first DDT-caused fish kill, on a beach in New Jersey where DDT spraying techniques were being tested.

COMMONER: Before I left the Navy, the last assignment I had was temporary additional duty in Washington assigned to the staff of a subcommittee of the Senate Military Affairs Committee. A subcommittee headed by Senator [Harley] Kilgore [Dem., W. Va.]. And for some reason, largely I think because Kilgore was a friend of Truman's, that committee got the job of writing the National Science Foundation Bill. In other words, of preparing legislation that would reflect the history of the role of science in the War. I guess that's why it was in the military, the Senate military committee. I was assigned to it, at the request of Senator Kilgore, by the Navy. I think the Navy was interested in doing it since they had the Office of Naval Research, which obviously was going to be affected by any creation of an NSF; they were undoubtedly interested in being in on the inside. In fact, I wasn't the only naval officer . . . and there were, I think there was somebody from the Army on the staff, too.

What was your rank then?

COMMONER: Lieutenant.
So, my job was liaison with the scientific community. In preparation of the bill, in testimony and so on, on the National Science Foundation Bill. That was the first time around on it; it got into a tremendous hassle over whether it [the Foundation] would be subservient to the scientific community or to the bureaucracy, and [Senator Warren G.] Magnuson [Dem., Wash.] was on one side and Kilgore was on the other, and the scientific establishment favored the Magnuson bill—which would have had the Foundation dominated by the scientific community—and Kilgore and Truman were on the other side. It was all very interesting.

Well, the way in which it got involved with atomic energy, is that I was spending a lot of time with members of the scientific community about the bill and testimony, and I ran across a botanist, I think his name was Griggs, whose son was at Los Alamos. Griggs—I think it was Griggs, I'm not sure—but anyway, from him I learned that the people at Los Alamos were dying to speak out on what had been going on, and I was brought in contact then with the whole crowd. With Ed Condon, Willy Higginbotham, Harold Urey, and so on. [These were the leaders of the Manhattan Project's scientific work.] The result was that I was able to organize the first—to help Kilgore organize—the first hearings at which they [the atomic scientists] spoke out. Which were in our subcommittee, ostensibly on the National Science Foundation Bill. . . . From then on, of course, I was in on a lot of the discussion.

When you say they were allowed to speak out for the first time, what were they saying?

COMMONER: Oh, that was the point where they came out and said that there were very serious problems, to do with the impact of the Bomb on military policy. There wasn't much talk about radiation. It was just the question of what a fearful weapon this was and how powerful, and that it was very important to, well. . . . The thing was, it was the first cry of anguish, really. [J. Robert] Oppenheimer testified.

What was the thrust of the initial testimony?

COMMONER: Very quickly, the issue became civilian control of atomic energy. And a split developed, as you know. Oppenheimer going off toward the military. So in other words, the testimony in the Kilgore hearings was almost accidentally connected with the National Science Foundation Bill, and the only importance of it is, that it was the first surfacing in Congress of the ideas and attitudes of the atomic-bomb project people. Very quickly, the scene shifted over to the hearings on the [Senator Brien] McMahon [Dem., Vermont] Bill [which called for a civilian atomic-energy agency].

The issue became one of civilian versus military. . . .

COMMONER: Very quickly that's what the issue was.

Why did that seem so important at the time?

COMMONER: Because I think the project scientists felt that the military didn't understand the dangerous potential of nuclear weapons, and they mistrusted. . . . They probably felt that it was too important a thing to leave in the hands of generals. From their own experience, with General [Leslie] Groves [head of the Manhattan District] and all that, they—many of them felt these birds wouldn't understand what the hell it was all about, in terms of danger, and political significance, and so on.

Well, was it their concern that the weapons would be used? Or that they would be improperly developed?

COMMONER: Oh, no. You want to remember that—when I left the Navy, the captain, or the admiral, or whatever he was, in charge of the corps to which I was attached, was blackmailing me in trying to make me stay. . . . He made two appeals to me: One was, that since I was Jewish I ought to recognize that they had been very nice to me, for being a Jew. In the Navy. That was one of the things that he told me. (*Laughs.*) That I owed them, to stay in the Navy because of that. You can imagine that didn't appeal to me very much. The other point that he made was, he said, "Lieutenant, World War III has begun, whether you know it or not." I looked at him, and he said, "We are preparing to confront the new enemy, the Soviet Union." And that was the substance of what was going on, and I think a lot of people, including myself, were worried about nuclear weapons being, you know, becoming the basis for an actual Third World War.

Looking at the written record of the time, it seems as if there was some argument that the military was paying inadequate attention to the civilian potential of nuclear energy.

COMMONER: There was very little discussion, as I remember, of civilian potential. As a matter of fact, I remember I knew, I had casual acquaintance with a very interesting guy who was head of the Bureau of the Budget at the time. . . . Brilliant, you know (*laughing*), unlike any subsequent heads of the Bureau of the Budget. I think he had been head of the University of Syracuse School of Public Administration. A really sound and humanitarian guy, and [he] really understood things. He asked me at the time to talk with him about the potential of uranium as a power source. He was interested. I recall those conversations as almost the only evidence, the only contact I had, with that question, although there was a lot of talk, you know, that a teaspoon of uranium would run a steamship across the Atlantic, and stuff like that, and [the budget director] was concerned about whether there really was enough uranium around to provide much energy. Obviously, the data were very crude at the time; I did a little looking into it, and [he] had some of his people look into it, and funny, we came to the conclusion that nuclear power was pie in the sky—or might well be pie in the sky—because there really wasn't that much uranium around. And I have a memorandum somewhere that I wrote . . . about it. . . . You know, there were discussions, sort of science fiction discussions, and so on, . . . and I am sure that the future importance of using reactors for power must have figured in the argument on civil control. But I think mostly civilian control was, you know, just like control over the Pentagon now, and the CIA for that matter. Who wants to leave them in charge?

The talk about the pea, or the teaspoon of uranium, that will do wonderful things, that was very common. . . .

COMMONER: Very common, very common.

Do you think it colored the discussion? Did it affect people's attitudes?

COMMONER: No, I don't think so. This was a sudden new thing, and there was a lot of talk about the Brave New World aspects of it, but nothing concrete. I don't think there was any sort of practical discussion about nuclear power. As I say, the only one I know was [the budget director], who was a very smart guy and, I remember very well, it was he who raised the issue with me and asked me to help him.

By the 1950s it was the existence of a civilian agency which permitted the development of the whole nuclear-power program?

COMMONER: Yeah. But that—you must know by now, you've been digging into it—that really was the whole Eisenhower "Atoms for Peace" thing, which was a political maneuver. In other words, I, certainly, absolutely am convinced that the government never developed nuclear power out of a coherent analysis of the energy question. Never. In the first place, it [energy analysis] wasn't done at all, in the beginning, because all they were interested in were bombs. And then when it was done, it was a political ploy on Eisenhower's part. You remember Operation Candor [the government's program to reassure the public about fallout]? And all that jazz. It was at the same time, when they decided they had to have some kind of nonwarlike excuse for continuing the development of nuclear energy. Let me put it this way. I tend to look at the whole nuclear-power thing as a kind of political Potëmkin Village. A very expensive one, constructed in order to make Eisenhower's political position credible. It's the most expensive charade in history.

[Admiral] Lewis Strauss [former member of Atomic Energy Commission] says that there was a big push for civilian power under the Eisenhower administration because they felt that it was very important that private companies, rather than the government, operate the electric powerplants.

COMMONER: This is in the fifties?

Yeah.

COMMONER: I don't remember much [about that aspect]. The thing that I have to remind you of, is that by 1952, certainly, my interest in all of this was focused on radiation hazards. And I must say I paid damn little attention to the question of nuclear power, in terms of energy, economic need, or anything of the sort. From then until quite recently. . . . Quite early in the fifties [nuclear fallout from weapons testing] surfaced, and that was all I paid attention to from then on. My knowledge of the nuclear energy thing in the fifties and sixties is nil.

A lot of liberal Democrats then were in favor of nuclear-power development.

COMMONER: Yes, yes. Undoubtedly.

And this was the mirror image of what the Eisenhower administration was afraid of.

COMMONER: Yeah, I think there was a sort of knee-jerk reaction to the notion that the Republicans wanted to have nuclear power developed by private industry. The liberal response was, "No, it's got to be done by the government," without asking what the hell do we want nuclear power for? And, I am sure that the ploy, Eisenhower's ploy, of wrapping the nuclear enterprise up in a peaceful guise took them in. I don't think anybody thought, really gave it any serious thought. Certainly there was no examination of the need for energy. None.

I'd like you to comment a little bit on the role of scientists in the early years of the discussion in the 1950s.

COMMONER: Well, okay. It's complicated. In the first place, apart from the flurry of immediate postwar talk, you know, by the atomic scientists, nothing much happened until a meeting [of the American Association for the Advancement of Science, the nation's largest professional society for scientists] in 1952 or 1953, let's say 1953. That was an interesting business; there were rumors [about fallout], and a number of us walked the issue out on the floor of the AAAS council, and there was a hard-fought battle as to whether it was appropriate for us to be barging into this secret military thing. That was one of the first issues. It was the issue of whether the scientific community had the right and, shall we say, the guts, to take up an issue which was formally secret. As I remember, we won a very difficult and complicated debate on the floor of the [AAAS] Council, which required the AAAS to look into the fallout situation. They were literally required to do it. As far as I know, that was the first, you know, official scientific-organization action. It was a brash, an antiestablishment thing to do, and the elder statesmen in the AAAS were quite worried about it. And what happened was very interesting. That resolution was never carried out by the AAAS. . . . Warren Weaver said that the National Academy [of Sciences] was going to do [a report on] radiation. . . . And the AAAS said (*laughs*), breathing a sigh of relief, "Blessings, it's going to be done." Now, that was the opening thing.

The response of the Atomic Energy Commission and its captive scientists was to say, "You don't know what you're talking about; we do, because we're well informed and you're not. And there is no danger." (*Somewhat to his visitor's surprise, Commoner takes out a scrap book of press cuttings and locates the first news story about him and radiation, a 1958 student-newspaper story about a paper he delivered at the AAAS's Christmas, 1957, meeting, in which he talked about fallout.*) That was, I think, Christmas, '57. Yeah. Interesting. (*Surprised at the date.*)

By 1957, all the important decisions about the civilian nuclear-power program had already been made.

COMMONER: Really? That happened without us noticing.

Private plants had been ordered, the [Atomic Energy Act] had been changed, companies had ordered the plants. . . .

COMMONER: Speaking for myself, we weren't paying any attention. We really weren't.

Was anybody paying any attention?

COMMONER: Probably. People like Ralph Lapp, who was probably talking about it. [Lapp, a Manhattan Project scientist, strongly advocated civilian atomic-power development in the 1940s and continues to do so.]

There was quite a lot written by some of the Manhattan Project guys, who were in favor of civilian nuclear power and opposed to military applications.

COMMONER: Right.

Such as Ralph Lapp.

COMMONER: Right. (*Laughs.*) So that's his position from way back. Yeah. I had nothing to do with that.

Some of the liberals who were once in favor of civilian development are now opposed to it, and the conservative and military people are in favor of it.

COMMONER: Yeah. Sure. Yeah, yeah.

Do you think that if you had–I don't mean you personally, but if people had paid more attention, the whole thing would have been done differently? About the civilian agency [Atomic Energy Commission] and the whole civilian development?

COMMONER: It's very hard for me to wrap my mind around that kind of question. My sort of optimistic approach to it is that the intense concern with the fallout problem and the military aspects—you know, civil defense, that whole thing—that that was a very important training ground for the scientific community. It gave the scientific community the experience of taking on, not only the government but the military establishment and, frankly, I think that the scientific community wouldn't have had the guts to get into the nuclear-power thing if it weren't for that background. So that, as far as I'm concerned, anyhow, I'm not that sure it should have been done differently, if we had it to do over again. Because clearly, the main battleground was the question of nuclear war and fallout; the scientific community played a tremendously important role in demonstrating its, shall we say, political value to the rest of the community [by taking on the government]. . . . I'll bet you that there are people now who are concerned about what scientists say about nuclear power, and so on, only because they remember the political service that the scientists performed in connection with nuclear war.

CHAPTER THREE

The Secret of the Atom Bomb

T HE ATOMIC SECRETS of the Second World War are now open to view. Any bright high school student today can assemble enough information— but not necessarily the skills and materials—to build atom bombs of the types which were exploded during the war. From what we now know, we can assess what then, in 1945, was secret and also what was known; there now seems little doubt that nothing of any scientific importance was then secret.

The explosion of a nuclear bomb over Japan revealed the most important single datum: such a bomb was possible. Much of the Manhattan Project work and research had been directed toward establishing this fact. If there were other facts of importance, most were revealed within weeks of the explosion by the Army's publication of the famous Smyth report, Henry DeWolf Smyth's detailed book-length account of the scientific developments which led to the atom bomb. The publication of these data surprised and angered many people who felt that protecting the secrets of the atom was the nation's first task. The Smyth report confirmed that uranium 235 and plutonium could be used for a bomb, and that plutonium could be made in a reactor; it discussed the various means by which uranium 235 could be extracted, and claimed that all four techniques tested by the US were successful. While the report stopped short of engineering details and remained silent on the actual design of nuclear weapons, it left nothing of scientific importance unsettled.

This being the case, it was difficult for many people, then and now, to understand why extraordinary measures were taken to protect the secrecy of the atom-bomb project. From the moment the atom bomb's existence was revealed, the public discussion of its future revolved around the need to protect the secret of the atom bomb. Military commanders resisted surrendering the atom to civilian control because, they claimed, of the difficulty of maintaining security in a civilian enterprise. And civilians, for their part, were suspicious of military atomic activities precisely because they viewed as excessive the secrecy requirements of the military.

The reasons for this preoccupation with secrecy may never be entirely explained. During the war years it was important to prevent our enemies from getting any hint that would assist their own supposed—but nonexistent—atom-bomb efforts. While Germany never mounted a serious effort to develop the atom bomb, and Japan never seems even to have investigated the possibility, wartime secrecy can be charitably attributed to this motive. After the war, the only remaining motive for secrecy seemed to be the need to keep the Soviet Union from benefitting from our atom-bomb development program. But with the explosions in Japan and the publication of the Smyth report, this consideration, whatever its initial importance, was much diminished. Details of atomic production and planning were sensitive, as any military information would be, but there was no *unique* hazard in atomic-bomb data; all the scientifically important material had been published.

This state of affairs was apparent to many of the scientists associated with the atom-bomb development project and deepened their distrust of the military. There seemed no rational purpose behind the Draconian secrecy measures taken by General Leslie Groves and the other military managers of the Manhattan District.

In retrospect, it seems likely that at least one motive for this extraordinary secrecy was the effort to conceal the illusory nature of our nuclear armament. At the close of the war, and until 1949 or 1950, we did not have any militarily significant stockpile of atomic arms; but the compartmentalization and strict secrecy of the Manhattan Project seems to have kept this secret even from those who were fighting so bitterly for control of the project.

During the war, three weapons had been built more or less by hand by some of the nation's most able scientists and engineers. The plutonium bomb, the only nuclear bomb available after the exhaustion of our meager supply of U-235, was a device of extreme complexity. Carefully machined high explosives were shaped into dozens of lenses, so that their explosive force would be directed inward; these lenses were detonated by a complex electronic system which was in turn set off by a fusing device, to allow detonation at a preselected altitude. Within this nest of high explosives and wiring was a sphere of finely machined plutonium, surrounded by layers of neutron reflectors and a steel shell. At the very center of the device was a tiny source of neutrons to initiate the chain reaction at the proper moment. All of this was encased in a spherical steel bombshell about six feet long, dubbed "Fat Man" in honor of its supposed resemblance to Winston Churchill.

This was not a weapon designed for production in numbers; it was simply the first and safest design which had occurred to the Manhattan District designers. No production lines had been set up; there was insufficient nuclear material to justify such preparations and insufficient time to design all of the units which a production line would require. The two plutonium bombs and the single, less complex uranium bomb were built on an *ad hoc* basis by teams of the talented inventors of these weapons.

But many of the hundreds of scientists who participated in this effort did not themselves know the importance of their role; secrecy was too tight. The various steps of design and assembly were carefully insulated from each other. The scientists in Chicago who protested the use of the bombs on Japan could hardly have known that those bombs represented the entire inventory of weapons assembled in New Mexico.

Chafing under security restrictions and military command, feeling without purpose once the war had ended, the scientists left the Manhattan District as quickly as possible. Some left with regret for the role they had played in the destruction of two Japanese cities; some left with pride and relief; but nearly all of them left. Behind them remained a meager accumulation of plutonium, some assistants, and masses of reports—literally thousands of documents—describing the technology of atom-bomb manufacture. But there was no production line, no machinery to grind out atom bombs at a steady pace.

General Groves was able to hold the last scientists together long enough to make two more plutonium bombs of the Nagasaki design. These two bombs were exploded at Bikini, in the Pacific, in July, 1946, once again exhausting our nuclear armory. By September of that year, the remaining scientists in the program had been demobilized, and many of them enthusiastically entered the political fight to wrest control of the nuclear program from General Groves.

Given the circumstances of the moment, the Bikini tests were remarkable. They destroyed a good part of our accumulation of plutonium and our entire arsenal of weapons. The tests left us without the ability to respond quickly with atom bombs to a military emergency, yet they were not tests of new weapon designs and thus did not furnish any new information beyond that secured during the war.

The reason for the tests remains obscure. The official historians of the Atomic Energy Commission [now two new agencies—see appendix C], maintain that the tests were carried out to impress the Navy. At that time, of course, atomic energy was exclusively the property of the Army and its Air Corps; the Navy was both jealous and skeptical. Fleet Admiral William Leahy, Truman's close friend and advisor, had consistently denigrated the importance of the atom bomb. Admiral Lewis Strauss warned of the effect the weapon might have on Navy appropriations if it proved to have made the Navy vulnerable to air attack.

In any event, the atom bombs were tested on warships, with representatives of the Navy and members of Congress invited in large numbers. Also invited was a shipload of reporters, including radio correspondents who described the test explosions to a breathless world, a representative of the Soviet Press and a "Russian scientist." As a public-relations event, the Bikini tests would have been hard to equal. So well publicized were they that a French entrepreneur was able to launch a new bathing-suit design simply by naming it for the test site.

Moreover, a few days before the Bikini tests, the United States had dramatically revealed its plan for international control of atomic energy. This plan was drawn up by Dean Acheson, then Undersecretary of State, and David Lilienthal, chairman of the Tennessee Valley Authority, and was presented to the United Nations' atomic-energy committee by Bernard Baruch, whose name it thereafter bore. Within the United States and around the world the Baruch Plan was greeted as a magnanimous gesture and a true step toward peace. The *Bulletin of the Atomic Scientists*—a magazine of enormous prestige published by a group of the scientists who had recently left the Manhattan District—hailed the Baruch plan and condemned the Bikini tests as an attempt by the military to sabotage negotiations with the Soviet Union in the UN.

Whether the Bikini tests were intended solely to demonstrate the potency of atom bombs to the Navy (which was not impressed) or to brandish a sword at the Russians will never be known with certainty. What is clear is that the tests were public-relations events with no military purpose; they in fact exhausted our small arsenal of atom bombs.

A curious question is whether Harry Truman knew, when he approved the tests within days after launching the Baruch plan in the United Nations, that we had no nuclear arsenal and that the atomic might we were offering to turn over to international control was still nonexistent.

Acheson and Lilienthal, who drew up the plan, apparently were kept in the dark as to the true state of affairs. The two laymen were given intensive briefings by the Manhattan District scientists on the fundamental principles of atomic energy. The scientists apparently communicated quite clearly the theoretical possibilities, but did not say or did not themselves know how far these were from practical realization. Thus misled, Lilienthal and Acheson both fought long and hard for civilian custody of our nonexistent weapons stockpile.

Lilienthal for a long time clung to the idea that the vast military capabilities developed during the war could be devoted to civilian purposes. As chairman of the Tennessee Valley Authority, he had come to be greatly impressed with the possibilities of government-sponsored electric-power projects, and he hoped that atomic power would make enormous new TVAs possible, particularly in the Columbia River Valley where military reactors had already been constructed. Many others seemed to share this hope or, in some quarters, this fear.

Throughout 1946, there were bitter fights in Congress over the nature of postwar control of atomic energy. Almost everyone who participated in the debate assumed two things, neither of which was then true: First, that we possessed a large military atomic force; and second, that the atom could be used for dramatic peacetime work in the near future.

Much of the fight revolved around a bill introduced by Brien McMahon, the freshman Democratic senator from Vermont, who sensed an opportunity to carve out a place in history. His bill provided for all nuclear energy work to be turned over to a civilian agency, the Atomic Energy Commission. General Leslie Groves, head of the Manhattan District—and his allies in the Army and in Congress—fought the bill bitterly. The very bitterness of the opposition enhanced civilian suspicion of the Pentagon's motives. When the McMahon Act was passed, Groves continued the battle by seeking to retain custody of all nuclear weapons. From the time of the bill's passage in July until his control ended on December 31, 1946, Groves refused to turn over even so much as an inventory of nuclear materials to the new five-member civilian commission.

Truman had announced the nomination of David Lilienthal, TVA chairman, to the post of chairman of the Atomic Energy Commission in mid-1946, but the

Senate for months refused to confirm the nomination of Lilienthal and his four colleagues. Various charges—that Lilienthal was a Communist or a sympathizer with Communists; that he had allowed Communists to work for the TVA; that he was unfit by character and temperament; that he would jeopardize the security of the atom-bomb project; that he would reveal its secrets—protracted the confirmation hearings. In the meantime, Groves sat tight. And the military nuclear-weapons program decayed still further.

The record of the Manhattan District from the end of the war until the assumption of control by civilians on January 1, 1947, is almost a complete blank. This period drops conveniently between the two volumes of the AEC's official history, the only authoritative source of information on then-secret activities. None of the many other accounts provide much light.

We do know that the gaseous diffusion plant at Oak Ridge, Tennessee, was expanded during 1946, and for the first time it began producing a trickle of relatively pure uranium 235, suitable for use in weapons. At the end of 1946, the electromagnetic separation plant (which produced the enriched uranium for the single uranium bomb made until that time) was shut down. No more uranium bombs were fabricated for some years.

The two plutonium weapons exploded at Bikini were apparently of the same design used at Nagasaki. No further plutonium bombs were immediately assembled (nearly all the scientists capable of assembling more weapons had abandoned the Manhattan Project), but a small quantity of plutonium was accumulating. No new sources of uranium ore had been developed; the United States' sole source was still the single Shinkolobwe mine in the Belgian Congo, then personally owned by King Leopold of Belgium, on whose whims we depended for further supply. At the Hanford Works, in the state of Washington, the United States' three plutonium-production reactors were in a state of slow disintegration. Swelling of graphite from radioactive bombardment, which we had not yet learned to understand and counteract, was slowly distorting the channels into which uranium was fed and through which cooling water passed. Large quantities of radioactivity were being discharged over the surrounding countryside, but little plutonium was being produced. To preserve their integrity as long as possible, two of the reactors were operated at only half power, and the third was shut down entirely.

There was no bomb-production system. Some components were still being made; others were not. Three wood-frame houses and a Quonset hut at the Sandia Air Force Base near Albuquerque constituted the nation's atom-bomb assembly plant. If we had had any weapons, they would have been stored in igloos in a dry arroyo behind the Quonset hut. But no weapons were being assembled.

This was the state of affairs while the struggle over control of the atom was waged in Congress and in the United Nations. This was the reality behind the hopes for a massive diversion of our supposed military resources to civilian benefits. When Lilienthal and his fellow commissioners slowly began assembling the information which Groves had, for two years, so successfully withheld from them and from the Congress, they were appalled. Lilienthal's diary records his disappointment that the civilian applications of nuclear power, for which he had held such hopes and which to some degree accounted for his presence on the Commission, were so very far in the future. On April 2, 1947, the five commissioners prepared a detailed report on the state of the atomic-energy program, and on April 3, they went to see Harry Truman.

How much President Truman knew about the weapons program before that meeting is uncertain. The only records now publicly available seem to show that General Groves had concealed the true state of our armaments even from his superiors. In any case, the new commissioners of the AEC felt obliged to lose no time in reporting to the president, and they chose the unusual measure of reporting as a group. A written summary of the previous day's detailed report had been prepared. In all written documents, however, the actual number of weapons in our possession was left blank.

On April 3, the commissioners handed Truman a brief typewritten report in his oval office and suggested that he read it aloud for the benefit of the military-service representatives who were present. The report began this way:

"After three months of authority over the American Atomic Energy enterprise, with access to sources of information and opportunity gradually to fit the facts together, the Atomic Energy Commission must report to the President certain serious weaknesses in the situation from the standpoint of national defense and security: 1. The present supply of atomic bombs is very small. The actual number for which all necessary parts are available is _____. None of these bombs is assembled. The highly technical operation of assembly hitherto has been effected by civilian teams no longer organized as such. Training of military personnel to effect assembly is not yet complete."

When Truman reached the blank in the report, the number of weapons for which components were available, Lilienthal supplied the number orally. It was never written down nor made public. That the number must have been very small indeed is evident from the context; Lilienthal recorded in his diary Truman's surprise on hearing it.

The remainder of the report detailed the sad state of the uranium-supply and weapons-production system as a whole; it recounted the need to design and test new weapons capable of systematic production, to obtain raw materials and build new plutonium and uranium facilities, to construct bomb-fabrication and assembly facilities. Within weeks, the president had authorized, and the Commission had undertaken, the nation's first actual nuclear-weapons production program. The designs for such weapons were tested in 1948 and, by 1949, we had begun accumulating atom bombs. At what point we acquired militarily significant or decisive numbers of weapons is still secret.

In August, 1949, the Soviet Union exploded an atom bomb of its own. Our nuclear monopoly, which had so little substance, but in which the world had believed for four years, was publicly ended.

But another fantasy, the fantasy of atomic power, had been created. Overwhelmed by what seemed to be the significance of atomic power, a significance carefully ballyhooed by General Groves through the use of the weapon itself and the artful release of the Smyth report, civilians had mobilized to take control of the new force. Democrats—particularly the more left-leaning, profoundly suspicious of the military—fought to gain control of the atom and its apparent potential for furtherance of the New Deal's programs. Groves and the Army, while unable to reveal the lack of substance in what was the center of our foreign policy, nevertheless discounted the potential civilian applications of nuclear power—quite correctly, but again arousing widespread suspicion of their motives. The Soviet Union, the American Communist Party, and most of those on the Left, publicly claimed that a military monopoly of the atom was stifling its enormous civilian potential.

The civilians won the battle, and Lilienthal, the champion of government powerplants, found himself in command of a program which was years away from even the simplest applications. For two years, Lilienthal and his colleagues labored mightily at their first order of business, the construction of an adequate military machinery. If Lilienthal was aware of the irony, he did not record it in his copious published diaries. But by 1949, with the military production programs finally well under way, and with civilian programs still far in the future, Lilienthal submitted his resignation. The champions of civilian control had succeeded, where the military had failed, in recruiting scientists and enlisting their enthusiasm for a large-scale military program. For the first time, as a result of victory in the fight for civilian control, a truly devastating nuclear-weapons force had been developed, and an arms race with the Soviet Union was launched.

In later years, civilian management did indeed produce civilian nuclear power, but not until those who originally had advocated civilian control of the atom had lost interest in the subject. Ultimately, the same protagonists would meet again on opposite sides of the question; Lilienthal and Barry Commoner now oppose nuclear power; the *Bulletin of the Atomic Scientists*, which still speaks for many of the physicists and chemists who worked in the Manhattan District and who fought so hard for civilian control of the atom, is now severely critical of civilian nuclear powerplants. But the conservative military and Republican proponents of military control in the 1940s, who then were pessimistic and doubtful about civilian nuclear power, are now among its strongest supporters.

CHAPTER FOUR

◆

Civilians

CARL WALSKE is an attractive and engaging man of evident intelligence who lives in the Murray Hill section of Manhattan, a short walk from his office on Park Avenue South, and who heads a remarkable enterprise through which he speaks for Exxon, Gulf Oil, the Chase Manhattan Bank, Westinghouse, General Electric, Allied Chemical, the Tennessee Valley Authority, the nation's largest power companies—and the National Wildlife Federation. All of these organizations are represented on the board of directors of the Atomic Industrial Forum, the trade association of the nuclear industry—the companies that build, buy and finance nuclear powerplants—and Carl Walske, a fifty-three-year-old physicist, is president and chief operating officer of the Forum.

As the head of a private trade association, Walske is singular in many respects. He had worked only briefly for a private company before becoming president of the Forum in 1973. He is one of the very few people in the world to have designed nuclear weapons. And he is also pleasingly frank about the problems and hazards of the industry for which he speaks.

The Atomic Industrial Forum was born, along with the nuclear industry, in the early 1950s and, through most of its life, had its headquarters in New York, where the financial institutions and industrial corporations which make up its membership are based. In the fall of 1975, the Forum moved its headquarters to Washington, DC, in part to carry out more vigorous lobbying, and perhaps because of the personal preference of Walske, who had lived in a Washington suburb while working for the federal government. The Forum continues to maintain a subsidiary office at the site of its former headquarters at 475 Park Avenue South, a new skyscraper faced, improbably, with brick and tinted glass.

In the summer of 1975, the Forum has not yet departed for Washington, but still there is no sign of its presence in the skyscraper's small lobby, where, presumably to protect tenants of the building from unwanted visitors, a uniformed guard presides behind a high counter. There is no visible directory of the building's occupants, perhaps in the interests of security; the guard directs a visitor to the Forum's offices on the fifteenth floor. Here again, there is no sign indicating the floor's tenants. Behind unmarked double doors is a small, expensively furnished reception area, where a bored young black woman with straightened hair acknowledges that this is the office of the Atomic Industrial Forum. "An appointment with Dr. Walske?" The receptionist is a temporary employee, and unsure; will the visitor have a seat?

The sofa, leather and chrome, is comfortable; a few copies of the Forum's monthly publication, *The Nuclear Industry*, are on a coffee table. The receptionist makes a call through her switchboard and shortly a Mr. Eugene Gantzhorn appears. Mr. Gantzhorn is a public-relations functionary who is irritated that a reporter has arranged an appointment with Carl Walske without his knowledge or assistance. Gantzhorn wears an open-collared shirt and a supercilious manner; he is a second guardian protecting the president's sanctum, a corner office overlooking Park Avenue, where Carl Walske receives his visitor.

Walske looks even younger than his years and has remarkable, intelligent eyes and a soft voice. He is a bomb designer, privy to the world's most closely guarded secrets, holder of the most dangerous knowledge ever found. In religions now long-forgotten, women were the custodians of unspeakable secrets and performed the rites of the great Goddess in a sanctum guarded by eunuchs and virgins. Now it is largely men who are the keepers of the terrible secrets and their offices are surrounded by women and publicists. Walske has spent most of his life within the inner recesses of the nuclear temple: after receiving his doctorate in physics, he went to work for the federal government and learned to design weapons, working in the same laboratory with Theodore Taylor, another bomb designer who has become well known for his public criticism of the lax manner in which plutonium is guarded by the civilian power industry. In 1966, Walske became Assistant to the Secretary of Defense and chairman of the committee that served as liaison between the Atomic Energy Commission, which designed and manufactured nuclear bombs, and the Defense Department, which would deploy and perhaps use them. From this position at the top of the nation's military nuclear programs, Carl Walske was selected president of the civilian trade association of the nuclear-power industry.

I gather you don't use the title "doctor"? People here have been calling you "Mr. Walske."

WALSKE: Generally, no sir, I don't.

Can you think of another trade association that's headed by a theoretical physicist?

WALSKE: Not offhand. I don't think of myself as a practicing theoretical physicist. Let me just say, I am *not* a practicing theoretical physicist. But it's a very useful background—educational experience—to have been exposed to.

Would there have been a civilian nuclear power industry of this scale [56 nuclear plants and an investment of more than $100 billion], do you think, without the military development?

WALSKE: Well, the industry of this scale, of course, derived a tremendous impetus from investment from the government for R&D [research and development] in the fifties. That would be, I would say, the crucial thing, although they certainly used what was available and useful from the military program. If you mean, Would we have developed everything from scratch, like the chemical separation, the mining, the milling, the enrichment, and all this sort of thing, not by now, we wouldn't have, no. Ultimately, maybe, yes, because ultimately we will be in such an energy crunch that it becomes compelling. But that had to be later. . . .

About the time you were studying for your doctorate, there was a struggle going on as to whether the civilian or military agencies would control nuclear power.

WALSKE: Well, that was really 1945–6. Nineteen hundred forty-six especially was when the Atomic Energy Act was passed, and it went into effect the first of January, 1947. So that was just, really, just before I got going on my doctorate—but anyway I was alive and I recollect it well enough.

Did it make any difference?

WALSKE: Civilian control?

Yes.

WALSKE: It made one big difference in terms of military programs. (*Smiling.*) It assured adequate funding for the military program and good continuity. (*Laughter.*) If it had been Defense Department, it would have been like all things in the Defense Department, subject to the cyclical nature of funding, and therefore instead of having the grade A-1 program, never short of dollars—at least in the earlier years, and in the middle years—it would have had its ups and downs, I suspect. Somebody that wanted to order a nuclear artillery shell would have had to compete with somebody who wanted to order new uniforms or something like that. And the result would have been that, once in a while, the new uniforms would have been ordered. Well, to be less facetious, it would have been, say, a playoff of whether you ordered new nuclear artillery shells or whether you had some additional men in the ranks. And, of course, in the military, personnel— soldiers, sailors and so on—are very near and dear. Without them you don't have any services at all. So it would have suffered from that.

Do you think civilian control really made any difference to the civilian nuclear industry, one way or the other?

WALSKE: Well, I think that having the technology, and the government expenditures on the technology, in a civilian agency certainly did make a difference. I would find it hard to think of a military R&D program advocating and sponsoring a commercial development, the way the AEC did with the reactor industry.

CHAPTER FIVE

The War with Russia

IN SEPTEMBER, 1949, United States aircraft over the Pacific Ocean picked up samples of radioactivity hot enough to trigger warning systems. The aircraft were part of the Army's recently created network to detect nuclear explosions by foreign powers. It had been established by the AEC over military objections that no other country was even close to developing a nuclear weapon; intelligence estimates put the first Russian test at least two years in the future. Yet the radioactivity collected by the aircraft clearly indicated an explosion had occurred. More samples were recovered; careful radiochemical analysis began to indicate the type of weapon and the date of its explosion—August 29, 1949. By mid-September, it was impossible to maintain secrecy any longer; President Truman announced that the Soviet Union had tested a nuclear explosive. He stopped short of calling it a weapon, since he was reluctant to trust the scientists' analysis this far. Nor did he believe that the industrial machinery for large-scale weapon production had been assembled in the USSR.

The nuclear monopoly which the United States had claimed since the Second World War was broken and the outcry was immediate. There were accusations of spying and of treachery. Above all, there were calls for increased armaments. In China, the nationalist government had fallen and a Communist regime friendly to Russia was taking power. Now the Russians had the Bomb. The United States was in peril, voices cried, and must mobilize for the coming struggle with Russia.

At first, the clamor did not affect the atomic-weapons program very strongly. The Congressional Joint Committee on Atomic Energy, which had responsibility for overseeing the AEC's activity, had so far declined to receive any classified material and was quite unaware of the true weapons situation. The committee pressed for expanded weapons-production efforts but, by and large, accepted the Commission's statements as to progress. No one in Congress was aware of our effective lack of atomic weapons in 1949; most believed that our aggressive foreign policy was based on a strong nuclear arsenal. In fact, Senator Brien McMahon, chairman of the Joint Committee on Atomic Energy, said as much when, following the Russian nuclear explosion, he called for redoubled efforts to expand military weapons programs—*and* to develop the H-bomb, which would preserve the nuclear supremacy on which he claimed our successful foreign policies were based.

The AEC assured Congress that every possible step was being taken to develop the H-bomb and to enlarge our nuclear arsenal; given the circumstances, this was true, since the Commission had started from zero only two years before, and had begun regular production of weapons only recently. The development of an H-bomb would be far in the future.

But McMahon was unaware of the low level of activity. He knew only that the AEC had a surprisingly small budget and seemed not to require more funds. From these facts he could only draw the conclusion that production of A-bombs and the development of the H-bomb were inexpensive matters and that the Soviet Union must be capable of doing both quite rapidly.

In fact, no efforts were being devoted to H-bomb development at this time. Scientists within the Commission were deeply divided over its theoretical feasibility; none of the schemes which had so far been advanced could be shown to be practical and computers still under development by John Von Neumann would be needed to carry out the necessary massive calculation. Instead, the Commission devoted most of its energy to enlarging the arsenal of existing weapons and to designing smaller, more efficient A-bombs. Atom bombs in 1949 were still bulky. They could be delivered only by strategic bombers and were suitable only for use on the largest targets: whole cities and industrial complexes. In reality, atom bombs were terror weapons, since only civilian populations provided targets large enough to justify their use. The military had only a mild interest in such weapons. There were no good means of delivering strategic weapons in a war against the Soviet Union fought from the continental United States, the only circumstances in which it was easy to imagine their use. (Military planners, then and now, assumed the USSR would quickly overwhelm western Europe in any new war.) Despite energetic efforts to develop a delivery aircraft capable of traveling twelve thousand miles at high speeds, this task seemed impossible; missiles, also under intense development, would not be available for several years. Until then, atomic weapons would be of only marginal interest to the military; and an H-bomb which would be a still larger version of the A-bomb was of still smaller interest. General Omar Bradley told the Atomic Energy Commission that such a weapon would be of use primarily for its psychological impact.

But this was precisely what the Congress now demanded. McMahon argued that our nuclear superiority had allowed us to take a firm stand against Russia in Europe; we must now maintain this superiority by developing the H-bomb.

In 1950, the Los Alamos scientists who had been recruited by the civilian AEC succeeded in substantially improving A-bomb designs. They developed schemes, quite different from the clumsy weapons set off in World War II, which would use plutonium or uranium efficiently and which accordingly could be applied to quite small weapons. The Commission was informed it was now possible to build tactical as well as strategic weapons: atom bombs could be used on the battlefield, in naval engagements, for troop support—for the whole range of military applications. And then, in June, 1950, war broke out in Korea.

The response to war in Korea bordered on hysteria. Harry Truman immediately transferred the nonnuclear components of atom bombs to forward bases in Britain, presumably to be ready for use against Russia. Shortly thereafter, bomb components were transferred to American bases in the Pacific. The United States and the world stood poised for general warfare.

It is still not clear what the actual state of our nuclear armament was at this point. For the first time, however, the Joint Congressional Committee on Atomic Energy ended its self-imposed ignorance and demanded information on weapon production and stockpiles. The members of the committee may have received a surprise like the one delivered to Harry Truman in April, 1947. Whatever the precise information they received, their response was to call for an immediate and drastic increase in weapon production, in which they were joined for the first time by the armed services. With the news that tactical atomic weapons could be made, the Joint Chiefs of Staff abandoned its old disinterest in atomic weapons and began to lobby vigorously for increased bomb production.

In Congress, the demand for increased production of battlefield weapons was confused with the wish for an H-bomb, although the AEC and the Joint Chiefs of Staff had both indicated the very limited military usefulness of such a weapon. Smaller, not larger, atom bombs were the military's principal requirement. In any case, no one knew how to make an H-bomb, and it would take years to do so once a design was invented.

But the clamor rose and Truman was forced to agree, not only to a vast expansion of military production, but to an all-out effort to develop the H-bomb.

Enormous sums were spent in this effort. As in the Manhattan District, no one knew what H-bomb design would work and so many efforts were made, most of which were unsuccessful. In 1950, it seemed likely on theoretical grounds that very large quantities of the radioactive form of hydrogen, called tritium, would be needed in an H-bomb, when and if one were designed; so five enormous reactors were built at Aiken, South Carolina (at a site chosen for its distance from transpolar Russian bombers) to generate plutonium and tritium, in case it should be needed. Although the records of the H-bomb effort are still secret, it seems from what information is available that large quantities of tritium were not, in fact, needed for H-bombs. Nor, indeed, were the H-bombs themselves particularly important. Today, as then, it is the smaller atom bombs which are most important in our arsenal, as tactical weapons and as the relatively small warheads delivered by intercontinental missile. The giant H-bombs which can be carried aloft by lumbering bombers are as much an anachronism now as they were superfluous in 1950. But nuclear technology continued to construct itself on fantasies.

In 1950, Truman submitted a request for a billion-dollar supplemental appropriation for the AEC in that year, and a total expansion program costing five billion dollars. Before this expansion was completed, the AEC's budget would rise to three billion per year and its capital investment would soar to nine billion dollars. At its peak, the AEC consumed ten percent of the electric power produced in the United States, in a vast effort to bring into being the overwhelming nuclear might which most Americans assumed we already had.

Admiral Lewis Strauss, a member of the Atomic Energy Commission during its first two years and then chairman during the Eisenhower years, and Senator Brien McMahon, then chairman of the Congressional Joint Committee on Atomic Energy, both spoke in tones which were common at the time in describing what seemed to them the importance of the H-bomb development effort; in a conversation McMahon said that "if the Russians should produce the thermonuclear weapon first, the results would be catastrophic, whereas if we should produce it first, there was at least a chance of protecting ourselves." Strauss entirely agreed. Later, looking back on the decision to proceed with H-bomb development,

Strauss said, "We were able to test our first hydrogen bomb in November 1952; the Russians tested their first weapon involving a thermonuclear reaction the following August. The President's decision [to proceed with development in 1950] was not only sound but in the very nick of time. By so close a margin did we come to being second in armament, not only in the eyes of the world, but in fact. Had we begun our development after the successful Russian test, there is no reason to believe that we would have been accorded time to equal their accomplishment."

In 1952, at Eniwetok, in the Pacific, we exploded a device with the equivalent force of slightly more than ten million tons of TNT, which the United States allowed the world to assume was the test of an H-bomb; it was far too large an explosion to have been an atomic-bomb test. In reality, however, it was the test of an experimental system, not of a weapon, and not until *after* that test was the United States sure it knew the principles on which to develop a weapon. The first operational American H-bomb was actually tested in 1954. In the interim, in August, 1953, the Soviet Union exploded a device which Gordon Dean, then chairman of the AEC, later conceded was "more advanced in some respects" than the American test of the previous year. Knowledgeable people have described the 1953 Russian test as one of an operational H-bomb, and one source described it as an airdrop. The United States had no H-bombs which could be carried aloft in aircraft at this time; the 1952 device tested at Eniwetok was housed in a special building constructed for this purpose.

Whatever the situation in the Soviet Union, in 1950 the United States was vigorously preparing for warfare in Europe or in Asia. These preparations involved small, tactical atom bombs, not H-bombs. The once-miniscule atom-bomb assembly facilities at Sandia Base in New Mexico gave way to elaborate fabrication and assembly plants scattered across the country. The rate of testing of new weapons skyrocketed: in 1947, we had tested no weapons at all—we had none to test; in 1948, we tested three, prototypes of new strategic atom bombs, the first production weapons for the US; in 1949, we tested no new weapons; in 1950, we tested no weapons; but in 1951, we conducted *sixteen* weapons tests, a new generation of tactical weapons designed for battlefield use. And, for the first time, we tested a dozen bombs at the Nevada test site; all previous tests—except for the single war-time test at Alamagordo—had been conducted at remote islands in the Pacific. In 1953, we conducted eleven further tests in Nevada; and year after year thereafter, the number of tests remained high, as tests of production models began to supplement tests of prototypes.

The weapons developed in those years now stand on display at the National Atomic Museum at Kirtland Air Force Base, near Sandia. The museum is a shabby building, a three-story warehouse in the midst of the sundrenched base. The Air Force has not provided enough money to maintain the museum, but has urged the city of Albuquerque to assume support of this odd tourist attraction, the exhibits of which are slowly decaying.

To visit the museum, one obtains a visitor's pass from a bored airman in shirtsleeves and blue beret at a guard post at the entrance of the base. One is warned to stay on the main road. The museum itself is surrounded by a low wire fence. Across the street is a building in much better repair: brass letters describe it as the Nuclear Weapons School, and the emblems of all the services decorate its doorway. This is where officers today are trained in the design and use of nuclear weapons.

The museum which faces the school holds only the shells of past weapons, the slowly decaying emblems of earlier national fantasies. There is a propaganda film shown every hour, telling the romantic and untruthful story of our development of the atom bomb during World War II, full of heroes and entirely without women (even Lise Meitner, who discovered nuclear fission, has been omitted), with Albert Einstein, looking uncomfortable in front of a camera, reading slowly from notes, as if the words were unfamiliar: "Ee equals em, cee, squared."

The first weapons on display are models of the bombs which destroyed Hiroshima and Nagasaki, jocularly titled by their designers "Little Boy" and "Fat Man." Beside them, looking like exact duplicates of the round, finned Nagasaki bomb and evidently much like it, are the nation's first true weapons. The MK-5, about eleven feet long, weighing 3175 pounds. The MK-6, described as an "improvement": length ten feet, seven inches; weight 8500 pounds. These weapons were to be delivered by B-29 bombers or equivalent large aircraft.

A few paces further, we reach the early 1950s, and suddenly there are ranks and ranks of weapons of every conceivable shape, size and purpose. The Mark-12, to be carried by supersonic fighter aircraft; the Mark-8, for the Navy, "for use against such targets as underground command and communication centers"; the Mark-91 penetration bomb, a long slender shape which "will penetrate up to twenty-two feet of reinforced concrete" before exploding; the Mark-105 Hotpoint Walleye air-to-surface missile with sophisticated in-flight guidance systems, yield in the subkiloton (equivalent to less than a thousand tons of TNT) range; the Mark-57 "general purpose tactical weapon"; the Gente AIR-2A unguided air-to-air missile, for nuclear dogfights in the sky; the GAM-83 guided air-to-surface missile for use by carrier-based aircraft; SRAM, the short-range attack missile to be carried by fighter-bombers; Little John, an "unguided surface-to-surface rocket that is rail launched, can be transported by helicopter. Its range is equal to medium field artillery." The Hawk surface-to-air missile, "for army and marine corps troop defense." The Honest John rocket, designed "for close support of ground troops, range twenty-two miles. It is twenty-seven feet long, mobile launched." The Lacross "all-weather guided missile that supplements conventional field artillery." The Corporal missile, "an artillery support weapon," range seventy-five miles. The Pershing solid-propellant guided missile, kiloton explosives. The 280-millimeter cannon, the nation's first nuclear artillery piece, an enormous self-propelled cannon described as "long-range artillery." The Mark-23, a Naval sixteen-inch gun which is to be "fired from shipboard." The Mark-49 artillery projectile, just over two feet long and about six inches thick, "provides nuclear capability for the 155 howitzer." The Mark-33 projectile "provides nuclear capability for the eight-inch howitzer." The Davy Crockett: "close-support antitank used by the infantry. The projectile is fired from a man-portable or a vehicle-mounted launcher." On display are Tactical Atomic Demolition Munitions, the landmines of nuclear war, which use the Mark-30 warhead and the XM-113 case, this last item looking for all the world like a length of corrugated sewer pipe braced with four-by-fours.

This extraordinary armamentarium gives mute testimony to the military plans of the 1950s. The visitor imagines infantrymen carrying nuclear antitank weapons into battle, nuclear antiaircraft shells exploding overhead, nuclear troop-support missiles being fired from fighter aircraft, nuclear artillery softening up the positions against which the infantry advances. The armed services' vision of World War III.

At the outbreak of the Korean War, we were not yet prepared to fight such a tactical nuclear war; but by its end, we had become so prepared. When the enormous buildup of the 1950s was complete, we had deployed roughly seven thousand nuclear weapons in Europe, nearly all of them tactical atom bombs. The development of nuclear weapons has continued to emphasize small, tactical bombs and shells. The United States now has about 3000 warheads on land-based intercontinental missiles, and about 6000 warheads on submarine-launched missiles. Estimates of the number of tactical weapons range from 10,000 to 50,000, deployed all over the world. The intercontinental missile warheads themselves are small and growing smaller, as accuracy increases. Some may be H-bombs but they need not be. A few strategic H-bombs with the power of millions of tons of TNT are still carried aloft in the lumbering and increasingly anachronistic B-52 bombers.

Many billions of dollars were expended in this rapid development of the means to conduct tactical atomic warfare. Fourteen enormous plutonium-production reactors (nine at Hanford, Washington, and five at Savannah River, South Carolina) were built. Three huge new gaseous-diffusion plants were constructed at Oak Ridge, Tennessee, Paducah, Kentucky and Portsmouth, Ohio. Plutonium- and uranium-processing facilities, the factories for fabricating bomb components of various kinds, the factories where the bombs were assembled, all these were constructed in Idaho, Colorado, New Mexico, Ohio and elsewhere. Mountains were hollowed out to provide storage facilities for the bombs.

CHAPTER SIX

Atoms for Peace

IN 1953, THE US Atomic Energy Commission owned and operated three towns; it employed sixty-five thousand construction workers, five percent of the nation's entire construction labor force; it consumed ten percent of the country's electric power. The construction program then underway was to bring the Commission's capital investment to nine billion dollars, more than the combined 1953 investment of General Motors, US Steel, Du Pont, Bethlehem Steel, Alcoa and Goodyear. The three vast uranium enrichment plants under construction would consume more electric power than the combined production of the Hoover, Grand Coulee and Bonneville dams. Moreover, this power would be provided in large part by another government-owned enterprise, the Tennessee Valley Authority, which would have to expand its own power-production capacity to accommodate the military armament program.

The atomic-bomb program had become, for the first time, a major enterprise. It had become, in fact, the country's largest industrial enterprise, but it was owned entirely by the government. The intensive effort to combat Communism had called into being the largest socialist enterprise in the nation's history.

David Lilienthal, that champion of federal power, resigned from the Atomic Energy Commission in 1949, after less than two years as chairman, unwilling to devote himself to the purely military development which he found facing the Commission. For four years thereafter, the Commission's chairman was Gordon Dean, a Truman appointee, but far removed from the New Deal. Dean managed the vast military expansion with considerable ability and when, in 1952, he found himself at the head of a huge government-owned industry, he began urging an end to the federal government's monopoly of nuclear power.

Turning the atom over to private hands would not be easy, however, for only military facilities had been built. The reactors and uranium-enrichment plants had been designed to give the nation armaments in as short a time as possible. There was neither leisure nor inclination to stop to examine what, if any, nuclear facilities might later be of use to a civilian nuclear-power industry. With more time, the Commission knew, it would have been possible to design reactors which produced *both* electric power and plutonium; meeting the enormous demand for electricity at the uranium plants *and* supplying military plutonium simultaneously. But there had been no time to develop such reactors. The plutonium-production plants built at Hanford were scaled-up models of Enrico Fermi's uranium and graphite pile; and at Aiken, South Carolina, the Savannah River plant was being built in pursuit of the presumed requirements of H-bomb production.

Even as the military program expanded toward its greatest effort in 1952, private industry and the Atomic Energy Commission began to consider the task of turning the vast atom business over to private owners.

As early as 1951, Charles A. Thomas of the Monsanto Company had begun pressing for steps toward eventual private ownership of the nuclear enterprise. In 1952, three groups of power companies—led by Pacific Gas and Electric of California, Detroit Edison, and Commonwealth Edison of Chicago, all early investors in nuclear powerplants—reviewed the top-secret designs which had been drawn for reactors. The three groups reported to the AEC that none of the reactor designs could produce economical electric power, but that some, if used for military plutonium production *as well as* power generation, might prove feasible. The most attractive design then appeared to be the sodium-cooled fast-breeder reactor, which promised to generate large quantities of plutonium and operate at the high temperatures which utility executives thought necessary for economical power production.

Of course, very little actual research into reactor building had yet been done. There had been essentially no development work outside the crash military programs; the private power companies were simply guessing from paper designs and theoretical considerations. But their guesses made the task of turning nuclear energy over to private hands seem difficult and complex.

In the fall of 1952, Dwight D. Eisenhower and a Republican majority in Congress were elected to office; Gordon Dean left the Atomic Energy Commission, to be replaced by Admiral Lewis Strauss, an investment banker who had served on the Commission during its first three years. Harry Truman went back to Missouri.

Within weeks after taking office, Eisenhower proposed a vast new Atoms-For-Peace program, an updated version of the Baruch Plan which, in 1946, had offered the United Nations control of our nonexistent nuclear armory, subject to our right to inspect Soviet nuclear development. In 1953, Eisenhower offered the United Nations control of a nonexistent civilian nuclear-power industry. Plutonium and uranium were to be diverted from our military program to civilian purposes throughout the world, Eisenhower proposed, as a means of effecting nuclear disarmament in stages, without abandoning the enormous investment which had already been made in nuclear technology.

In the spring of 1953, the Atomic Energy Commission proposed legislation which would allow private industry to own atomic facilities (existing law would not permit this); a domestic counterpart of Atoms-For-Peace got underway immediately, with the proposal that private power companies owned by men named Dixon and Yates should build the new electrical generating plants required by the Atomic Energy Commission and which the TVA would otherwise construct.

In 1954, a new atomic-energy act was passed; as amended it gave the Atomic Energy Commission responsibility for fostering the development of a private atomic-energy industry. The law passed with little debate and with little opposition, except from a few stubborn public-power enthusiasts. The World Bank began organizing meetings among industrialists and members of foreign governments to expedite the financing of exports of nuclear technology; AEC Chairman Lewis Strauss began a strenuous effort to find private power companies willing to invest in nuclear power.

The first to be persuaded to do so was the Duquesne Company of Pennsylvania. The company expected its investment to be a loss, but agreed for patriotic reasons to contribute five million dollars toward the construction of the nation's first nuclear powerplant for the generation of electricity at Shippingport, Pennsylvania. The plant, which was owned by the government, consisted of a scaled-up version of the Navy's submarine propulsion plant. (Rumor still asserts Hyman Rickover had intended the reactor running this installation for the first nuclear aircraft carrier, which Congress then was stubbornly refusing to pay for.) In the following months, the Detroit Edison Company was prevailed upon to lead a group of utilities in building the Enrico Fermi Power Plant, the "breeder" reactor which, it was hoped, by virtue of its production of copious quantities of plutonium for the military, would be the prototype of the first economically viable nuclear powerplants.

Edward M. Spencer, an official of the Atomic Industrial Forum and of the Detroit Edison Company, at a 1956 meeting in Belgium, held to assist the development of private nuclear power in that country, remarked, "The passage of the 1954 Atomic Energy Act has made it possible in the United States to apply private-enterprise methods to a new and essentially full-grown business. This interruption, and I hope complete reversal, of a most disturbing trend towards socialism presents history's greatest challenge to our free-enterprise system."

The challenge lay in the fact that no profitable civilian use for nuclear power had yet been found. Power companies were markedly reluctant to invest their own funds in further development, and the existing technology seemed unlikely to be profitable. But the danger of government ownership overrode such considerations; and a reluctant industry invested its stockholders' money in yet another struggle to stem the tide of socialism.

II. GOVERNMENT POWER

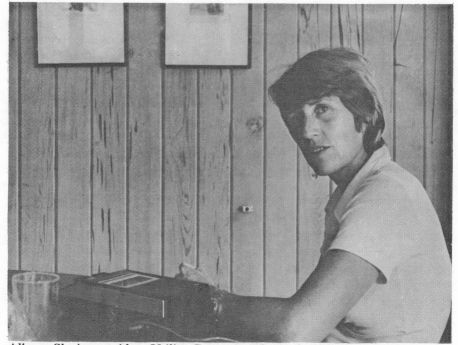
Alberta Slavin, president Utility Consumers Council of Missouri (Chapter 8)

Reverend Eleazar Wheelock, founder and original charter-holder, Dartmouth College (Chapter 9)

Daniel Webster (Chapter 12)

Former Chief Justice John Marshall (Chapter 14)

David Newburger, legal counsel to the Utility Consumers
Council of Missouri (Chapter 17)

William Anders, former chairman, US Nuclear Regulatory
Commission (Chapter 19)

Russell Train, administrator, US Environmental Protection
Agency (Chapter 21)

Frank Zarb, administrator, Federal Energy
Administration (Chapter 22)

CHAPTER SEVEN

Heartland

JEFFERSON CITY, capital of Missouri, is a small town in the center of the state. Approached from the north, across the Missouri River, the capitol building seems to rise abruptly from the cornfields. Visitors often stay at the old Governor Hotel, a square brick building near the governor's mansion, which charges seven dollars a night for a single room. The leather furniture in the hotel lobby and the old-fashioned steam radiators in the rooms speak of a time when women guests were more likely to be paid than paying. The Ramada Inn is now the fashionable place to stay and dine, however, and there the rooms are modern and somewhat more expensive.

Between the two hotels stands an office building that houses many of the state agencies, including the state Public Service Commission. The Commission's large hearing room is on the tenth floor and would have an attractive view of the surrounding countryside if the windows were not blocked with heavy drapes. The room is very much larger than the hotel rooms of the Governor—it holds benchs for perhaps a hundred spectators—but its wooden panelling somehow seems to conceal the smoke-stained walls of Missouri's steam-boating days. The ceiling has been lowered to make room for elaborate artificial-lighting facilities to replace the sunlight from blocked-off windows. Across the far end of the room stretches a raised judicial bench, behind which is a row of high-backed leather swivel armchairs for the five commissioners. A witness stand is at the left of the bench, and arranged in courtroom fashion before it are tables for opposing counsel and, behind them, the ranks of wooden benches for spectators. The lowered ceiling and the draped windows make the room oppressive, and somehow the setting fails to convey the august judicial atmosphere that was intended. Smoking is forbidden, and so spectators frequently step through the wide doors of the hearing room into the gray hallway to smoke or talk.

For several weeks during the fall of 1974, the Missouri Public Service Commission heard testimony in this hearing room on the question of nuclear power. The occasion of the hearings was the application of the Union Electric Company for a permit necessary to construct a nuclear powerplant.

The Union Electric Company, with headquarters in St. Louis, sells electricity to people and businesses over an area of nineteen thousand square miles in Missouri, Illinois and Iowa; through a regional network of utilities it buys and sells power to companies throughout the south-central states. Its assets in 1975 were well over $2 billion, and its operating revenues were about $500 million during the year. Even for a company this large, the nuclear power station it was proposing was an ambitious undertaking. There were to be two nuclear plants with common transmission facilities at a total cost of $1.7 billion. These would be by far the most expensive powerplants the company had ever built: their combined cost would be almost equal to the company's entire existing capital investment. But the plants would be capable of generating only about one-third of the company's power output in 1975. Union Electric hoped to offset this enormous capital expense by savings in fuel cost over the life of the powerplants.

The task of designing and building the plants was beyond Union Electric's own resources. The two reactors had been ordered from the Westinghouse Electric Company as part of a package, in which four utility companies had joined to order six identical nuclear systems. The design was to be done by Westinghouse, which would also handle the complex task of preparing environmental and safety documents; construction was the responsibility of an architectural-engineering firm, Bechtel; and still another private firm had been retained to oversee the work of equipment manufacture and plant construction. The General Electric Company was to supply the turbines and generators—the nonnuclear portion of the plants—and still other contractors and subcontractors would be engaged for various parts of the task. The over-all design of the six nuclear plants, including only generic items, such as the general design of the reactor itself, filled ten fat loose-leaf volumes. The environmental statement, describing factors and design items which were unique to the two plants planned by Union Electric, was equally long and detailed.

The nuclear power station was to be called the Callaway plant and, if built, it would be the first nuclear power station in Missouri. The chosen location was in a rural area just north of the Missouri River, about twenty-five miles east of the state capital and eighty miles west of the state's largest city, St. Louis. Because it would be outside Union Electric's usual service area, the plant could not be built without a certificate from the Missouri Public Service Commission stating that the plant would serve public convenience and necessity.

When Union Electric first announced its plans in 1971, there was little public response; the plans were both vague and distant. In 1973, the company said that it had already acquired the land for such a plant in Callaway County and had ordered two reactors from the Westinghouse Company. The cost of the plants was estimated at slightly over one billion dollars—a figure that would increase rapidly in months to come. The plants were to be capable of generating 2.2 million kilowatts of power—about one-third the company's capacity at the time—and would go into operation between 1981 and 1983. The plants would be large enough to supply the domestic power needs of a city of two million people—roughly the size of the St. Louis metropolitan area.

The enormous size of the investment and the imminence of construction focused attention on the company's plans. Local news media had in the previous few years given increasing attention to the national debate over the safety of nuclear power, and while this debate had not yet become a matter of local concern or general awareness in Missouri, there were a few people and organizations that immediately expressed opposition to Union Electric's plans. The first protest came from Kansas City on the Missouri-Kansas border. The Kansas City Power and Light Company was one of the utilities which had joined Union Electric in purchasing nuclear plants and Kansas had been sensitized to the issue by earlier aborted attempts to locate a radioactive waste-disposal facility in the state. The 1973 powerplant announcement was therefore followed by formation of the awkwardly named Mid-America Coalition for Energy Alternatives, a small group led by a slender young woman named Diane Tegtmeier and a bearded young man named Ron Hendricks. While Hendricks accumulated scientific information and support, Tegtmeier quickly built a membership list and began urging citizens' groups throughout Missouri to join in opposition to powerplants being built there and in Kansas.

One of the groups that talked to Tegtmeier was in St. Louis: the Utility Consumers' Council of Missouri, led by Alberta Slavin, who had achieved wide publicity in a variety of disputes over consumer issues. The UCCM devoted much of its organization's modest resources to contesting rate increases by the electric, gas and telephone utilities which serve eastern Missouri. Slavin's group had long discussed its attitude to Union Electric's nuclear plans and, early in 1974, announced opposition to the plant on the grounds that nuclear power was expensive, unsafe and unnecessary. Rising power prices and increasing conservation efforts would eliminate the demand for power which Union Electric was planning to meet, the group argued.

Some months later, the Coalition for the Environment, a loose alliance of conservation organizations in St. Louis, announced its own opposition to the plant, on similar grounds, but emphasizing the radiation hazards and environmental problems which the Coalition feared the plant would create. The two organizations met informally to discuss a joint strategy for opposing the plant and agreed that their limited resources would not permit duplication of effort. The UCCM, because of its experience in state proceedings, was to oppose Union Electric's application to the Missouri Public Service Commission; the Coalition was to oppose the application for federal licenses.

CHAPTER EIGHT

A Protagonist

ALBERTA SLAVIN lives with her husband, a physician who has been active in environmental work, their children and pets, in a very comfortable house on a large wooded lot near St. Louis. She is president of the Utility Consumers' Council of Missouri, a vice-president of the National Consumers Congress, and active in a variety of consumer and environmental organizations. She is a strong-featured, suntanned woman in her forties, with dark blonde hair cut straight and short. She smiles often and talks easily about herself and her work. She is undecided as to a political career; later, in April 1976, she will enter the race for lieutenant governor—a post whose only function is to serve as a step to higher office. While this first hastily organized effort is not expected to be successful, Slavin is seen as a likely candidate to eventually become the first woman to hold state-wide office in Missouri. She has not yet abandoned her nonpolitical role as a consumer activist, however. At the time of this interview, the Utility Consumers' Council of Missouri claimed about a thousand members and vigorously opposed the construction of a nuclear plant.

When was the Utility Consumers' Council of Missouri organized?

SLAVIN: In December of 1970.

You have been the head of it?

SLAVIN: *(Laughs.)* I don't believe in democratic structures.

UCCM isn't the only group you work with?

SLAVIN: One is "Housewives Elect Lower Pricing". . . . Our survey, not even knowing what the term meant, uncovered "price discrimination" in food in five stores located in the city. Our data [helped to bring] a congressional hearing to the St. Louis area in price discrimination. That was the first area that I was into. Then . . . we began to realize that food was becoming almost minor [compared] to people's utility bills. In the seventies we began to see what had happened in that area. And that people were not really represented by the regulatory agency.

We announced the formation of the group [the Utility Consumers' Council of Missouri] and then, our first formal action was shortly after we were started. Now, I don't really, I can't even call it an intervention. We were involved in the proceedings in opposition to the rate increase for the power company, but we were not able to have our own lawyer represent us, so the Public Service Commission attorney represented us. We put in my testimony and depended on them to cross-examine. That one was our most successful effort. That case was for twenty million dollars and they were awarded five [million] and, I think, it was the first time we brought up the interests of *residential* customers as a group in a rate case.

You say "we" when you're talking about all this. Who was involved?

SLAVIN: Well, at that time, our level of organization involved sort of a core group of volunteers who were set up when we principally moved into handling complaints, so that I'd have a woman who would handle the phone company complaints, and a woman who'd handle gas complaints, a woman who handled the electric complaints. We used the complaint process to become aware of the structure, and it is a very effective way. There's no better way to learn how a utility operates than to start taking complaints.

In each utility we had a contact person, and we continue to have a contact person in the utility, where we take a complaint and intervene on behalf of the customer who's having the problem.

Well, what we began to discover is that there are some things we can correct and some things that we can't. We bumped up against the fact the filed tariffs allow a certain *modus operandi* on behalf of the utility. The filed tariffs have been approved by the Commission and they are the working regulations of the company. Let's take a, you know, like a deposit, a security deposit. The tariff filed before the Public Service Commission permits every utility in the state of Missouri to charge a deposit which is twice the amount of the highest monthly bill. It doesn't say the *highest* monthly bill on the tariff, but that's the one they choose. All right? Not only that, they can terminate a person's service whose bills may be paid in full, over lack of payment of deposit. They can refuse to reinstate service over a deposit. Well, when I learned this I couldn't believe it at first, you know. You run up against the hard fact that you either have to change the tariff, or operate within it and try and get them to be somewhat sensitive to the problems. . . . The Commission now has a billing practice or procedure before it, now, this is how many years later? They have a billing practices procedure before it, in which they will make some changes or propose some changes. Not all acceptable to us, but we will at least have an input into them. The best we were able to work out was an arrangement of deferred payments of the deposit: work out a system of installments for them so that they aren't suddenly zapped with their bill, plus a hundred dollars, which they're supposed to come up with overnight and it's usually a person who may be unemployed who has a most difficult problem to come up with the cash.

The UCCM represents a class of consumers. Who are they? How do you visualize your constituents?

SLAVIN: Well, we carved out a goal which said that our constituency is residential customers of utilities. I'm getting more and more requests, as the economy faces its crunch, to represent smaller commercial interests that are really getting hit hard, but we've continued to focus on the [residential]. . . .

I'm very sensitive to the problem the poor customer has, and a lot of our testimony has directed itself to the fact that the poor customer is hit the hardest by this [rate increases]. I have been unwilling to say that we are only in it for the poor low-use customers, because I think residential customers are treated by the utilities as a group and we should consider them our constituents. If we don't, we run into the divisiveness that the utilities are very quick to [exploit] in our economy. You know, if you really are sensitive to the problems of the poor, [you will hear that] they're deadbeats, they don't pay their bills and you, you white middle class out there, are paying your bills and subsidizing them. So you set up a friction between the classes which is very nonproductive. I know, when I get a break for the first 250 kilowatt hours, that I'm helping the guy who lives in the old house in the city who needs a fan, a refrigerator and a lightbulb. And I also just helped Mr. [Charles] Dougherty who is the president of the company [Union Electric]. Who's earning $180,000 to $200,000 a year. Right? He didn't want my help, but he got it! The other thing is, we know we can't help each person with a complaint, so we take the complaint, and there are massive complaints in certain areas, to change things which hurt a whole lot of people.

So the Utility Consumers' Council really began very much as a consumer-service organization, oriented toward prices.

SLAVIN: Yeah, I would say that that's really true, although really it would be unfair to separate my own interests in environmental issues. I was under the impression, at one time, that in some respects we could have more impact through the consumer front on issues than environmentalists have.

We started out on a pretty much dollar-and-cents-relief-for-the-residential-consumer kind of approach. In just reading the testimony, the first case, on Union Electric, I saw what had happened for years with no representation in a rate case, in which no one was really worried about the residential customer as a class. We realized that the industrial customer had gotten an enormous break in prices and is continuing to do so.

When you looked at the rates, you found that industrial consumers weren't carrying their share.

SLAVIN: It leads you first to the notion that we're building a lot of power-plants to serve the industrial customer, and that they're very costly. Now I don't know when that realization began to hit me. You realize the tremendous capitalization of [electrical] generation which is going to a very small number of large customers, and there hasn't been much discussion of efficiency. . . .

From looking simply at rates and trying to bring down rates for some residential customers, you began examining the way the company was managed?

SLAVIN: That's correct. In their decisions on rate increases and so forth. And their decisions on what their demand was, what their growth picture was, and you know, we now have a set of three rate cases behind us and another one pending. And essentially, they could almost refile the same material year after year after year with a different set of figures. And that's almost what happens. And if you defeat one thing, it'll show up in the next round. So that there has been what I would call just a minor shift, despite our efforts in trying to educate the Public Service Commission to these matters, and to try and reach the company with some idea of change; they have been pretty well unresponsive in the filed material.

We have gotten, what I would say [are] minor victories, which is, no increase for the first 250 kilowatt hours and a change in the rate structure itself. They used to have a minimum of 40 kilowatt hours, now they have a minimum of 100 kilowatt hours. That was beneficial to the poor customer, the small user, the fixed-income customer. We got a small victory for low-use customers, and on the last two cases, not the very last one, we actually had no increase up to 250 kilowatt hours. Of course, there are proposals now being tossed around the country, the Lifeline proposal which attempts to do the same thing. A certain charge for a certain number of kilowatts. We didn't call it a "Lifeline rate," but that's essentially what it is.

Would it be difficult to decide on rate increases without evaluating the management of the company?

SLAVIN: Or thoroughly cross-examining and finding out whether their projections for demand are accurate. The whole system of utilities, electric, gas and so on, has been a promotional business. You have to sell, right? And I think that public-service commissions all over the country have endorsed the concept of growth and sell. They have never questioned. . . .

They have not inserted themselves into decisions on where powerplants are built, what kind of powerplants they are, where they're located. It was, you know, extraordinary for us to be able to get a full-blown hearing on the nuclear-plant siting. We got that through a settlement of a rate case.

As a result of this process of getting further involved in the management of the company, you looked at their decision to build a nuclear powerplant?

SLAVIN: Uh-huh. I should point out that I'm also a stockholder.

You're a stockholder?

SLAVIN: Union Electric. And you find that when you go to a stockholders' meeting, that you get a lot of different information than you get in rate proceedings, and I would urge anybody who wants to get into this to spend a little money to buy some stock in the company, because it really is interesting. The first stockholders' meeting I went to was back in '71 and Mr. [Charles] Dougherty [president of Union Electric] at that time said, "Well, we're in really bad shape," he said, "but when the Public Service Commission authorizes our twenty-million-dollar increase we will be able to provide better earnings for our customers".

In my testimony I pointed out that Mr. Dougherty didn't say *if* he was awarded an increase, but *when* he was awarded an increase, implying he had no question the Public Service Commission would award him everything he asked for . . . .

I first learned of the intention to build a nuclear powerplant at the stockholders' meeting which occurred in, I guess, 1971. The next year I came back to the stockholders' meeting and suddenly there wasn't any word about the nuclear powerplant. It just, like, disappeared. And then everybody seemed to be dodging whether they would or wouldn't build a nuclear powerplant—that's when they were involved in their land acquisition. And a month and a half later they announced the acquisition of the land in Callaway County, despite the fact that in answer to a direct question from myself and other stockholders, they denied any definite decision at that point. So it's been moving sort of inexorably since that time. . . .

Let me ask you very bluntly, who elected you to represent the general public?

SLAVIN: Nobody.

The things that you do affect the whole populace, not just the public of Missouri, but people all over the country. You're trying to interfere with a multibillion-dollar company and change its management.

SLAVIN: I have been not so graciously called a "self-appointed watchdog" by people who don't like what I'm doing.

Well, that's accurate, isn't it?

SLAVIN: That's correct.

Don't you feel . . .

SLAVIN: It was a dumb thing to do, it brought a whole lot of work! Well, does it bother me? It bothers me only from a standpoint that I feel an obligation, a big sense of obligation to be accurate in what we're doing and responsible for what we're doing. Anything that we've done has not been based on emotional appeal, but [on] an effort to sort through the materials and the facts to arrive at a conclusion that we think is accurate.

If there were to be an election in Missouri or St. Louis, as to whether Union Electric should build a nuclear powerplant, how would it come out?

SLAVIN: Right now?

Yeah.

SLAVIN: It would probably come out in favor of building the plant.

CHAPTER NINE

Dartmouth College

D ANIEL WEBSTER was thirty-six when he argued the Dartmouth College case before the United States Supreme Court. He had already been in Congress for five years, where he had begun to acquire a reputation for oratory; for four years he had practiced before the Supreme Court. A portrait from that period shows him as a thin man who looked even younger than his age, with a high forehead and unruly black hair. As a young man he seems not to have been remarkable in appearance (although he was tall and broad-shouldered), but he was possessed of an unusual personal magnetism on which most later descriptions dwell. He said of his origins, "Gentlemen, it did not happen to me to be born in a log-cabin; but my elder brothers and sisters were born in a log-cabin; raised amid the snowdrifts of New Hampshire, at a period so early that, when the smoke first rose from its rude chimney, and curled over the frozen hills, there was no similar evidence of a white man's habitation between it and the settlements on the rivers of Canada."

But this was said after Webster had campaigned for the presidency, in an era in which a log-cabin birth had already become a considerable political asset. In fact, Webster's boyhood was not entirely rustic. His father, Ebenezer, had served in both houses of the state legislature and was appointed Judge of the Court of Common Pleas. The salary attached to this position enabled the elder Webster to send his son Daniel to Exeter Academy, where he apparently learned little, for a further course in the classics was necessary under the tutelage of the Congregational minister of Boscawen, New Hampshire. But at last Daniel was prepared for college and, in August, 1797, he entered Dartmouth; the requirements for admission in those days were "six books of the Aeneid, a part of the Greek Testament, and arithmetic."

Dartmouth College, in whose defense Webster was to become famous, began in 1754 in Lebanon in what was then the colony of Connecticut. The Reverend Eleazar Wheelock in that year had established, at his own expense and on his own land, an Indian Charity School, at which he "maintained and educated a number of children of the Indian natives, with a view to their carrying the gospel in their own language, and spreading the knowledge of the Great Redeemer among their savage tribes," as the Supreme Court described it later. The school prospered and Wheelock was encouraged to seek more substantial support than was then available in the provinces. A friend, the Reverend Nathaniel Whitaker, evidently an able fund-raiser, travelled to England for Wheelock, taking along as a kind of exhibit the Reverend Sampson Occom, an Indian minister educated at Wheelock's school. The trip was successful; Whitaker was able to secure the patronage of the Right Honorable William, Earl of Dartmouth, and a parcel of jurists and solicitors.

The substantial contributions made by these worthies were put into a trust by Whitaker, acting under a power of attorney given him by Wheelock; with the financial well-being of the projected college assured, he was able to obtain the promise of extensive tracts of land in western New Hampshire, given on the proviso that the school be settled there. The governor of New Hampshire, as a further inducement to moving the well-endowed school from Connecticut, offered a charter of incorporation which would put the college on an equal legal footing, at least, with the great colleges of Britain.

The charter of incorporation was granted by the governor of New Hampshire, in the name of King George III, on December 13, 1769. To a modern eye, it is a curious document. While the college so incorporated was a private institution established with private funds, it had a strong resemblance to a governmental body. Its twelve trustees, who together constituted the corporation, included the governor of the Province of New Hampshire, the president of the provincial council and the speaker of its house of representatives; an official of the colony of Connecticut; Dr. Wheelock; and seven ministers of the gospel, presumably chosen by Wheelock. This group was declared by the charter to be "henceforth and forever, a body politic" and was given the power to own and sell property "in as full and as ample a manner . . . as a natural person, or other body politic or corporate. . . ." The trustees were required to take the oath of allegiance to the king which was required of public officials. The trustees were empowered to make "such ordinances, orders and laws, as may tend to the good and wholesome governance of the said college. . . ." Furthermore, ". . . such ordinances, orders, and laws . . . we [the king] do for us, our heirs and successors, by these presents ratify, allow of, and confirm, as good and effectual to oblige and bind all the students, and the several officers and ministers of the said college."

The trustees of the college had the power of a government, so far as the inhabitants of the college itself went; this was perhaps natural enough for a college in a wilderness. But the charter also was made reflexively binding on the king and his heirs and successors; the king gave up the power to alter the charter of the college or to meddle in its governance. Furthermore, no future king could, under the law, alter or change or withdraw the charter of the college.

The college, now renamed to honor its patron, the Earl of Dartmouth, moved across the Connecticut River to New Hampshire and prospered. From educating the natives the college turned to training missionaries and, by slow degrees, to providing a proper religious education to the children of the colonists. Revolution swept the colonies but left the college untouched. New Hampshire became a state and joined the United States under the new Constitution, but the college went on as before, governed by its trustees and the successors they appointed for themselves. The Rev. Dr. Wheelock, as the charter provided, served as president until his death and was succeeded by his son. The self-perpetuating government of the twelve trustees took on the look of a self-enclosed barony within the larger republic. The college had become a kind of metaphysical being, incarnate as the trustees of Dartmouth College. The education the college provided—a smattering of the Latin classics, some Greek, and exercises in a debating society—was that designated by the trustees in continued imitation of the landed gentry of England. But the charter of the college, inviolate to the king and his successors, was held to be protected from any change by the new state legislature.

The situation was irritating to the Republican sentiments of New Hampshire inhabitants and to others. Thomas Jefferson, who assumed the presidency in the same year that Daniel Webster graduated from Dartmouth College, wrote to his friend and fellow Republican, the governor of New Hampshire, William Plumer: "The idea that institutions established for the use of the Nation cannot be touched nor modified, even to make them answer their end, because of rights gratuitously supposed in those employed to manage them in trust for the public, may, perhaps, be a salutory provision against the abuses of a monarch, but it is most absurd against the Nation itself. Yet our lawyers and priests generally inculcate this doctrine."

The charter of Dartmouth College continued to rankle until, in 1816, the New Hampshire legislature finally gathered its courage and passed a statute which attempted to alter the charter of the college. The bill purported to enlarge the college to a university, to add schools and funds, and to enlarge the number of trustees from twelve to twenty-one. This was most politely and hesitantly done. The twelve existing trustees were retained, although Federalists all. The nine new trustees appointed by the governor, and presumably Jeffersonian Republicans, would not have voting control. But the bill provided that future vacancies among the trustees were to be filled by the governor alone rather than by the trustees, so that gradually the new university would come under state control.

Daniel Webster, in the meantime, had risen rapidly. His extraordinary personal presence had secured him a good position as an apprentice with a leading Boston law firm. Admitted to the bar, he returned to New Hampshire to practice law and was soon elected to Congress.

Washington, DC, the new seat of the government, was then still largely a matter of mud and forest; few attorneys could or would travel from distant states to appear before the Supreme Court (which met in a shabby room in the basement of the Capitol), and so it was the custom that a state's representatives in Congress would argue a state cause before the Supreme Court. Webster, as congressman, joined the small circle of attorneys in Washington who practiced before the Supreme Court and began to make a name for himself.

The trustees of Dartmouth College, his alma mater, were to serve as his first really important clients. They did not choose to accept New Hampshire's efforts to dislodge them, no matter how gently. At first they simply obstructed. Then they fought. They hired attorneys and brought suit in the state's courts, demanding that the college's books and premises be turned over to them. And so began "The Trustees of Dartmouth College v. Woodward," a lawsuit which had the most profound and unexpected effects on United States history. The final decision in the Dartmouth College case established the corporation as the preeminent form of business enterprise in the nation and clothed it with extraordinary constitutional protections. Sir Henry Maine called the decision in the case "the bulwark of American individualism against democratic impatience and socialistic fantasy." One and a half centuries later, the arguments in the case would still be of considerable interest.

CHAPTER TEN

States' Rights

THE MISSOURI PUBLIC SERVICE COMMISSION hearing on nuclear power in the fall of 1974 was curious in many ways. All of the participants seemed to acknowledge that the hearing was a kind of ritual whose outcome was predictable. The members of the Commission repeatedly complained that they were being asked to decide matters outside their competence; the chairman was visibly impatient with the hearings and kept referring to other important business which was being delayed. At the end of one evening session, he shook his head and said that "Fifty million dollars in rate cases are pending," as if these overshadowed the two-billion-dollar proposal before the Commission.

At first glance, the proceeding before a Missouri agency seemed to make little sense. Union Electric was one of four utilities in four states, buying six nuclear powerplants from a manufacturer in Pennsylvania. The nonnuclear portions of the plant would be built by a company in New York, and the construction would be conducted and supervised by two firms based in still other states. Financing would be obtained in national markets. The US Atomic Energy Commission would approve the design and license the construction and operation of the facilities. If built, the nuclear plants would be part of a six-state power pool, and would supply power to a company that served customers directly in three states and indirectly in all six. The Missouri Public Service Commission's staff was admittedly inadequate to evaluate the design of the plant in even its most general aspects and the Commission was forced to rely on the utility company—which in turn was forced to rely on Westinghouse—for all information concerning the design and cost of the plant. The Commission was also dependent on Union Electric for the most elementary information about the demand for electric power in the region. The contracts with Westinghouse for the reactors and uranium fuel were secret and were never given to the Commission.

Yet, the Missouri Public Service Commission had the power to say "yes" or "no" to this two-billion-dollar plan of Union Electric. The decision conceivably could mean success or bankruptcy for the company, with vast repercussions throughout the Midwest and elsewhere. A refusal would have affected the power generating capacity of the entire Midwestern power pool and might have drastically affected the value of billions of dollars of bonds and stock held by investors throughout the country. (Earlier in the year, the Consolidated Edison Company of New York had omitted the payment of a single quarterly dividend and utility securities across the country had immediately declined by ten billion dollars.) The five men of the Missouri Public Service Commission seemed to have extraordinary power over matters affecting the nation as a whole. And yet, there was something in the manner of the hearings and the attitude of those present which belied this power. Even those in opposition to the company's proposal seemed to feel that the Commission simply *could not* refuse the request before it.

There are a number of familiar influences, of course, on a state agency. Some of the members of the Commission may receive favors from utility companies;

Commissioner William Clark was forced to resign during the nuclear hearings because of revelations that he had been entertained by the Southwestern Bell Telephone Company, which was regulated by the Commission. But such venality belongs to an earlier day; the bright young men appointed in more recent years to the Commission, in reward for their campaign efforts, have generally not been tainted by such suspicions. They do, however, share some community of interest and background with those they regulate; they are all, of course, white male Protestants of local upbringing and education. The commissioners are not men of wealth, but neither are most of the utility employees who appear before them. They have ample opportunity to meet and discuss mutual problems.

The staff of the Commission is severely hampered. It relies entirely on the companies it regulates for information. The limited salaries of state employees are another difficulty—able attorneys and technical staff find that their only route to advancement is through the private companies they face in a supposedly adversary setting. There is thus a great temptation to soften one's attacks during the time of government service and, for the most able individuals, to move on to private companies as quickly as possible. But the day of the outright bribe to the Commission or its staff seems to be past, and even the prospect of future employment with a utility firm is less important than it once was; government service at the state and federal levels is increasingly rewarding.

At a deeper level, the Commission is chosen by a governor and confirmed by a senate, politicians who are unable to campaign or win elections without business contributions; and the utilities, through their officers and directly with corporate funds, are heavy contributors to state elections. Company lobbyists maintain the relationships established at campaign time. Early in 1975, Missouri Governor Christopher Bond nominated a man to replace William Clark, but the senate refused to confirm him—ostensibly for his lack of a law degree (although non-lawyers had often sat on the Commission), but in truth because he had once headed a federally-funded organization which had fought, quite modestly, for the rights of the poor. It is highly unlikely that anyone critical of the business community in general, or of utilities in particular, could be nominated or confirmed to the Public Service Commission in Missouri.

But fundamental sympathy does not preclude an independent exercise of judgment; the Public Service Commission was called on by the law to make its own decisions as to the necessity for a new nuclear powerplant, and the importance of the decision was certainly not lost on the members of the Commission. Despite their limited access to information and their necessarily limited grasp of the issues involved, the five members of the Commission were required to arrive at an honest and independent decision. In Missouri, as elsewhere, the campaign laws have been reformed. Campaign contributions by corporations have been exposed and presumably limited. Those like William Clark who had a more casual attitude toward accepting private favors have been encouraged to leave public life, and the members of the Public Service Commission and its staff were as honest as one finds men elsewhere. They were aware of the importance of the decisions they were to make and the intense scrutiny to which these decisions would be subject.

But there was still no practical likelihood that the Commission would depart from the decisions the utility companies asked it to ratify. The relationship between them was and is far more intimate and complex than that created by campaign contributions and lobbyists. Like all other such state agencies, the

Missouri Public Service Commission is in large part a creation of the utilities, and, unless this fact is understood, the utility industry and the manner in which it is regulated will be incomprehensible.

CHAPTER ELEVEN

The Judges

HENRY KENDALL is a tall, slender, blond professor of physics at the Massachusetts Institute of Technology who speaks in the patrician accents of Boston and who has come to Jefferson City, Missouri, to appear in opposition to the nuclear powerplant which is proposed by the Union Electric Company. He has flown from Boston at the request of the Utility Consumers' Council of Missouri, despite his own reservations concerning the usefulness of a hearing before a state public-service commission on the national issue of nuclear power. But Dr. Kendall devotes his spare time to criticism of the nuclear-power industry; indeed, he often seems to be the only physicist not employed by that industry who has taken the trouble to learn enough about it to criticize the technology in use.

Kendall's credentials as an experimental physicist are impeccable. As a physicist he has also worked on many secret military projects; he has served as an advisor to the Department of Defense and as a member of the JASON group whose assistance on weapons development has been much criticized by liberals—he is by no means a radical critic of military uses of nuclear energy. However, he is a man of independent means, which apparently allows him to travel around the country and to appear as an expert witness in various proceedings without receiving any payment beyond his expenses. With an associate, Dan Ford, Kendall carried the burden of developing evidence in a months-long Atomic Energy Commission hearing which resulted in a modest tightening of the regulations affecting some key safety features of nuclear plants.

The Missouri Public Service Commission rarely receives out-of-town witnesses at its hearings. Kendall was invited by UCCM's attorney, David Newburger, a professor at Washington University in St. Louis. The Union Electric Company has hired a Washington law firm which specializes in such proceedings to conduct the cross-examination at this hearing, in part because of the appearance of out-of-town expert witnesses, and in part because Union Electric's own counsel seems to know very little about nuclear power. The cross-examination is handled for the company by Gerald Charnoff, a former employee of the Atomic Energy Commission, who now represents the firms whose activities he once helped to regulate. The five members of the Commission and the staff of Union Electric have watched with some bemusement as this exotic imported attorney exercised a ruthlessness rarely observed in the placid hearing room of the state agency. With Kendall's appearance, however, the commissioners take a more active part in the questioning. William Clark, a bearlike man from the southern bootheel of the state, a former slave-owning region that is still planted to cotton, rolls in his reclining chair at the far end of the Commissioners' bench from the witness and mutters to himself and to the reporters who sit at that end of the room his various dark suspicions about the nature of Dr. Kendall's organization, the Union of Concerned Scientists; Clark audibly suggests that a locally famous hunter of Communists would be useful in the present circumstances. Other members of the Commission are more restrained but apparently feel that the real issue of the hearings is finally before them.

CHAIRMAN JAMES MAUZE: Let's go on the record. This is a continuation of Case No. 18,117. As of last evening Mr. Clyde Allen [of Union Electric] was on the stand, and so I presume—

MR. DAVID NEWBURGER: Your Honor, we have Dr. Kendall here this morning. He was supposed to take the stand.

CHAIRMAN JAMES MAUZE: Oh, that's right. I apologize. Are there any preliminary matters to come before the Commission before we start our examination of Dr. Kendall?

MR. GERALD CHARNOFF: I would like to raise one, either on or off the record. These lights [for television cameras] are going to produce an awful lot of heat in this very small room. I don't know if it's customary to have these kind of lights in your hearings or not, but they are burdensome in a small room.

CHAIRMAN JAMES MAUZE: If they were here all day, I would agree. But, I don't think they're going to be on very long.

CAMERAMAN: I would hope to have them off in ten minutes.

CHAIRMAN JAMES MAUZE: Under the procedure that we have adopted, there is one witness who has to testify somewhat out of order. Mr. Newburger, would you call your witness, please?

MR. DAVID NEWBURGER: Dr. Kendall, would you please take the stand.

CHAIRMAN JAMES MAUZE: Would you raise your right hand to be sworn? (*Witness sworn.*)

MR. DAVID NEWBURGER: Would you please state your name and address for the record?

WITNESS KENDALL: My name is Henry W. Kendall. I live at 221 Mount Auburn Street in Cambridge, Massachusetts.

MR. DAVID NEWBURGER: Are you the Henry Kendall who prepared testimony and had it filed in this case, a document of twenty-three pages?

WITNESS KENDALL: Yes, I am.

MR. DAVID NEWBURGER: Do you have a preliminary statement that you would like to make for the Commission?

WITNESS KENDALL: I have a short statement, yes.

MR. DAVID NEWBURGER: Would you please proceed with that?

MR. GERALD CHARNOFF: Is this in the form of a summary, or is it in the form of additional direct?

MR. DAVID NEWBURGER: It's in the form of a summary, as I understand it. Is that right, Dr. Kendall?

WITNESS KENDALL: That's right. The general purpose of my testimony goes to the question of whether it's presently prudent and whether it's a good idea to continue construction of nuclear powerplants in this country. There is at the present time a major controversy over a number of issues with regard to safety in the nuclear program. They generally fall into three categories, and there is detailed discussion of these in my written testimony. They are, first, the question of plant accidents and the risks that are posed to the general population and to the economy from some kind of an inadvertent event at an operating nuclear plant.

CHAIRMAN JAMES MAUZE: How many inadvertent events have you discovered or have occurred?

WITNESS KENDALL: Well, there have been a very large number of accidents of various kinds. I can describe those at more length. But, what they really relate to is the possibility of an inadvertent massive release of radioactivity. This can occur by a large number of different mechanisms.

CHAIRMAN JAMES MAUZE: I didn't mean to interrupt you. Why don't you go back to your summary.

WITNESS KENDALL: The second area has to do with the storage of radioactive materials, the so-called wastes, after the fuels have been removed from the reactor. These materials are biologically enormously toxic, and they have no further use to society, and they have to be sequestered away, kept out of the biosphere, for absolutely prodigious lengths of time—time measured in multiples of a hundred thousand years. The technology at the present time is simply not able to deal with the question of this kind of archival storage. It's an open question, and it remains a very nagging question.
Lastly is the area referred to generally as the safeguards area, and this encompasses a number of different items. But, it's generally related to the fact that there are materials in the nuclear program from which you can make nuclear weapons. And it's now been established by weapons designers and people who are familiar with the field that the problem of a terrorist group constructing a nuclear weapon hinges principally upon whether they can get the materials. . . . This, essentially, means stealing them in some way from the nuclear program. Then the ease with which they could construct a nuclear weapon is relatively high. And this poses and it's believed to [pose], in the future even more, a very considerable threat.
All of these issues hinge on one very special and unique character of nuclear energy, and that is the fact of radioactivity. Radioactivity is inescapably connected with the production of electricity by nuclear fission. And in the normal

course of a reactor operation, a single reactor will accumulate very large quantities of these materials. A typical large reactor of the sizes that are proposed for this area and are generally being constructed in the country, each one will contain something like a ton and a half of radioactive material after it's run for a while. This compares with the approximately two pounds of similar stuff that was produced in the Hiroshima explosion. An inadvertent release [of radioactivity] in accidental circumstances from the plant itself can pose the threat of [an] exceptional-size accident with very long-range consequences in addition to very devastating short-range consequences.

By "long range" I mean not only in distance, because the lethal range can extend for tens or many tens of miles, but also in time; because the materials can't be detoxified. They can, simply, only be diluted. And they persist in their radioactivity according to their own schedule until they've undergone radioactive decay. For the materials in the reactor, the most hazardous ones [the decay period] is measured in many tens of years; [for] a few of the components it's in thousands or tens of thousands of years.

Now, my general feeling about the nuclear program is that it never could be considered prudent to construct devices like this with these very large quantities of radioactivity in the areas where they could harm people if there was something else you could do instead of that. It's just a question of prudence. It's a spectacular potential hazard. And a thoughtful approach would be to see if there would be some other way around it. So, the real question is: how severe is the need? Is there a rock-bottom need for nuclear plants, and is there some way around it?

These matters have been addressed by a number of studies. It has to be looked at within the context of the risk that is presented, and this in turn depends on studies of accident modes, accident experience, the way the reactor program has been run, the whole question of how fragile the devices are against accidents.

And that particular situation, on which our group [Union of Concerned Scientists] has devoted a great deal of time and effort to reach an understanding of it, is generally regarded as rather glum. The numbers of important people and groups who have looked at the whole question of reactor safety are coming now, increasingly more and more, in making recommendations not to continue in nuclear programs.

And riding in on the plane last night, I found that another group had checked in with this kind of a recommendation; this is the kind of group whose views should not and really cannot be ignored. This was the Ford Foundation's three-million-dollar energy-policy project, which has spent the last couple of years looking at the question of what the country should do with its energy difficulties. And according to *Business Week*, they have now recommended that there should be a total halt to the construction of new nuclear reactors, and this is the opinion I hold.

CHAIRMAN JAMES MAUZE: Was that the Freeman—

WITNESS KENDALL: That's S. David Freeman's study.

This is a very, very brief description of . . . it's the tip of an enormous iceberg, if you like. The safeguard problem is taken very seriously, as I have said. The waste storage raises important moral questions which haven't really been addressed by many of the utilities, for example, who plan to install nuclear reactors. And, yet it is very troublesome.

That, I think, is the end of my statement.

MR. DAVID NEWBURGER: Thank you, Dr. Kendall. We would now tender Dr. Kendall for cross-examination. . . .

COMMISSIONER FAIN: (*After two hours of cross-examination.*) Dr. Kendall, I would like to ask you a question, and it's one that has disturbed me a great deal as I've sat here and listened to you. You, undoubtedly, have more knowledge about the nuclear-power program than I could possibly put together in my lifetime. As a layman I have followed it for the last twenty years. Now, let me ask you: does it make any sense at all, sitting where you are, that laymen, such as myself, with just a very, very limited knowledge, that you would come here and ask us to make a decision as to the safety of a nuclear reactor when you know that I couldn't possibly have an nth of a degree of the intelligence and knowledge about this matter that you do? Does that make any sense at all in our system? Does it make any sense at all? Now, I'm just asking you for your analysis of that. It disturbs me greatly. Does that make any sense?

WITNESS KENDALL: The question which I think is posed to you people is a very troubling question, which certainly your question is evidence of. One of the great difficulties that we all have is that these enormously complicated technical issues are now becoming part of society, because we're implementing these nuclear reactors. And I have given testimony before other committees, particularly one I remember which was the legislative committee in the California legislature dealing with California's energy policy. They, too, had to deal with—

COMMISSIONER FAIN: Confine yourself to the thing that I'm asking you. Does it make any sense to you as a scientist with tremendous knowledge in this field to let babes—[to] put an issue before a baby, and say, "Make a decision whether you should eat the candy or not"? Does that make any sense to you?

WITNESS KENDALL: I think it has to happen that way, difficult as it is.

COMMISSIONER FAIN: Doesn't it make more sense that we rely upon the AEC which is made up of experts such as yourself, even though you say you're not a part of that community? Wouldn't you trust them making a judgment on this matter when they have the knowledge, the technical know-how, rather than individuals who have absolutely no way of evaluating the things that you're saying?

WITNESS KENDALL: First of all, on the basis of a very deep study that I personally carried out with my colleagues, I no longer can confirm that the AEC is able to support their position adequately, and that is the whole brunt of my testimony.

COMMISSIONER FAIN: Do you think that I as an individual, a layman, that has merely followed the nuclear-power program in the twenty years as a layman would, do you think I would be able to make a better decision? Do you really believe that, as a scholar? . . .

WITNESS KENDALL: I think that your board cannot resolve the actual technical issue of whether the reactor that is proposed or any other reactor is or is not safe on the basis of the board's own technical review of the situation. It is a matter, because of its intricate character, which has to be carried out at least in part by technologists, such as myself and the Atomic Energy Commission. Those are the principal inputs to the situation. The decision which boards such as yours have, is not to resolve the safety issues explicitly but to decide whether in view of the controversy and the quality of the participants and the weight of the evaluations . . . whether it is prudent to endorse the installation of a nuclear reactor or not. . . .

COMMISSIONER PIERCE: Do you have any measure of those risks that we could apply to this plant, that we could make any use of?

WITNESS KENDALL: Well, as of today the risks are not quantitatively established by our study because of great difficulty in doing that. But the body of information we have is very disturbing.

COMMISSIONER PIERCE: As yet there is no other measure other than that report [referring to an AEC report on reactor safety—the Rasmussen Report—that consumed "one hundred man-years" of effort and cost three million dollars to produce].

COMMISSIONER JOSEPH REINE: May I interrupt there? Has anybody at MIT ever spent a hundred man-years on how to cure cancer or how to stop automobile accidents? Or are you just going to spend all your thinking time involved in all this theoretical stuff? If that's true, I'm glad I didn't go to MIT.

WITNESS KENDALL: I am spending my time on—

COMMISSIONER JOSEPH REINE: Answer my question. Is anybody at MIT going to spend a hundred man-years on worrying about the carnage on the highways that happens every day, thousands and thousands of times over this nation? Answer it.

WITNESS KENDALL: There are groups who are doing that.

COMMISSIONER JOSEPH REINE: At MIT?

WITNESS KENDALL: So far as I know.

COMMISSIONER JOSEPH REINE: Have you ever thought about it?

WITNESS KENDALL: Yes. I've thought very much about it.

COMMISSIONER JOSEPH REINE: Are you going to spend a hundred man-years on that?

WITNESS KENDALL: Well,—

COMMISSIONER JOSEPH REINE: Would you devote your free time? Or do you think your fellow scientists would devote a hundred man-years free on something that's very practical today?

WITNESS KENDALL: The auto-safety situation I find very unhappy, but it is not in my opinion a largely technical issue, and I don't believe it could be resolved by people like me.

COMMISSIONER JOSEPH REINE: Are people, who are as deep a thinker as you are, working on that at MIT?

WITNESS KENDALL: I believe so. But, they would not be physicists, because it's not a physicist's problem.

COMMISSIONER JOSEPH REINE: Obviously. I wonder whether or not the campus at MIT or any of these other Eastern universities are involved in trying to solve practical problems that this country has.

WITNESS KENDALL: They're very deeply involved in it, and I—

COMMISSIONER JOSEPH REINE: Well, the results have been very poor.

WITNESS KENDALL: Well, I agree with that, and I'm sorry to say—

COMMISSIONER JOSEPH REINE: I hope the results that you come up with aren't as poor as the results that came out on that thing if that's what MIT is doing and Harvard and Yale and all those places.

WITNESS KENDALL: The very sad and touching thing about the automobile accidents is that it seems to us, as we look at it, that it's not really a technical problem. I don't believe I could contribute anything to that.

COMMISSIONER JOSEPH REINE: I'm not asking you whether you would. But, you're so interested in the public welfare of this nation that you and your people are going to spend a hundred man-years. Is anybody interested in the public welfare of this nation enough to spend that much time on highway accidents or cancer or heart attacks?

WITNESS KENDALL: Cancer and heart attacks—

COMMISSIONER JOSEPH REINE: They're very practical problems that we have.

WITNESS KENDALL: Cancer and heart attacks are getting as much devotion from the technical community as I think it can take. It's an enormously complicated problem.

COMMISSIONER JOSEPH REINE: I think you see my point.

WITNESS KENDALL: Yes, sir. I see it very much, and it's very sad that it is a very live issue, and that I am agreeing with you—

COMMISSIONER JOSEPH REINE: Sometimes I don't think you guys ever reach reality. That's what I'm trying to say, and I have a series of questions on that later. I have my doubts whether anybody out of an Eastern college reaches reality.

WITNESS KENDALL: Well, this is not—this reactor-safety issue I could not agree is a theoretical problem. It's a problem of the greatest practical importance.

COMMISSIONER JOSEPH REINE: Is he going to be back on the stand at 1:30?

MR. GERALD CHARNOFF: I'll be through in about five minutes, and then you can have him.

COMMISSIONER JOSEPH REINE: I'll get you right now. Are you familiar with the operation or the duties of a regulatory commission? You're not talking to a board or a committee. You're talking to a commission, a regulatory commission. Are you aware of that, or did you know that?

WITNESS KENDALL: Yes, I did.

COMMISSIONER JOSEPH REINE: Are you aware of what our duties are?

WITNESS KENDALL: In part. I haven't looked at your specific—

COMMISSIONER JOSEPH REINE: Tell me what in part you do know, or do they teach that out East?

WITNESS KENDALL: My understanding is that, as a regulatory commission, you have responsibilities to see that these programs are implemented—

COMMISSIONER JOSEPH REINE: What programs?

WITNESS KENDALL: Programs of power implementation in the state, of electric generating capacity.

COMMISSIONER JOSEPH REINE: Do you understand our duty to see to it that utilities provide electricity and adequate service, which means enough electricity to serve this state at reasonable rates?

WITNESS KENDALL: That's generally my understanding.

COMMISSIONER JOSEPH REINE: Is that what your understanding was of the duties of a regulatory commission?

WITNESS KENDALL: Yes.

COMMISSIONER JOSEPH REINE: All right. You make a lot of theoretical assumptions. Let me make one. Assuming that there is a need for an electric plant in the Union Electric territory to serve their territory in Missouri, you're advocating us disapproving the building of a nuclear plant; right?

WITNESS KENDALL: That is right.

COMMISSIONER JOSEPH REINE: What would you do then? What would you advise us to do?

COMMISSIONER WILLIAM CLARK: Go to Harvard and take a postgraduate class.

COMMISSIONER JOSEPH REINE: That would be the last place I'd go, or MIT, either one.

WITNESS KENDALL: The question of the need for the plant has to be studied in detail, and in the—

COMMISSIONER JOSEPH REINE: Huh-uh; huh-uh. I'm asking you a question. Let's assume that there is a need for a thousand-megawatt generation plant to serve Union Electric's customers, and they have come in here, and they have asked us for permission to build an atomic-energy plant, and you are saying to this Commission, "Don't let them do it." Right? That's what you're saying and giving your reasons.

WITNESS KENDALL: That's correct.

COMMISSIONER JOSEPH REINE: Obviously you think they're sound, and I'm not going to argue that point. If we say, "no," then what happens?

WITNESS KENDALL: If the plant is genuinely required, which I'll take as an assumption in the conversation,—

COMMISSIONER JOSEPH REINE: Take that first assumption.

WITNESS KENDALL: —then the plant could use a fossil fuel.

COMMISSIONER JOSEPH REINE: Now, let's take a further assumption. Let's take an assumption that the fossil-fuel plant is substantially higher in operating costs and would cost substantially greater sums to construct, would you still give the same advice?

WITNESS KENDALL: In view of the economic consequences of an accident, which has to be a part of any economic thinking—economic thinking can't just be on the operating costs, it has to include consideration of accidents—on balance, the fossil plant would be preferred. It is not likely, in my opinion, given the present costs of constructing nuclear plants, that the nuclear plant would be anything but much more expensive to construct than a coal plant.

COMMISSIONER JOSEPH REINE: You have to answer my question on the assumption, because the Union Electric people indicate that it's cheaper to build this nuclear plant, construction and operating costs, a combination of the two. That is their allegation, which we'll have a determination on. But, do you think that the people of this state or any other state would be willing to pay the higher rates necessary to have a coal plant versus an atomic plant if the coal-fired or fossil-fuel plant is substantially higher in cost?

WITNESS KENDALL: I certainly do.

COMMISSIONER JOSEPH REINE: You really do?

WITNESS KENDALL: Yes, sir.

COMMISSIONER JOSEPH REINE: Have you ever attended any hearings by any regulatory agency that has been held in any large city regarding a rate increase where they've heard testimony from the public?

WITNESS KENDALL: I'm very much familiar with the difficulties, but—

COMMISSIONER JOSEPH REINE: Have you ever attended one of those hearings?

WITNESS KENDALL: No, I have not.

COMMISSIONER JOSEPH REINE: Do you think the people on welfare and the low-income groups in this country would agree with the statement that you've just made?

WITNESS KENDALL: I do, if they had a general appreciation of the hazards, an appreciation which has not yet been communicated to the public.

COMMISSIONER JOSEPH REINE: You think they would be willing to pay more money to have a coal-fired plant?

WITNESS KENDALL: Yes.

COMMISSIONER JOSEPH REINE: One further question. Do you advocate the theory that regulatory commissions should hold utilities down from constructing plants to create a shortage of electricity because of environmental considerations?

WITNESS KENDALL: I would hope that a commission was never in that position.

COMMISSIONER JOSEPH REINE: That's not what I asked you. I don't want to be in that position either. Are you an advocate of that theory?

WITNESS KENDALL: I would never give a global answer to that problem. It depends very much on the circumstances. It depends on the risks, the environmental damage, the relative cost, how badly electricity is needed and many factors which would be specific to particular situations, and I cannot give a general answer.

COMMISSIONER JOSEPH REINE: I would suggest that, in your thinking process that you carry on in these ivory towers, that what you do, some time, is go to downtown Boston when the public-service commission there is holding a hearing for a rate increase and listen to the people. I think you would learn a lot more than you're ever going to learn in some theoretical books. I also think you would have learned a lot more if you would have gone to school in Missouri.

CHAPTER TWELVE

Private Property

DANIEL WEBSTER began the argument for Dartmouth College before the Supreme Court of the United States at eleven o'clock on the morning of March 10, 1818. With him was Joseph Hopkinson of Pennsylvania, then forty-eight years old, appearing for the first time in a constitutional case. Their opponents were John Holmes, a politically agile congressman from Maine, then forty-five years old, and William Wirt, attorney general of the United States, a man of forty-six years, retained in his private capacity as an attorney. Webster was the youngest counsel there by a substantial margin; he was also, by far, the best orator of the group. Holmes, the leading attorney for the other side, was not impressive. During the arguments, Webster wrote to a friend, "Holmes did not make a figure. I had a malicious joy in seeing Bell [a New Hampshire judge] sit by to hear him, while everybody was grinning at the folly he uttered. Bell could not stand it. He seized his hat and went off."

The Dartmouth College trustees had asserted a right to the college's property and records under the old charter granted by the governor of New Hampshire in the name of King George III; the case therefore turned on whether the New Hampshire statute altering the royal charter was valid. The Supreme Court of New Hampshire had upheld the statute and dismissed the trustees' claim.

The defeated New Hampshire lawyers for the trustees, Jeremiah Mason and Jeremiah Smith, were by all accounts unusually able men, and Webster later acknowledged that much of his own argument before the Supreme Court was based upon the reasoning developed by these two attorneys in their unsuccessful suit before the state courts. Mason and Smith had proceeded on two main arguments. The statute, they claimed, was invalid because it deprived the trustees of their property in a manner contrary to the New Hampshire constitution; and, they argued, the statute was rendered invalid by the Constitution of the United States.

For the second argument, Mason and Smith had fastened on a provision of the Constitution which prevented the states from passing laws which would "impair the obligations of contracts." This phrase had been added to the Constitution by its draftsmen to protect creditors; in the widespread bankruptcy and disruption that followed the Revolution, some state legislatures had simply abrogated private debts; the constitutional phrase was intended to prevent such actions. To avail themselves of this provision, the attorneys for Dartmouth had argued that the charter granted Dr. Wheelock in 1769 by the late king was somehow a contract with the present trustees, which the New Hampshire legislature was impotent to disturb. The New Hampshire state courts rejected this argument.

On the morning Webster began argument for his client, there was only a small audience of attorneys (although some later accounts claim a large audience of important citizens). This was not surprising: the case did not seem in itself to be of great importance, and Webster was not yet widely known. Opposing him was, apparently, a political hack.

The room must have been dominated by Webster; tall, broad-shouldered, deep-chested and an orator of compelling gifts, Webster spoke for most of the first day, and his speech has been much praised over the years as one of the best orations made by one of America's greatest orators. As a legal argument the speech was also remarkable, in that three-fourths of it dealt with matters that were outside the jurisdiction of that US Supreme Court before which the argument was made.

The case had been lost in the state courts; on appeal to the US Supreme Court, only one question could be considered—whether the New Hampshire statute contravened the federal Constitution. But Webster treated this issue almost as an afterthought. He devoted the bulk of his time and eloquence to attacking the New Hampshire statutes under the *New Hampshire* constitution, issues that had been settled and could not properly be before the court.

That this argument not only was permitted but was ultimately successful suggests that the Supreme Court was not concerned solely with narrow questions of law.

Webster began his now-famous oration with what amounted to a defense and praise of private property. Dartmouth College, he argued, had been created by, and was the property of, Reverend Wheelock. It had been the reverend doctor's own, legal, private property. This property Wheelock had consigned to trustees, who now held legal title to the college. The New Hampshire legislature, Webster insisted, simply had no power to destroy this private property by putting it under the control of the state.

On this point the New Hampshire court had ruled that there was no difficulty. The state legislature had passed a law—which was, by definition, the law of the land—amending the college's charter. The state constitution clearly provided that private property could be taken by the government so long as this were done in accordance with the laws. Against the apparent plain meaning of the New Hampshire constitution, upheld by that state's highest court, Webster opposed the Magna Charta: the great English charter which established the property rights of medieval barons in the face of depredations by a king, at a time when Parliament was still unimagined. Webster then followed with a long series of quotations from more recent writers which raised the specter of an unbridled legislature, trampling on property rights, throwing off any restraints imposed by the courts and taking private property at will for political purposes.

Webster then returned to a recurrent theme: not even the king could overturn the charter of a corporation. This was, indeed, a well-established principle of the common law. Corporations were not yet very often business enterprises. The corporation had originated in what one historian (Henderson, *The Place of Foreign Corporations in US Constitutional Law*) called the "century-long struggle between the English boroughs and the kings and feudal barons, and more immediately from the struggle for Charter rights and privileges between the colonies and the king. These struggles had come to invest the standard terms found in all corporate charters with a peculiar sanctity, as things the [Americans of the early 1800s] and their forefathers had fought and bled for."

For the colonies themselves had been corporations; some were trading companies granted exclusive trading rights in the New World, while others were chartered in forms closer to that of Dartmouth College—charitable corporations with governmental powers. These corporations were created by grants similar to the municipal and borough charters which embodied rights that had been wrested from the kings and barons.

Charitable corporations had also come to have inviolable protections, ever since the efforts of James II to interfere with the governance of the private colleges at Oxford and Cambridge. These struggles had invested the college corporate charters with strong protections from royal encroachment. And this protection for religious and charitable enterprises had in time come to be used by the wealthy to protect their bequests.

But Parliament, of course, could and did alter the charters of the corporations. An act of Parliament had altered the charter of the giant East India Company which governed India. It was the *king* who was prevented from altering corporate charters, although only he could grant them; this was a long-fought matter of protecting individual rights and property from royal despotism, once rights and liberties had been wrested from the king's grasp.

Webster marshalled all of this precedent in favor of Dartmouth College, by placing the New Hampshire legislature in the role of the king, and the trustees of Dartmouth College in the position of the embattled municipalities of England.

Dartmouth College, as Thomas Jefferson had noted with distaste, was a remaining enclave of the landed gentry, or their emulators, surrounded by a Republican government. The legislature admittedly had the power of altering its own political creatures—the municipal governments. Webster was claiming the privileges and protections of the ancient public and municipal corporations, established to protect them from the power of the king, while at the same time insisting that Dartmouth College was private property. The result was the argument that private property must be protected from the despotism of a popularly elected legislature.

All of this was the point and purpose of Webster's long diversion into New Hampshire law, which in theory could not be considered by the Supreme Court. Without ever having to make the naked argument that property rights needed protection against democracy, he was still able to drive the point home in the strongest language possible, comparing the New Hampshire legislature with Charles II's and James II's "illegal acts . . . open piece of burglary. . . . This act of violence—arbitrary interference with private property." Over and over again he pounded home the point that this was an act of despotism which exposed the college to the mob; the corporation is possessed of "liberties, privileges and immunities, [which] being once lawfully obtained and vested, are as inviolable as any vested rights of property. . . . Colleges and halls will be deserted by all better spirits and become a theatre for the contention of politics. Party and faction will be cherished in the places consecrated to piety and learning. These consequences are neither remote nor possible only. They are certain and immediate." Having put the argument on this footing, there was little need to deal in detail with the constitutional issue.

The real question was plain: the Court was being called upon to protect private property from the mob. As a conservative and admiring biographer of Webster put it at the turn of the last century, "The Supreme Court of New Hampshire decided . . . that the provision of the New Hampshire Constitution that no man could be deprived of his property but by the law of the land, meant any law the legislature might choose to enact. This was the most dangerous feature in the original decision. . . . Webster's argument . . . called attention to this fundamental error. . . ."

The only provision of the US Constitution which could be used to strike down this pernicious law of New Hampshire was the clause which prohibited the states from impairing the "obligations of contract," the phrase intended to cover private debts. Webster asserted that the charter of Dartmouth College was a contract, between the Reverend Wheelock and King George; that the trustees had succeeded to Wheelock's interest and the legislature to King George's; and that, hence, the legislature was prohibited by the Constitution from altering and so impairing the contract.

If this alteration was permited, Webster warned, all of the colleges of the nation would be subject to the ravages of unbridled legislatures, and private property would no longer be safe.

The opposing attorneys were evidently not prepared for such an onslaught. Holmes argued badly and is barely reported in the official records of the debate. Wirt, the attorney general appearing in a private capacity, presented a clear and simple argument. He addressed himself to what should have been the only issue before the Court—the constitutional prohibition against impairing private contracts. The charter of Dartmouth College, Wirt pointed out, was simply not a contract. Wheelock did not own the original property, except in the most meagerly legal sense. The donations of the Earl of Dartmouth had been put in trust to be used for a college; Wheelock had nothing to give the king in exchange for the corporate charter; he contributed no funds and could make no promises beyond those already made to Lord Dartmouth. Hence the corporate charter was a gift and no contract at all. (Later commentaries all agree, in fact, that the charter was not a contract in any sense of the term which was then accepted.) Even if it were a contract, however, the trustees had no rights or property to be impaired. And finally and most importantly, the college was not private property at all; it was a public institution set up for governmental purposes for the benefit of the public at large, and the legislature could alter its charter just as it could alter any municipality's charter.

All of this was most certainly correct, but it was extraordinarily dry and unimpressive after Webster's performance. His colleague, Hopkinson, delivered the reply for the trustees, again driving home that it was of no importance whether the college were Wheelock's property or that of the Earl of Dartmouth; the important thing was that it was *private* property and hence "cannot be revoked." This, once again, brushed aside the only issue which could properly be before the Court and reemphasized the claim for the protection of property.

At the close of argument on March 13, a newspaper reported, "The Chief Justice observed that the Judges had conferred on the cause. Some of the Judges had not come to an opinion on the case. Those of the Judges who have formed opinions do not agree. The cause must therefore be continued until the next term." The following day, the court adjourned. Webster wrote to a friend that he believed he would win the case; that Chief Justice John Marshall and Justice Bushrod Washington were definitely with him, and Justice Story probably with him as well.

But Webster did not rely on the deliberations of the justices. During the Supreme Court's recess he and the supporters of the college arranged as much publicity for their cause as possible, in the hopes that it would influence the judges. Webster also had a hand in bringing several lawsuits against Dartmouth, suits brought in federal court which would directly raise the issue of protecting private property from the New Hampshire government. Webster discussed the case with Justice Story and concluded that the private-property issue, far more than the artificial legal question of whether the corporation was protected by the Constitution's contract clause, would sway the Court.

CHAPTER THIRTEEN

Management Questions

ALBERTA SLAVIN is good-humored about her activities and her chances
for success. She has won some modest victories for the consumer with
UCCM, helping to protect the household rate-payer from dramatic increases in
the price of his rates, and forcing the Missouri Public Service Commission, which
must decide on rate increases, to operate in a glare of publicity. By providing the
kind of dramatic contest which the news media can easily report, Slavin has
allowed them to give much more attention to the details of rate-making and
regulation than would otherwise be afforded these dull proceedings in a small
town, Jefferson City, far from the camera crews of the major television stations in
St. Louis and Kansas City. Slavin has slowly begun to use public support, which
news coverage makes possible, to influence a variety of management decisions in
the utility companies which serve Missouri. In her own view, Slavin is trying to
help the managements of these companies avoid mistakes which would be damaging both to the companies and to the customers they serve.

*In that first rate-increase proceeding that you won a partial victory, the company
appealed. What happened?*

SLAVIN: That appeal was still pending when they filed their next [request for
a] rate increase, very shortly thereafter, three months later. And when they filed
their next rate proposal, in arriving at the amount that was awarded, they agreed
to drop the first appeal. So, they not only have the possibility of going for an
appeal, but they also have the capability of filing three months later, after they
get a bad decision, for an increase. And, of course, you realize that all of this
[filing and appeal] is supported by the rate-payer. Any costs that they incur.
We're still paying off rate cases that go back to '72. They cost "we the rate-
payer." They're very costly—they spend a lot of money on it. They spent
between $180,000 and $200,000. That was back in '71. We know that in the
nuclear-power case they've spent at least half a million dollars, and that's probably a low amount. So they don't have any of the decisions that we face, in terms
of whether we have the capability, or the personnel, or money, to see something
through.

*How does that affect the Public Service Commission? Knowing every decision will be
appealed? And, that there are essentially unlimited resources available to fight them in
court?*

SLAVIN: You know, I've never even thought about that, but that must have
some impact. That really must have some impact. And, of course, I think—we've
just combed all of the [court] cases in the state, going back to 1900, and you
can—there are maybe five which you can quote, which have been brought by
intervenors in the case. All appeals have come from the companies, themselves.

So, you can see that over the years either no one was there, or they [parties opposing rate increases] really didn't have the capability. And usually, the industrial customers have worked out their piece of the pie satisfactorily, so that they really had no need to appeal. But you know, that probably is a very good point. They're going to award enough that the company . . . I think that that first decision [against the power company] was really a, "I'm gonna show you guys a little bit." That hasn't happened since.

Three months later the company filed for a new increase, which was how big?

SLAVIN: I don't remember that . . . seems to me it was in the range of thirty-nine million. Each one has gotten larger.

And they–

SLAVIN: Yes, and I think they got twenty million dollars.

Which was the amount they had originally asked for.

SLAVIN: Uh-huh.

And in exchange, they dropped their appeals. Now you're trying to get the state Public Service Commission to do more than revise rate increases, you want to get into the management decisions of the company.

SLAVIN: Well, the decision, of course, in this case was nuclear. The decision is very clear-cut. You've got a very expensive plant to build and you're holding it out as a cheap source of power and a clean source of power. . . . We, the environmentalists who are yelling about coal and dirty air, find ourselves in the uncomfortable position now of preferring coal to nuclear. So it's a kind of a peculiar switch that's occurred there, in my view, and, of course, there's a great disagreement [among utilities]. I just was in Kansas City, and Kansas City Power and Light claims great success in using scrubbers [to remove sulfur oxide from stack gases]. They have absolutely no problem with scrubbers and you take Union Electric and they say they wouldn't work. You see, I think that coal has now been labelled a dirty word by the utility. Of course, it's the coal-burning utilities that have survived in the energy crisis better than any other utility in the country, as you know. So they [the utilities] hold out the decision and I think it's been really a decision that has been almost foisted upon them by a federal policy and their own trade association, which now extracts large amounts of money from each one of them to conduct their continuing nuclear research. . . .

Now here's a situation in which you haven't had much success in getting rate changes. And you are going in, asking the Public Service Commission not only to do what they traditionally do, to change rates, but really go right in and alter the fundamental management decision, which they don't claim they have the power to do.

SLAVIN: I know.

How would you evaluate your chances of success?

SLAVIN: Tends to look suicidal, doesn't it? Tends to look like this is not the route to go. Well, I think there's enough concern about the possibility [of success]—that . . . one of these regulatory commissions around the country, may, in fact, turn down a powerplant on the evidence—that they [the utilities] have gotten busy changing the laws of the states. We had a major turnover in our Commission in the last three months. Now, if the courts look upon our appeal in a different way than the Commission does, and sends it back, at least we have a shot at another set of people. Another set of men who are going to make this decision.

When the governor [Christopher Bond, a Republican] nominated a purported consumer-advocate commissioner, the state senate turned him down.

SLAVIN: Rapidly.

Do you want to say anything about that?

SLAVIN: Well I, first of all, did not lobby nor have I ever asked for a consumer advocate on the [Public Service] Commission, nor would I. I would like to have somebody on the Commission who has the intellectual capability and background and the expertise to sort through this incredible mountain of evidence and deal with each case in an objective, intelligent way. That's what I would like to see. . . . And I don't think that they necessarily have to be attorneys. Now, I am not convinced that the appointment that was suggested by Bond was that kind of a person. I think the person who was appointed was a person who would view matters perhaps more from my perspective than from industry's and, from that standpoint, was a treat. But beyond that, I don't know what his capability [is] . . . we've never asked to have a Commission stacked by consumer advocates. . . . I just think that there are so many important decisions to be made in state regulatory agencies today; we really should try and put the most capable qualified [people] into those spots and we're not getting that.

Do you think the utilities have been heavily influencing those decisions?

SLAVIN: It's not clear to me that they have or haven't. I think where they probably have been most effective is in really curtailing the power of the Public Counselor's Office by keeping that budget very low, and in continuing to have a situation where underrepresentation of the public occurs in every proceeding. I mean, I think and I have stated that I think it's a tragedy, that the public has to be defended by me. We're a volunteer organization! That is a tragedy that our government cannot put together a better proceeding or a better possibility of evening the odds. It's David and Goliath, the little leagues against the big leagues and we certainly saw it in the nuclear proceeding. I don't think the utility would have been in very good shape on the nuclear proceeding if they hadn't had their hired gun from Washington.

I don't mean to underestimate our people. I think we have incredibly talented legal assistance at this point. What becomes difficult is that they are attempting to work in this effort within the range of their other duties, be it academic or a practice. And if you get into a hearing that may go three weeks, we're juggling attorneys, we're bringing in one attorney one day and he's supposed to cross-examine one day and you know we're staying at the cheapest flea-bag hotel next door, and you know, the contrast is so remarkable. It became really clear to me when we brought in all of our records, first in an apple box, then in a pear box, then a beer container, beer case and another cardboard box of some sort; and [counsel for Union Electric] came in with leather, enormous briefcases, each one with a secretary ready to leap to attention and bring forth the exact document that was needed at each moment for the cross-examination. We're a ragtag group.

CHAPTER FOURTEEN

Tyranny of the Mob

THE CHIEF JUSTICE of the Supreme Court in 1818, when Webster presented his argument on behalf of the Dartmouth College trustees, was John Marshall, appointed to the court by John Adams in 1801. Marshall was a Virginian. His family, farmers of modest means, descended from a captain of cavalry in the service of Charles I who had come to Virginia in about 1650. In 1818, when the Dartmouth case was heard, Marshall was sixty-three years old and had been chief justice for seventeen years; it was the precise midpoint of his extraordinary term, for he would serve yet another seventeen years as chief justice.

Marshall was a tall man like Webster and was largely built, but he had an unusually small head, with a low forehead; his only attractive feature is said to have been his brown eyes. His speech was slow except when he was aroused, and he was somewhat clumsy. He was one of the many Federalist judges appointed in the last hours of Adam's presidency, before the Jeffersonian Republicans swept into office.

Webster and Marshall would, in effect, collaborate in several important cases to extend the reach of the federal judiciary, but the Dartmouth College case was the first such instance. In 1818, it was still an unusual and extreme step to ask the US Supreme Court to strike down a state law which had been properly enacted; the situation was made the more delicate because Marshall (like Webster) was a Federalist, while the government of New Hampshire (and a majority of Marshall's colleagues on the Court) were Republican. A decision in favor of the Dartmouth trustees would not only strike down Republican laws, but would overrule the New Hampshire Supreme Court as well.

The United States Supreme Court's decision was announced on February 2, 1819, on the second day of the term; the Court met for the first time in the "splendid" new room provided for it in the Capitol. Marshall read the opinion of the majority. He was joined by his fellow Federalist Bushrod Washington and three of the five Republican members; one justice took no part, and only one dissented.

Marshall began with great care. Early in his opinion in favor of the Dartmouth trustees against the New Hampshire government, he said, "This court can be insensible neither to the magnitude nor the delicacy of this question. The validity of a legislative act is to be examined; and the opinion of the highest law tribunal of a state is to be revised . . . . On more than one occasion this court has expressed the cautious circumspection with which it approaches such consideration of such questions. But . . . on the judges of this court . . . is imposed the high and solemn duty of protecting, from even legislative violation, those contracts which the constitution of our country has placed beyond legislative control; and, however irksome the task may be, this is a duty from which we dare not shrink."

Irksome or not, the duty was clearly self-imposed, for Marshall had concluded, without discussion, that the charter of Dartmouth College was, in fact, a contract protected by the Constitution of the United States: "It can require no argument to prove that the circumstances of this case constitute a contract." Marshall took this for granted, and so gave the Supreme Court authority to review the state's laws. Later in his opinion, he indicated awareness that there were some obstacles to believing that a contract had been formed between the Reverend Wheelock and King George III; but he insisted that there was a contract. "This is plainly a contract to which the donor, the trustees, and the crown . . . were the original parties." This conjures up an absurd picture of Wheelock, the Earl of Dartmouth and King George III entering into a three-cornered arrangement, each receiving something of value from the other, whereas the scheme for the school had already been committed in the earlier exchange between Wheelock and the Earl. The corporate charter was merely a gift, or a grant, and so it was ordinarily viewed.

In short, Marshall accepted the dubious point which allowed the federal courts to intervene and which formed the only possible basis for a constitutional challenge; most later commentators agreed that he was simply wrong as the law then stood.

Justice Story, who concurred with Marshall in the decision, evidently found the contract question a serious obstacle and devoted most of a very long concurring opinion, of which Marshall must have been aware, to a consideration of this point. This point may have been, to Marshall, a minor matter. Or perhaps the justices could agree only on the conclusion, and not on the proper reasoning; perhaps Marshall wisely saw that a weak argument would be far more vulnerable than no argument at all.

In any case, once having established a contract, Marshall's reasoning followed quite simply: the Constitution protected contracts respecting private property; the corporate charter is a contract; if the college is private property, it is protected by the Constitution.

Nor did Marshall have much difficulty in finding that the college was private property. It was founded by the Earl of Dartmouth's money; it remained, in effect, an extension of the Earl's will. No matter that it was a college for public education founded on land given in part by the state government, with all the powers of a local government. The college was private property.

But whose property was it? The Earl was long dead, and his heirs were not in court. The trustees had only bare legal title and no real ownership, for the college was, in fact, limited to performing its public services. The students and teachers were not before the Court, nor was the public through its representatives. There must be someone in whom the property rights are vested; if no one was being injured, then the Court had no reason and no authority to intervene.

This, Marshall says, is the only question of real difficulty before the court. The surprising and revolutionary answer to this difficult question is that the *corporation itself* owns the property and is possessed of constitutional right: "An artificial, immortal being, was created by the crown, capable of receiving and distributing forever, according to the will of the donors, the donations which should be made to it . . . . In this respect their [the donors'] descendants are not their representatives. The corporation is the assignee of their rights, stands in their place, and distributes their bounty, as they would themselves have distributed it, had they been immortal . . . . [T]he body corporate . . . completely representing the donors . . . has rights protected by the Constitution."

There is something faintly chilling about this language. The wealthy, it seems, are to have immortality on earth by creating corporations, which are metaphysical extensions of themselves. The corporations, as extensions of their creators, may own property and, in every other way with regard to property, are the equivalent of human beings. But they are immortal. As if they were actual human offspring, the legislature had no power to bring them to an end, for their immortal rights were protected by the Constitution of the United States.

In this passage the modern corporation was born. And from the beginning, it was clothed in the armor of the Constitution. Like an infant Hercules, it was from birth something superhuman, mantled in the privileges and immunities which generations of Englishmen had wrested from the king, to protect the helpless individual and his property from royal tyranny. The corporation, protected not only by the rights created to secure the permanence of municipal governments and the freedom of institutions of learning, was clothed as well in the constitutional protections for which the nation had fought a long and bloody revolution. The corporation was shielded by the instrument meant to protect the individual from the overpowering strength of the state. In this way the accumulation of wealth that is a corporation would be protected from the tyranny of the people.

As Hopkinson wrote to the president of Dartmouth College, announcing their victory, the Court had established "principles broad and deep, and which secure corporations . . . from legislative despotism and party violence for the future . . . . " In Webster's words, the decision would be a "defense of vested rights against state courts and Sovereignties." Marshall's decision fixed the corporation as the preeminent form of business activity; it was to become the principal weapon in the battle between industry and democratic government.

CHAPTER FIFTEEN

The American Way

CHAIRMAN MAUZE: Are there any further preliminary matters? Mr. Birk [house counsel for Union Electric], you may call your first witness, please.

MR. BIRK: Call Mr. C.J. Dougherty [president of the Union Electric Company]. (*Witness sworn.*)

MR. BIRK: Please state your name and address for the record.

MR. DOUGHERTY: Charles J. Dougherty. I live in St. Louis County, Missouri.

MR. BIRK: And you are the same C.J. Dougherty whose prepared testimony has just been copied in the record; is that correct?

MR. DOUGHERTY: I am.

MR. MCCABE   [Public Service Commission staff attorney]:   Mr. Dougherty, I wonder if you would tell us whether or not you participated in the final decision to construct the Callaway 1 and 2 [nuclear] units?

MR. DOUGHERTY: I did.

MR. MCCABE: And could you give me an indication in what capacity you participated in that decision-making process?

MR. DOUGHERTY: I participated in that decision-making process, Mr. McCabe, as the chief executive, and I had the final determination. And I made the determination that we should go ahead with the Callaway project.

MR. MCCABE: I wonder if you could indicate to me on what basis you made the decision to go nuclear rather than coal.

MR. DOUGHERTY: Well, it was an over-all decision, recognizing the obligation which we have to provide electric service, to provide that electric service at the lowest possible cost consistent with sound business principles, and necessarily in doing that we must provide as reliable and as efficient a service as is possible. With that objective, we estimate that there will be a need to supply a greater load in the future; it then becomes a question as to how do you supply that load. And, frankly, we considered all possible alternatives. And all things considered, and I guess you could probably umbrella all things in a descriptive term of reliability and acceptability, both environmentally and economically; on that basis we decided that nuclear came out as the preferred method of providing the expected load growth.

MR. MCCABE: Let me define the parameters of that admittedly very, very broad question a little more. You mentioned considerations of cost, I wonder if you could indicate to me what data or information you utilized in making the determination that nuclear would be more favorable than coal from a cost standpoint.

MR. DOUGHERTY: Well, we used—

MR. MCCABE: Assuming, of course, that you did, in fact, make that determination.

MR. DOUGHERTY: Which we did, that is correct. We figured the annual cost of nuclear versus fossil. I might say, as I indicated, we considered a number, Mr. McCabe, a number of—as I said, I believe, all possible methods of providing the load growth. You have narrowed, I believe, your reference to coal and nuclear, and I want to say that that in reality is what it came down to in the final analysis.

It was a question of coal versus nuclear; and in considering those costs we considered the annual costs, and [by] the annual costs I mean the servicing of the capital costs as well as paying taxes, all operating, maintenance expenses, depreciation and whatnot.

MR. MCCABE: Are you aware that Mr. Allen [Clyde Allen, vice-president, Union Electric] assumed an eighty percent capacity factor for this plant [i.e., that it would generate eighty percent of the total power theoretically possible—a measure of reliability]?

MR. DOUGHERTY: Yes, I am.

MR. MCCABE: Are you confident that this plant will perform at that capacity factor?

MR. DOUGHERTY: No, I'm not. And I will tell you, Mr. McCabe, when this decision was made we used eighty percent because at the time—I guess this was 1971, I believe—there were some of the newer nuclear units which were achieving an eighty percent capacity factor. So we thought it prudent to use an eighty percent capacity factor for the fossil-fuel plant, although, frankly, we weren't aware at the time of any fossil-fueled plants which had ever achieved that degree of capacity factor; but, as I say, I think we leaned over backwards in order to get a fair comparison. The truth of the matter is that eighty percent might not be a realistic capacity factor, but we have looked at this on a seventy percent capacity factor, on a sixty percent capacity factor, we have even taken the most recent capacity factor figures released by the AEC and the Edison Electric Institute, which showed—I don't want to be held to this . . . Well, I'm quite sure this is right—a fifty-nine percent capacity factor on nuclear units and a sixty-seven percent capacity factor on fossil units between three hundred and six hundred, I think, megawatts. We have compared our nuclear unit on a fifty-nine percent capacity factor with the competing fossil capacity at a sixty-seven percent capacity factor, and in all instances the nuclear is still the preferred.

MR. MCCABE: Is this all other things being equal, sir?

MR. DOUGHERTY: All other factors being the same except where they would be changed.

MR. MCCABE: If it were shown at this time from a cost standpoint, based on your latest information, [that coal] was, in fact, more economical than nuclear, would you reverse your decision?

MR. DOUGHERTY: I think we would, it would depend upon—we would have to evaluate the amount of money we have in the plant already, but I don't think there is any question that if it comes up—Mr. McCabe, we are not wedded to nuclear generation, we are wedded to providing economical and reliable electric service to our customers, and it just so happens that the result of all our studies indicate that nuclear is the preferred way to generate the necessary power and energy today.

MR. MCCABE: Then, I take it, you are confident that the capacity factor of that plant will be sufficient to make it more economically viable than a comparable fossil-fuel plant?

MR. DOUGHERTY: That is correct.

MR. MCCABE: All right, sir. I wonder if you would indicate whether [Union Electric] would agree that, if the Public Service Commission orders that a nuclear plant in this case could be constructed, but in the event that in 1982 coal-fired generation by plants begun in the years 1977 to 1978 is a less expensive method of producing electricity, the revenues allowed to your company would be equal to those allowed if the means of production was coal?

MR. DOUGHERTY: There would be no basis for such an order whatsoever, because you must bear in mind that the decision must be made now.

MR. MCCABE: Well, then, you are confident that the nuclear would be more economically viable than the coal at that period of time?

MR. DOUGHERTY: On the basis of what we know now, absolutely, and that's what we have to make the decision on, and that's the basis upon which we are seeking a certificate. If we want to wait until 1982 to decide what kind of a plant to build, it's too late.

MR. MCCABE: Well, I wonder if you could tell me, then, if your decision is wrong—that is, it does prove more economical for coal—and when we look at this retroactively, if you will, in 1982, who will bear the burden of that incorrect decision having been made in 1974?

MR. DOUGHERTY: The customers will bear the burden, as under the American economic system that's the way it is supposed to work. [*Pause.*]

MR. MCCABE: What I am trying to say is this: is it not a fact that this regulatory commission, like any other regulatory commission in this country, makes a determination of what is allowable to the company as a fair rate of return on your investment after allowable operating expenses, and would not make a determination as to the operating efficiency of your nuclear unit as opposed to a fossil-fuel unit that might have been constructed?

MR. DOUGHERTY: Well, I would like to think that that was true—

MR. MCCABE: All right, sir.

MR. DOUGHERTY: But I wouldn't attempt to prognosticate what this Commission may do in 1982, I wouldn't be so presumptuous.

MR. BARVICK [public counsel for the state of Missouri]: There is some testimony, I believe, that a nuclear powerplant has a potential for having an accident which would render it useless. Now the terms of how—are you aware of that at all?

MR. DOUGHERTY: I have heard such discussions, yes. I have heard such discussions.

MR. MCCABE: Well, you say you have heard them. You don't put any credence in them?

MR. DOUGHERTY: Not much.

MR. MCCABE: If such a thing did happen and render the property no longer capable of being defined as used or useful for purposes of your rate base [the value of the utility's physical assets upon which utility rates are based], what would happen to your company?

MR. CHARNOFF: I think we need some clarification. What would happen to the company in what regard, sir?

MR. BARVICK: If the Commission took out of the rate base the $1,700,000,000 [the cost of the nuclear plant] of rate base, or whatever the figure is ultimately invested in this thing, and computed your rate of return on some other figure, a figure which would be $1,700,000,000 less than that.

MR. CHARNOFF: In other words, if for one reason or another a nuclear plant or a coal plant went down, and the Public Service Commission saw fit to take the invested amount from rate base, what would happen to the rate base?

MR. BARVICK: What would happen to the company?

MR. CHARNOFF: What would happen to the company from a financial standpoint?

MR. BARVICK: Right.

MR. DOUGHERTY: I would say that we would go to court pretty quickly under the deprivation of property. I think it would be an unconstitutional deprivation of property if such a thing happened.

CHAPTER SIXTEEN

Railroads

A DOZEN YEARS after the Dartmouth College case was decided, railroads were beginning to spread with astonishing speed throughout the country. This industry rose at a pace which is difficult to understand for us who imagine a leisurely rate of change a century ago. The first steam locomotive was sold in 1830, and ten years later, track had been laid in twenty-six states. By 1860, there were thirty thousand miles of track throughout the US. Four great trunk lines had been built in the East: the Pennsylvania, the New York Central, the Baltimore and Ohio, and the New York and Erie. These companies were the first modern industrial corporations.

The new corporations grew in a condition of near warfare with state governments. Andrew Jackson's radical Democrats had succeeded the moderate Republicans and conservative Federalists in national office. Chief Justice John Marshall gave way in 1834 to Roger B. Taney, Andrew Jackson's close friend. Until the Civil War reestablished federal supremacy, Taney's Supreme Court struck at the structure of protection for private property against state depredations. In defense of state rights—and slavery—against the power of the industrial and banking wealth of the northeast, the Taney court partly reversed the decisions of John Marshall. Corporate charters were opened to state attack by a decision permitting states to enact statutes reserving in all newly issued charters the right of the state to alter or abolish the privileges granted. (All states now have such statutes.) The railroads, many of which operated under charters granted before this liberalization, were only lightly hampered in their expansion.

During the 1850's the railroads wrestled with complex engineering problems and grew from scattered, short-haul companies into large trunk-line roads connecting the major cities of the East. In the process they developed the administrative structure needed to operate a large and complex organization spread over hundreds of miles. Financial affairs were separated from operational concerns, a decentralized structure was developed to deal with the problems of large geographical areas, and cost-accounting and cost-control methods were developed.

Soon there appeared buccaneers like Erastus Corning, a former hardware clerk, the creator of the New York Central. His knowledge of railroading was minor. After amassing a modest fortune in land speculation and finance, Corning bought the Albany Iron Works which, among other things, made iron spikes for railroads. Largely to assure a market for his product, Corning began to buy the stock of local railroads and, through stock ownership and proxies, ultimately gained control of several lines; these continued thereafter as faithful customers of the Albany Iron Works.

These railroads were modest affairs, each running from one small New York town to the next, and their stock was not of great value. Having acquired control of several, however, Corning and a group of fellow financiers and New York politicians discovered a gold mine which to this day has not yet been exhausted. They consolidated ten small roads, two of which existed only on paper, as the

New York Central Railroad, running from Albany to Buffalo, in 1853. The combination of these separate lines into a single trunk line greatly enhanced their combined value, for a single long-haul railroad could attract far more freight and passenger travel than ten separate and competing lines.

The creation of this single trunk line was quite simple: a new corporation, the New York Central, was formed, which exchanged its new stock for the stock of the separate railroads. To stimulate the owners of the small roads to accept the stock of the new Central corporation, some inducement was needed. The stock of the Central ultimately would be far more valuable, the owners were told, and told rightly; but to strengthen the inducement, they were also offered a "premium"; interest-bearing bonds worth from 17 to 55 percent of the original holdings. This was a bribe to the stockholders, to induce them to exchange railroad stock for holding-company stock, and its effect was to "water" the New York Central—the holding company's—stock. In effect, the future income of the railroad was mortgaged to its organizers. Erastus Corning, the principal stockholder in the roads being combined, received $77,500 in bond premiums, a tidy sum at the time (and, of course, the Central, like its predecessors, bought iron spikes from the Albany Iron Works).

This is a brief example of how a utility holding company was created and why it was so profitable to organize one. Holding companies would come to dominate railroads and, later, the electric power utilities.

Large sums were also to be made on the construction of new railroads. Within a generation of the New York Central's formation, the last of the great Eastern trunk lines, four transcontinental railroads had been built, far more than could be justified economically; their purpose was to secure the fabulous land grants and inflated construction contracts which accompanied a great railroad scheme. Members of Congress and the Cabinet were tainted by the Crédit Mobilier, the greatest of these scandals; the courts and legislature of California were hopelessly corrupted, and the governor was himself a railroad man.

The embarrassing oversupply of railroads lent a good deal of volatility to their stock values and a great urgency to speculations and struggles for control of these stocks. An extraordinary series of battles was enacted; the most colorful and best publicized incidents occurred in the struggles between the New York Central and the Erie.

In the 1850s, former cattle-trader Daniel Drew, the supposed originator of the term "watered stock," gained control of the Erie's stock and drove it up and down in various grand speculations. Drew acquired control of the Erie when, as a director of the corporation, he drove down its price in one of his famous raids, selling the stock short—that is, selling at present prices for later delivery—and therefore profiting from every decline in the stock. When the stock price had reached a satisfactorily low level, Drew then bought enough to assure himself a controlling interest. Later, of course, the price of the stock would be driven up again with a program of buying; after it had reached a satisfactory peak Drew would profit once more from its decline by selling short.

The consequences of this sort of maneuver were dramatic. In its early years, the Erie had been a great trunk line by the standards of the time, with nearly five hundred miles of rail joining the port of New York to the Great Lakes. It had been built at a cost of fifteen million dollars, an immense sum at the time, and was considered an engineering marvel. But the Erie quickly became simply an object of speculation, and, by 1854, when Daniel Drew became a director, its track was in exceedingly poor condition: the wrecks on the line became so frequent that newspaper headlines would read, "Another Erie Disaster." The revenues of the railroad were so heavily mortgaged to its organizers and bond holders, and so largely diverted into stock speculation, that nothing remained for maintenance of the track and rolling stock.

The enormous profits being reaped by Drew tempted Cornelius Vanderbilt into the railroad business, and he and Drew quickly became antagonists. Cornelius Vanderbilt—called "The Commodore" because of his extensive shipping interests—was a man of modest beginnings who had amassed a fortune in the coastal shipping trade, a fortune which he multiplied during the Civil War by supplying the Union Army with allegedly shoddy goods. To Drew's tactic of selling stock short, Vanderbilt opposed the tactic of the "corner"—gaining control of all the stock in circulation. In a short sale, one may sell stock one does not own, with a promise to deliver it later, but on the day of delivery one must then buy stock to deliver. If all of the stock is held by one's opponents, this can be awkward indeed. Drew and Vanderbilt struggled for control of two rail lines running from New York to Albany, the Harlem and the Hudson lines, whose competition posed a potential threat to the Erie. In the first struggle, the New York state legislature announced its intention of removing Vanderbilt's franchise for the Harlem line, whereupon the legislators gleefully began selling the Harlem stock short. Vanderbilt quietly acquired all of the freely circulating stock in the line, so that when it became time for the legislators to deliver, they went scrambling madly to buy shares and drove the price up higher and higher. Vanderbilt made a million dollars on the transaction, half of it, according to popular rumor, extracted from Daniel Drew.

Drew launched a second raid against all the Vanderbilt railroad holdings. The old Commodore managed to organize a pool of five million dollars to buy up all of the outstanding stock in the Harlem line once again. This time, in their aggressiveness, the Drew forces, which again included much of the state legislature, managed to sell short more shares than existed. When the time came to deliver, the scramble to buy shares was frantic. The price of stock began to rise, apparently without limit, since the penalty for selling shares one could not deliver was a jail term. Vanderbilt finally permitted a settlement at roughly three times what he had paid for the stock in constructing his corner.

After these early successes, Vanderbilt entered the railroad business on a larger scale. Already in control of the rail traffic from New York to Albany, he managed to buy up the New York Central, which ran from Albany to Buffalo. In control of the Central, Vanderbilt proceeded to acquire other short lines in New York, with each acquisition inflating the value of New York Central stock. On May 20, 1869, the ever-pliant New York legislature granted Vanderbilt the right to consolidate all of his railroad acquisitions with the New York Central system. Vanderbilt thereupon issued a stock dividend of 80 percent, almost doubling the capitalization of the company. The stock dividend was worth forty-four million dollars, and then as the principal stockholder, of course, Vanderbilt was also the principal beneficiary. One night, at midnight, he carried away from the office of Horace F. Clark, his son-in-law, six million dollars in greenbacks—a part of his share of the profits; he had twenty million dollars more in new stock.

The truly spectacular struggles were over control of the Erie, which remained the great speculative plum. Securely in control of the Central, Vanderbilt launched a series of raids on the Erie, in the control of which Drew had been joined by Jay Gould and Jim Fisk, the famous Erie Gang. The struggle was for control of the stock rather than over management of the company, in which none of the principals was interested. It was a long and complex war, waged in the full glare of publicity to the huge entertainment of the public. The Drew faction had the advantage, for they were selling their own stock short and could print as much of it as they liked. Vanderbilt tried to buy up control of the company and Drew, Gould and Fisk issued millions of dollars of new convertible bonds and then issued stock against them, flooding the market and diluting Vanderbilt's holdings. The Commodore had a compliant judge issue an injunction against the new stock issues at one point, but Fisk circumvented it by stealing a hundred thousand shares from his own messenger and dumping them on the market.

Vanderbilt lost millions in this maneuver, but, undaunted, he had an order issued for the arrest of Drew, Fisk and Gould, for violating the injunction against further worthless stock issues. These gentlemen then withdrew across the Hudson River to New Jersey, beyond the reach of the arrest order, taking with them the company's cash reserves and securities. Two directors were arrested, but most of the Erie Gang arrived in Jersey City intact and set up headquarters in a hotel named Taylor's Castle, renamed "Fort Taylor" for the occasion. A force of railroad detectives guarded the hotel, and three twelve-pound cannon were mounted on the piers. Shortly thereafter, presumably at Vanderbilt's behest, a gang of New York toughs laid siege to the Jersey City fort, but retired in the face of superior forces. The cannon were not fired.

The long battle was concluded after Jay Gould visited the legislature in Albany, carrying a suitcase stuffed with five hundred thousand dollars in greenbacks. Under the leadership of state senator William Tweed, the legislature cultivated both sides in the dispute. Gould outbid Vanderbilt, being able, as he was, to draw on the captive treasury of the Erie. The books of the Erie showed "business expenses" of this episode amounting to one million dollars and payments to individual legislators went as high as seventy thousand dollars.

The legislature passed a statute lifting the ban on sale of Erie stock, allowing the gang's return to New York. A compromise was reached with Vanderbilt, who recovered some of his money and abandoned his attempt to take control of the railroad.

The condition of the Erie naturally continued to deteriorate during this period. A confidential letter from the superintendent of the railroad to Jay Gould, later submitted to the legislature in justification of a new bond issue, read in part: "The iron rails have broken and laminated and worn out beyond all precedent, until there is scarcely a mile of your road, except that laid with steel rails, between Jersey City and Salamanca or Buffalo, where it is safe to run a train at the ordinary passenger train speed . . . . "

The New York Central was in similar condition, its revenues having been diverted into stock speculations, with the result that none of the railroad's income was left for maintenance. According to one often-quoted estimate, by 1873 the railroad had fifty thousand dollars worth of inflated stock values for every mile of track between New York and Buffalo.

These early incidents will give something of the flavor of the process by which the railroads were built and these events were hardly unique. In a later, more successful struggle for control of the Erie, J. Pierpont Morgan battled Jay Gould for control of the Albany and Susquehanna line. The Morgan forces took control of the Albany end of the line, and the Gould troops occupied the western Binghampton end, using local police for reinforcements. Traffic on the line came to a halt, and there was a temporary standoff. Each side then sent a train loaded with armed men along the line toward the opposition camp. The two trains collided midway and a battle was fought, with the Gould men retreating. The struggle was then removed to the courts and the political arena; finally Gould allowed himself to be bought out on generous terms, issuing his famous dictum, "Nothing is lost save honor!"

While existing railroads were allowed to decay, new lines were built to fuel the speculative frenzy. In 1860, there were thirty thousand miles of railroad; between 1865 and 1873, thirty-five thousand new miles of rail were laid. While in 1860 there were 1026 people in the country for each mile of track, by 1873 there were only 590. There was no way of deriving sufficient revenues from the traffic of the time to pay for this massive overbuilding and for the even more massive watering of the resulting stock. The inevitable result was a crash of market prices. The stock market crashed in 1873, and then, fuelled by another round of railroad speculation, the market rose again. There followed a depression in the mid '80s, concluded by a brief market rise and finally the panic of 1893. In this panic, 155 railroads, capitalized at $2.5 billion, collapsed. Between 1890 and 1895, a third of the nation's railroad trackage was in receivership.

Such catastrophic booms and busts could not be allowed to continue, and the disastrous competition for control of railroad stocks was finally abated by the construction of vast nationwide trusts and holding companies, in which rival investors would have a share. In the period following the crash of 1893, financial institutions, acting as receivers for bankrupt railroads, took control and, in essence, abolished competition. The most successful of these groups was led by J.P. Morgan, who had entered the railroad business twenty years before to contest with Jay Gould for the Albany and Susquehanna. Morgan proved himself a successful reorganizer and consolidator of bankrupt companies; he said, "We do not want financial convulsions and have one thing one day and another thing another day."

But just as the great railroad holding companies were emerging from their greatest crash, a new pyramid of holding companies was being built with the country's newest utilities, the power companies.

CHAPTER SEVENTEEN

Historical Circumstances

DAVID NEWBURGER is a balding, bearded, thirty-one-year-old law professor at Washington University in St. Louis; he is also counsel to Alberta
Slavin's Utility Consumers' Council of Missouri and has represented that group
in its battle against the nuclear powerplant proposed by Union Electric. He is not
paid for this work, but he has been able to combine it, to some extent, with his
teaching duties; students who assist him in the case are earning credits toward
their degrees, and other faculty members participate in the legal work and in
supervising the students.

Do you see this as a traditional liberal issue, the opposition to nuclear power?

NEWBURGER: I think that's where it came from. As you know, the original
concern about nuclear power came with the Bomb, and the problems of radiation
exposure and fallout, which were closely tied to the peace movement. The peace
symbol means nuclear disarmament, I think . . . . I think that what happened
was that the people who were in the liberal causes happened to pick up the
issue. The other thing is that basically the nuclear-powerplant fight is now an
antiestablishment fight, because nuclear power is so much a convention within
the establishment. It takes the liberal folk, who are used to being antiestablishment, to oppose nuclear power. But what we see in a lot of different places—
witness the recent Common Market Battle in Britain—is the Right and Left
coming together, and it's not at all surprising to see people on the far Right also
opposing nuclear power for the same reasons that people on the Left might do
that.

*Union Electric is clearly doing business in many states and buying and selling power in
many states and in some ways is very nearly a national, certainly a regional, organization.
Doesn't it seem curious that a state agency should be making such important decisions about
it?*

NEWBURGER: It seemed curious to the state agency, when we went before it,
for us to argue that the state agency should be deciding the nuclear-safety issues,
which it was our position that they should.

*But, it was, in fact, Union Electric that went through the Public Service Commission,
they were required to get permission from the state . . . .*

NEWBURGER: What they are required to get is a permit to build the plant.
There was considerable debate at the hearing about the scope of that—what is
necessary to be established in order to obtain that permit. Traditionally, a permit
of this sort considers the financial advisability of building the plant, the public

need for the plant and things like that. We infused in the proceeding not only those financial issues but the issues of nuclear power.

Now, I think maybe the question you're asking is, Well, why is the state even deciding the financial questions? First of all, I've got to say that when you're in a federal system there are a lot of quirks about jurisdiction that are not necessarily explicable. It might be that a more rational, orderly way might be to put a decision some place where it isn't, but when you're a lawyer dealing in that system, there happens to be jurisdiction there, so that's where you happen to have the fight. I do think that somebody needs to regulate utilities and a decision was made, well, it was made at the turn of the century, and again in the thirties, to give the states, or to retain for the states, a very substantial role in that regulation. So the fact that a state has jurisdiction is not surprising from the historical point of view. The rationality of it, I guess, I really don't consider.

Union Electric has a permit, a general blanket permit, to build in its service area anything it damn well pleases, so that it wouldn't need to get a construction permit if it were building a nuclear plant in St. Louis or St. Louis County. Because it has to build [a nuclear plant] in an area of low population density, and because there were no such areas available within the company's service area, they had to get an extension of their blanket construction permit for the site in Callaway County. One of the commissioners, in the course of the closing argument asked me, Isn't it fortuitous—essentially the same question you're asking—isn't it it somehow silly that the Commission should be deciding all of these issues that are so hard to decide, when it turns out that, had they planned the plant either for Illinois or had they planned the plant for their own service area, the Commission would not have had the question come before them at all? And I said, Yes, isn't that surprising. The way the law is written is that when you have these certain facts fall together, the permit has to be granted, on a record, and therefore, fortuitous though it may be, that doesn't limit the responsibility of the Commission to make the decision in this context. I feel that there's an awful lot of irrationality whenever you're talking about jurisdiction, they're arbitrarily drawn lines, historical in origin, philosophically not put together very soundly. You can have fortuities in which a given act in one case may be able to occur with no regulation at all, and in another case it may have full-dress regulation. But that simply happens to be an incident of a federal system. So you sort of take it from there, rather than try to explain it.

Insurance has been protected from federal regulation by a long tradition, and as a result that's one [industry] in which you will see a lot more active state regulation than federal regulation. The auto industry is an industry that is rapidly becoming a regulated industry, *de facto*, and certainly the oil industry is in the same kind of position. That's coming out of need, it's coming out of the problems that we're faced with, and it's going to be natural for those industries as they become regulated to be regulated at the federal level, rather than at the state level, because the problems are perceived as being at that level when they first develop, when the need for setting the regulation first becomes clear. Union Electric used to be a little company that sold steam heat to some consumers at the downtown office buildings. Who the hell would think to put federal regulation on that?

The state agencies that have set out to regulate these companies are not exactly massive entities themselves.

NEWBURGER: Regulation is a complex of different things that have to happen all at once. There has to be a statute that directs somebody to do the regulating, and there have to be administrators to do the regulating, and there have to be ways of enforcing the directives for control, that these administrators implement under the authority of the statute. So that really, when you're thinking about regulation, the best way to think about it is not whether regulation exists but does an effective regulation exist. That leads to a question of what are the standards for control, what is the ability of the agency to implement those standards, and the question of what [is] the ability of the agency goes not only to matters of expertise, or practicality of the standards themselves, but also goes to questions of budgets and other factors which effect the *practical* ability of the agency to do its job. Now, it's my view, that much of the regulation in this country—that there's a pattern of regulation in this country. And that is, that we state either national or state policies of intending to have effective regulation, which sound good on the books and, if [the regulation] did occur, it would achieve the populist values or the progressive values that are its underpinning, and that makes for good public relations. Then the people who are authorized to implement the policy, in fact, turn out not to be authorized in that way, and as a result the regulation, in fact, doesn't have the effect that it appears it would on the books. That's very true of utilities commissions. There's a problem of cooperation and there's simply the problem of understaffing. It is ridiculous to think that a commission made up of five lawyers who have not been in the utility business and maybe don't know much about it, who are paid a modest fee and who are not given a very large staff, whose staff is basically very underpaid, so it's only young people getting experience who can afford to work with the staff; it's not reasonable to expect that kind of commission to be any match at all for the utilities. But then, there are very few agencies that really are matches for industry and generally you see them on the federal level. But there are some, you know, Herbert Denenberg in Pennsylvania, in the case of insurance—he's no longer there, but he was a match for the insurance companies in Pennsylvania. And then there are a few agencies on the federal level.

So you think the relative disparity in the staff size [and] budget is central to the problem—

NEWBURGER: I think staff size and budget are very central to the problem, when you're dealing with utilities—I mean, that's the only justification I have for going into a rate case. I know I can go into the rate case with a bit more experience than some of the people on the [Commission's] staff who are working on a rate, and I can see questions that ought to be asked that they just aren't seeing. But that's because I've got—I'm ahead of them in experience. And, everytime one catches up with me he goes on to another job because he becomes marketable someplace else. So that I think it is natural that the starved agency is simply not going to develop the expertise that is needed to regulate effectively. But I also believe that a person can go into an agency of this nature, if he is pretty effective and if he is a reasonably politic person—these agencies are really very responsive to political pressures—if he's fairly capable of handling political problems, I think that he can be much more effective than the agencies usually are, according to my standards. I think the real answer is that the people who are commissioners in most cases do not share my attitudes about the problems with utilities. I have never heard one commissioner say that, in a time of recession, a utility should be feeling the squeeze of the recession along with everyone else. Instead, what happens is the utility comes in for a rate increase, and the commission tries to give rate increases to make the shareholders as rich as they had been in times of plenty.

Is the biggest disability your group—UCCM—labors under a lack of money?

NEWBURGER: That's hard. I mean, I think that if we had all the money in the world, as I said earlier, I'm not sure that we would succeed in implementing the policies we have, because those policies are not the policies of most of the decision makers that we're dealing with. But I think that if we were able to pay for a staff of attorneys, and if we were able to use all the procedural advantages that can be used in a litigation, I think we would probably show up with more successes than we do.

You'd like to see the state agencies strengthened with staff and money and what else? Would you like to see them explicitly authorized to consider radiation questions and the nuclear-safety questions at the state level?

NEWBURGER: Well, I think that it's nonsense to think that they can make the financial decision without deciding the radiation question. Nuclear hazards are directly relevant to the financial analysis of whether to build the plant, if everything could be quantified in the kind of cost-benefit analysis that an agency goes through in order to decide whether financially it is a good idea to build the plant. The Public Service Commission is not, by the Northern States Power cases, not permitted to set new regulations on plants for low-level radiation release and things like that. But they were still left with the responsibility of deciding the financial feasibility of the plant, and to the extent that certain risks and costs related [to] fuel cycles, and so forth, are part of the costs of the plant, [then] those matters should be taken as a whole. I'll bet one of these days that question is going to hit the Supreme Court and that they're going to say that the Public Service Commission can look at the whole financial picture, even though they can't set different [radiation] standards.

Is there something else I should have asked you?

NEWBURGER: There's one other thing that I think you ought to think about. That is, I see an institutional problem that one ought to think about: energy alternatives to nuclear power. Clearly, energy alternatives include solar energy. One of the things that's very clear is that the strategies that are easiest to finance are the ones that tend to be pursued, or the alternatives for energy that are the easiest to finance are the ones that tend to be pursued, even though they may not in the long run be rational. Barry Commoner, in our nuclear case in the Public Service Commission, testified that, according to some data that he had collected in the St. Louis service area, the Union Electric service area, had the company, instead of building a nuclear powerplant, purchased efficient air conditioners and traded them in for the less efficient air conditioners that are presently in use in the market area, that they would be able to save—to reduce the long-run demand for peak capacity by about twelve hundred megawatts, which was the equivalent of one nuclear plant. Now, not only that, but the cost per kilowatt of the air conditioner was not more than a couple of hundred dollars, maybe three hundred dollars. The cost per kilowatt of the nuclear capacity was something like $765. So that you can solve the problem [by moving] in a different direction and clearly save money. The Commission thought that that was a joke. We presented the argument that solar energy used as heat collectors to solve heating and air-conditioning problems in private homes was clearly a superior alternative to nuclear power. What we are attempting to do is to reduce the increase in demand. But the Commission repeatedly says, That's not an alternative that we can handle, because what our question is, is *will* the public have those solar collectors. If they will, then the demand isn't there and we shouldn't be building the plant. But, we [the Commission] cannot force the alternative of putting the solar collectors there as opposed to building a plant.

That's where the mistake is. There are some things that are amenable to central problem resolution. There are other things that aren't. We can finance the building of a massive nuclear-power unit—we do *not* have devices at this point for financing the building of solar collectors on all the homes of the service area. That leads to a very unfortunate inability to take advantage of alternatives that may in the long run be preferable, simply because we don't have mechanisms. That suggests that really what you do need, in that kind of problem area, is a broader regulatory function. There are lots of ways that financing can be done, through subsidies, through tax incentives, through all kinds of programs, through government funding with long-term amortization rates, and so forth, such that the consumer would end up paying the same amount of money for the solar collector that he ends up paying for the nuclear powerplant.

What sort of structure do you imagine?

NEWBURGER: Well, the problem—part of the problem is, of course, that a major company can borrow a large amount of money in a short period of time and pay it back over a very long period of time. The individual consumer is the one who ultimately pays for that financing in his rate bill. If that money could be redirected to capitalize the cost of the solar collectors, instead of being directed in through the rates into the capitalization of the nuclear plant, then the solution is possible. A company that builds these solar collectors and that rents them to the public. . . . That's one mechanism, just doing it privately like that; it can be done through subsidies, as I said, it could be done through a tax incentive program. It seems to me it could be done through excise taxes on uses of other energy like gas or electricity or oil—there's quite a wide range of ways that you could do it, but nobody has really done that. The electric industry, of course, has the mechanism already in place for doing what it does, and no one has the kind of dominating self-interest to construct that alternative mechanism. No one has nearly the vigor or money at stake, or anything like that, compared to the electric industry.

CHAPTER EIGHTEEN

Judicial Socialism

T HE NIXON AND GRANT administrations had some characteristics in common: each was the most corrupt the nation had theretofore suffered, and each president had great difficulty finding among his friends men of minimal ability to appoint to high office. When the office of chief justice of the Supreme Court became vacant in 1873, President Ulysses S. Grant first offered the position to Senator Roscoe Conkling of New York, a man elected by the corrupt legislature dominated by the Tweed Ring; Conkling refused the appointment. (He is perhaps the only man to have done so twice, for President Chester A. Arthur also offered him the post fifteen years later, and he again refused; perhaps the emoluments of private law practice were more than he could bear to give up.) Grant then nominated Mr. Caleb Cushing, who was seventy-three years old, a Democrat and a sympathizer with the late rebellion. Mr. Cushing's nomination was withdrawn when the confirmation hearings in the Senate uncovered a friendly letter addressed by Mr. Cushing to the president of the Confederacy, Jefferson Davis. Nothing daunted, Grant then appointed another old crony, his attorney general, George Williams. This nomination, too, was greeted by a storm of protest and withdrawn.

Like Richard Nixon faced with a similar situation a century later, Grant then nominated a little-known Midwestern attorney of mediocre reputation, and this appointment was confirmed, at least in part for fear of what nominations might follow if it were not. Morrison Remick Waite, an attorney from Toledo, Ohio, took his seat as chief justice of the Supreme Court of the United States on March 4, 1874. The *New York World* heralded his appointment thus: "We suppose nobody doubts that he is up to the level of a highly respectable mediocrity, and we have all an interest in wishing that he may prove himself very far above it." Waite's new colleagues on the Court received him very coolly, with a manner bordering on discourtesy, as he told a friend at the time. The other justices insultingly suggested that perhaps Waite would let one of them preside until Waite had learned the procedures of the Court. Nor did acquaintance improve his associates' opinion of Waite's abilities; after five years, Justice Samuel Miller wrote to his brother that Waite "is much more anxious to be popular as an amiable, kind hearted man (which he is) than as the dignified and capable head of the greatest court the world ever knew. Of what is due to that court, and what is becoming its character, he has no conception." Upon Waite's death, the *American Law Review* could manage no more flattering obituary than, "He did nothing to lower the dignity of the great office which it was his lot for fourteen years to fill, and this is, perhaps, as much as will be said of him in future times."

He was a handsome, strong-featured man with a full head of gray hair and a square beard when he joined the Supreme Court at the age of fifty-seven. His opinions for the Court were not remarkable for their language or scholarship, but Chief Justice Waite was present during momentous events.

The corporation had become, preeminently, the railroad. The profits in rail-

roading were not to be made through transportation, but through the manipula-
tion of securities. Bonds—corporate IOUs—were issued in vast quantities as
self-awarded bonuses to the organizers of holding companies, or as payments for
graft-ridden construction contracts. There were more railroads, and more bonds,
than could easily be supported; freight and passenger rates rose and rose, until
they were all the traffic could bear, and more.

In the time of Marshall, the federal courts had been protective of the great
corporations. The Supreme Court declared their charters to be contracts protected by the Constitution from state impairment. The states had fought back by
writing reservations into all new charters. Corporations had nevertheless become
beings which, while they had many of the powers of governments, were themselves protected by the Constitution from the tyranny of the ballot box. And the
federal courts and the federal government were obliged to protect these corporate
rights with force of arms, if need be.

The battle against corporate depredations therefore became entwined with the
battle for states' rights; state legislatures tried to gain control of the corporations
which they found in their midst, but time after time the federal courts and, for a
while, federal troops defeated them.

The Civil War settled nothing in this regard: the railroads continued their
expansion as Congress overchartered transcontinental railroads, which operated
as monopolies through thousands of miles of the West; farmers who settled along
the railroad lines were dependent on the railroads for their livelihood and so could
be counted on to supply the funds needed to pay the interest on the steadily
mounting pyramids of bonds issued by the holding companies. Only four years
after the close of the Civil War, these farmers began to rise in a revolt which was
at first political; the Granger movement swept through the West, seizing control
of state legislatures. The states began to limit the rates which railroads might
charge, to prohibit them from discriminating in their rates and to lay new taxes
on the railroads. The railroads invoked the power of the federal government and
the protections the federal Constitution gave their charters and their rights. In the
South, this battle coincided with the states' efforts to expel the occupation troops
and end Reconstruction. And Chief Justice Waite, by virtue of that very mediocrity which had assured his selection to the Supreme Court in 1874, was at the
center of this national struggle.

Only a part of the outcome concerns us here. On March 1, 1877, the Supreme
Court announced that the Illinois legislature could limit the rates charged by
grain elevators in that state; and on the next day, Rutherford B. Hayes, the
Republican candidate, was declared president of the United States.

The two events were related. The Hayes-Tilden election of 1876 was in dispute because of the uncertain status of many of the votes. Each side was probably
guilty of widespread fraud, and each claimed victory. The electoral college was
unable to choose a president; consequently, the decision was put to Congress,
which in turn appointed an electoral commission of fifteen—including five members of the Supreme Court.

It now seems that the Democrats agreed to allow Hayes the victory if, in
return, the Republicans would abandon their program of Reconstruction and
withdraw federal troops from the South; a promise to share the wealth being
distributed by the corrupt Grant administrations may also have been given.

It could not have been coincidence—with five members of the Court sitting on
the electoral commission—that, at the same time the compromise over Recon-

struction was reached, the Court handed down the decision in the most famous of the Granger cases, allowing Illinois to regulate the charges made by local grain elevators.

This was a very radical decision at the time. The Granger movement was avowedly radical, and the efforts to control the rates of railroads and grain elevators, on which the Granger farmers depended, were seen as socialist programs. In Illinois, Minnesota, Wisconsin and Iowa, the cry of the Granger was "The state must absorb the railroads or the railroads will absorb the state." The contest was quite clearly between the privileges of private property and the asserted rights of the elected government. As the governor of Iowa later recalled, "Sheltering themselves behind the Dartmouth College decision [the railroads] practically undertook to set public opinion at defiance . . . . In other words, they got it into their heads that they as common carriers were in no way bound to afford equal facilities to all, and indeed that it was in the last degree absurd and unreasonable to expect them to do so."

The *American Law Review* expressed the conservative view of the regulatory laws enacted by the Granger movement; the Supreme Court should "strike down these laws" in order to restore public confidence in the rights of private property now severely shaken. . . . [No] movement would succeed in America which was really directed, not against abuses, but against the rights of property . . . . When the Grangers had once proclaimed that their object was to fix rates . . . it was perfectly clear that the Granger movement was rank communism."

Laws which fixed rates both for railroads and grain elevators were at issue in the several cases which the Court decided on March 1, 1877. In a movement analagous to the withdrawal of federal troops from the Southern states, the Court conceded to the states what were then seen as radical new powers to regulate private enterprise. The Court gave this rationale: it was not really overruling the Dartmouth Case and other decisions which granted extensive protections to property; but in the case of monopolies like those of the railroads and grain elevators, performing a needed function of great importance to the community, private property had become "affected with a public interest," and hence must be subject to some public control.

It seemed that the states had finally succeeded in regaining what they had lost in the Dartmouth case, and even more. For as Justice Stephen Field pointed out in his angry and vigorous dissents, it was difficult to say why a grain elevator is "affected with a public interest" and any other trade or calling is not. The courts would never succeed in drawing any reasonable limits around this doctrine, for every large productive enterprise naturally serves a social function, whether or not it is privately owned. The inherent contradictions in such a state of affairs could hardly be resolved by judicial fiat. Chief Justice Waite hardly seemed aware that he had touched on such profound questions. Commentators, however, were not shy in pointing out the apparent reach of his decision. The *American Law Review* carried many such comments as these: "The Grain Elevator Case strikes at the stability of private property, at rights which lie at the very foundation of modern society and civilization. . . . By the demagogues who are conducting the agitation now going on throughout the country, it is confidently appealed to and relied upon to sustain the yet more communistic and destructive legislation which they demand." The Dartmouth decision had been unseated, it seemed, and the private corporation was now at the mercy of the electorate.

CHAPTER NINETEEN

The Boy Scout Oath

A VISITOR expects William Anders to be taller and blonder than ordinary people: William Anders is an astronaut and he has been to the moon. The achievement is so extraordinary in every sense that one expects some unearthly characteristic to be visible. But the former Colonel Anders is a slight, dark-haired man whose short haircut seems merely conservative and not military. The pilot of the lunar module of Apollo 8—which went to the moon in 1968—has undertaken a new career as chairman of the Nuclear Regulatory Commission. It is difficult to imagine a person who more thoroughly embodies modern technology or its problems.

Chairman Anders is a friendly, blunt, frank person who is accustomed to long hours at work; he prolongs an interview with a reporter well through the lunch hour and then orders sandwiches to be eaten in his office while the conversation continues. The sandwiches are not very good; Anders says loudly for the tape recorder that he hopes the champagne and caviar lunch will not mar the reporter's objectivity. He talks with unabashed pleasure of his meeting with President Ford a few weeks before, a meeting for which he feels he was very well prepared. His pleasure in having been well prepared and consequently commended is almost boyish. There is a casual mention of Richard Nixon, who appointed Anders to the Commission in 1973 (Ford promoted him to chairman in 1975); Anders wants President Ford to be free to carry out nuclear foreign policy without being subject to veto by a regulatory agency or the Congress, but the suspicions awakened by Nixon threaten to hamper the President:

"I would think that the legislation was going to be restructured rather than harping on this. . . . You know, visions of Nixon tripping around the world throwing reactors off like rosebuds. We really ought to be thinking about responsible things. Think about the next Democratic president—if there is one—and couldn't at least he be responsible for implementing our foreign policy. . . . Congress has to decide who they're putting their foreign-policy trust in."

There is a good deal of this engaging talk. Anders is scornful of his predecessor, Dixie Lee Ray, and of the personal animosities which plagued her command of the Atomic Energy Commission. (Just before departing, Ray referred a file she had compiled on a fellow commissioner to the Justice Department for possible prosecution involving a conflict of interest; no observer has believed the charge to be substantial.) Anders freely predicts that he will manage—has already managed—better, and a visitor is inclined to believe him.

William Anders is a team player, accustomed to situations in which a great deal depends on cooperation and obedience. In the formless, directionless Atomic Energy Commission of 1973, he stepped into what should have been the chain of command. President Nixon had just announced that the nation's new first priority in the energy field was development of the "breeder" reactor—a new plutonium-fuelled powerplant that would generate more fuel than it burned.

"And they'd just gone through a real trauma of hiring a breeder-reactor-project guy and you know, all of that, and there was some great suspicion that the cost might be slightly out of line. And so here I was, the dumb new guy who didn't know any better, I said, What do you want me to do? And they said, Why don't you go down and look at the breeder reactor. That's great, you know. I just leaped into the buzz saw. Well, in that regard the management was screwed up. The cost wasn't slightly out of line, it was grossly out of line. And in doing that, [I was] really doing a noncommissionerlike thing, I mean getting down and doing management reviews, and the kind of stuff that if I were in NASA I would have said I was an assistant administrator for a manned space flight, or something like that. In other words, almost in a line capacity. Which was a much easier thing to fit into in the [space] Agency."

Anders uncovered the remarkable multibillion-dollar overruns in the breeder program, and now feels pleasure both in his role and in his evident independence in exercising it. He makes no secret of his feeling that there are serious difficulties in the plans for extensive use of plutonium fuel in the near future. He refers to India's nuclear "weapon" and then corrects himself ostentatiously: "Now excuse me—a nuclear explosive device."

But despite all this badinage, he is firmly committed to the development of nuclear power. William Anders is a team player. The team he is on right now is the nuclear-power industry and its governmental components. In the past, he explains, the Atomic Energy Commission was able to perform a kind of national executive function, overseeing, fostering and directing the development of the nuclear industry (as in the times when Lewis Strauss as chairman strong-armed utilities into buying reactors). But in the reorganization of energy agencies at the beginning of 1975, the Atomic Energy Commission was divided in two, as its critics had long urged. The huge research and development activities, primarily military but including the half-billion-dollar-per-year breeder development and plutonium-fuel programs, were shifted to an Energy Research and Development Administration. The AEC's former regulatory role was put in a much smaller Nuclear Regulatory Commission, which is headed by five commissioners, independent of the president, serving staggered five-year terms.

In the 1950s and 1960s, as Anders pointed out, the combination of development and regulatory authority had allowed the AEC to act as a kind of national executive agency for the nuclear-power industry; after 1954 it carried out the strongly-expressed mandate of Congress that the industry be developed rapidly, and the powerful Joint Congressional Committee on Atomic Energy sat in close and almost daily supervision of its activities; one of the commissioners was a former Joint Committee staff director, and it was taken for granted that the Joint Committee was fully and particularly informed of the activities of the Commission. This curious national executive for the industry came to an end with the division of the AEC. The research and development agency now has no influence over the existing industry and is directed solely toward the development of new technology; the Nuclear Regulatory Commission, which does have great authority, is supposed to be an independent and neutral regulator, without any responsibility to foster the growth of private nuclear power.

Anders feels the executive function still must be performed; that industry is unable to coordinate its activities sufficiently. Perhaps nuclear facilities should be consolidated into regional "parks," to reduce transportation and theft problems.

There is also the problem of securing trained manpower for the power companies:

ANDERS: They [the utilities] have certainly come to me and I've heard in discussion that qualified manpower is high on their list. . . . I've gone to meetings at the secretary of labor's office focusing on construction manpower training, and training of welders and training of pipefitters, electricians to build these more technically challenging power devices. The requirement [is] to insure that trained operators are available at the same time the plant is ready to go. They have to be trained. If they're not trained, ready to go, they can't operate the plant. So the utilities damn well better have the man at least a day before the plant is on the line.

Is this the kind of issue that makes you think there should be some kind of executive agency that can fit into the—

ANDERS: Yes, yes.
[In a speech to the electric-power industry's trade association] I kind of finally said in public what I've been saying, for as long as we've been in business, in private. That somebody in this government needs to do the executive function.

Unlike most people in Washington, however, Anders does not seem to be advancing his ideas as a means of advancing himself. The executive function should be performed, he says, not by his own agency, which is to perform an essentially judicial role, but by the Federal Energy Agency, headed by Frank Zarb, President Ford's closest advisor on energy. Anders has, in fact, urged Zarb to take on the role of directing the national nuclear industry. But Ander's evident esteem for Zarb does not hamper his differences of opinion in some matters.

Would you like to comment on the energy ant?

ANDERS: Energy ant?

That is the symbol, the Federal Energy Agency's symbol. The symbol of the national campaign to conserve energy. Generally to represent the agency the way Smokey the Bear represents—

ANDERS: When I see an ant I step on it.

William Anders, in short, is a team player who finds he must first put together a team and find a captain. He has thought a good deal about this and about the role the Nuclear Regulatory Commission will play. He shifts from the metaphor of a team to viewing the Commission as a link in a chain; in fact, his image of the Commission's function is a mechanical one. It is to make decisions, on a case-by-case basis, like a kind of computerized court.

One of the Commission's early tasks was untangling the regulations concerning routine emissions of radiation from powerplants. Emissions are entirely within the control of the power companies and can be reduced to any arbitrarily low level, given the expenditure of enough funds. The difficulty was to decide how low the emissions should be; AEC scientists John Gofman and Arthur Tamplin had forced a downward revision by publicly calculating the numbers of cancers which would have been produced by the allowable emissions; but how much money should be spent reducing them toward zero? The Atomic Energy Commission, before its dissolution, had dodged the problem by announcing in Delphic fashion that emissions should be made "as low as practicable." Anders cut through the difficulty with an arbitrary administrative decision.

Almost all radiation releases from a powerplant are intentional. Utilities dispose of their radioactive wastes in the least expensive fashion permitted by regulations: Mildly radioactive or very dilute wastes are simply dumped in a nearby stream or released to the air. The NRC could have prevented all such emissions, but the utilities complained that the costs of alternative means of disposal would have been prohibitive. Radioactive xenon gas, for instance, would have to be collected by freezing exhaust gases to near absolute zero, and stored compressed into steel cylinders under very high pressure (or so the manufacturers claimed). The NRC was therefore forced to decide at what point the costs of alternate means of disposal would outweigh the health benefits to be gained by stricter regulation. The Commission under William Anders' direction finally decided to set a fixed value for radiation exposure. Every exposure, no matter how small, is conservatively assumed to cause an increase in cancer and other damages from radiation. Presumably, therefore, a price can be set on radiation exposure—each unit of radiation causing so much damage—and a corresponding price can be placed on control measures. As soon as control measures begin to be more costly than the radiation damage, a balancing point is passed, and the additional controls are not justified.

The result of this approach, of course, is that some radiation releases are permitted because control efforts would be too expensive: some cancers and mutations are permitted to occur, because it would be too expensive to prevent them. By setting the correct value for radiation exposure, therefore, the NRC ensures the optimum level of control of radiation without issuing any complex directions or setting fixed standards. The operations of the market place, in theory, see to it that the best and most efficient controls are employed, and that the net benefit to society is greatest.

By fixing standards in dollar terms, however, the NRC is also ensuring that there will be no incentive to develop cheaper or more efficient radiation controls. Companies will be required to spend the same amount of money regardless of how much radiation they succeeded in controlling. But this is the fundamental difficulty of external regulation: a government agency cannot require ingenuity or innovation. Nor can it substitute its own judgment for that of a private company; a government agency can only set down fixed standards, and supply incentives or punishments to induce compliance with those standards.

ANDERS: We took a forced march and made a decision, which made a fundamental difference in the way we're regulating. We set a thousand dollars for a man-rem [utilities will be required to control radiation releases until their expenditures reach a rate of one thousand dollars for each man-rem—a unit of radiation exposure—eliminated]. We've set up, for the first time, not only an objective mechanism to weigh cost and benefit in a very subjective area, but we have also told them where to put the fulcrum. . . . There's a thousand dollars, and we've further said that we don't believe in that number, but that's the best one we've got right now. And we have a rule-making procedure to let you, the interested

public, and the industry, tell us where that fulcrum ought to be. Never before
have we done this.

Anders is knowledgeable about the other matters which still remain unre-
solved, and he takes a businesslike attitude toward them. There is no question in
his mind, it seems, that the nuclear program will proceed and that there will be
more nuclear powerplants built. He indignantly protests that "none of us,"
meaning the commissioners, "would be here if we didn't think that nuclear power
had the potential for economic and environmental benefits for our citizens." But
within the assumption that there will be a nuclear industry, Anders feels it is
important for the Commission to be a strong link in the chain, or a good team
player, by attending carefully to health and safety regulations, which are its
allotted responsibility.

Despite repeated questions, Anders declined to say what he thought the fundamental policy of the over-all enterprise should be or how it should be set; he
denied that he would be subordinate to the president—by law the Commission
must maintain its ability to make independent judgments that can halt or reverse
executive-agency programs like the breeder reactor; but the NRC does not look to
Congress for over-all policy—Anders did not even refer to Congress in this
regard.

One suspects that much of the NRC's policy evolves in its dealings with industry. The Commission cannot monitor all the activities of industry; it can only
insist on an effective system of industrial self-policing.

*NRC is now though moving toward the traditional position of a regulatory agency. A
relatively small body dealing with a very large industry, and the pattern in the federal
government, as you know, has been for the small agencies to simply give up and try to deal
through negotiation and compromise rather than really to regulate.*

ANDERS: I came away [from a meeting of agency heads with the president]
with the conclusion that most regulatory agencies are considered bureaucratic.
Some of them are considered somewhat captured by those they regulate. That
the safety and consumer-oriented agencies tend to be up against the stops. . . .
Well, these are some thoughts that occurred to me before I came here, or before I
even thought about coming here, and . . . I really thought that it would be very
dangerous if this agency was sort of taken over by the traditional regulatory
mentality, you know. An army of lawyers marching in lockstep with all their
judicial procedures.

*Is it going to be possible for the NRC to supervise the design and construction of each
powerplant, in the way it has in the past? Isn't that the fundamental—*

ANDERS: There is somewhat of a misunderstanding, and I must admit I had
it to a degree, as to the degree that the federal government—either the AEC or
the NRC—gets involved. Now, people do submit preliminary safety reports to
us. We do inspect operating nuclear powerplants. I think if I had to describe what
our main function is, it is based on the theory—and I think a reasonably valid
one—that the person who is ultimately responsible for the safety will be the
applicant [the power company]. We will hold him responsible. Okay. Our main
activity is to insure that the system, both the engineered system and the management system, the reactor itself, and the management procedures as such, are set

out in a general way, that he can handle the technology he is dealing with. Okay. Now specifically in the safety area. We don't inspect the product *per se*. We inspect the process. Now, indeed, you can go back to our inspection reports, you will find guys down there inspecting a valve. Well, that's what I call the ice-pick approach. What they do is jam an ice pick into this blob, and then they see all the way down deep, right to the handle and the color of that valve, if everything is okay. Just as the operator would. And if it's not okay, that's not so much an indictment against the valve as it is against the operator's system to inspect the valve. Okay. And we really club them hard when that happens.

It's rare that we find a case of what seems to be a deliberate attempt to circumvent the system. Because the penalties involved with utilities in any down time [the loss of revenues and the cost of buying replacement power when a reactor is disabled] are fantastic. And so their incentive is to play it straight and play it professionally. And I must say that the utility industry, since it began, has had a very deep professional ethic. The professional engineers, the ASME [American Society of Mechanical Engineers] boiler code which they developed themselves, have been very heartening. And so it is. If you don't want to be flippant about this sort of self-regulation, there are a lot of natural checks and balances built into it which we take advantage of.

The fire underwriters, for example, who give the utilities holy hell. Because they do the fire insurance. . . . It seems to me the chances of a reactor standing on the line when there is a safety question are considerably less than a reactor being off the line because you know somebody didn't do a good job of resolving that question. So the first question. The first rule I think you're generally following is, Let's do it right. Okay. It pays off in the long run.

The utilities have a good record in terms of standards of responsibility and so on, but they have rather a poor reputation for maintaining scientific staffs or well-trained people.

ANDERS: It's like the Chevrolet agencies don't do much automobile research. It's really, it's really not their bag. The manufacturers of their cars, Chevrolet, or, in this case, Westinghouse, General Electric and the others have done quite a bit. That has been focused somewhat off the real problems of the utilities. The utilities realize, there's been sort of a developmental—I don't want to use the "gap"—but lean developmental years and they have gone and formed such things as an Electric Power Research Institute. That's the R&D kind of thing. They recognize, and the federal government has also recognized, that it should do more in these sort of utility-oriented things.

But they're the guys who have to do the on-the-spot regulating, now.

ANDERS: Well, they have to interpret the rules and execute them and their licenses are a charter that's pretty damn severe. If they break the nuclear boy-scout oath, they're gonna get in serious trouble.

There is a fundamental conflict between Anders' insistence that nuclear power be governed by free enterprise, and his clear perception that the industry requires a national executive agency to plan and direct its development. As an able manager he seems out of place, a team player with no team. Throughout 1975 and in the early months of 1976, Anders was under steady attack from industry for his energetic efforts to enforce NRC's regulations. The Boy Scout oath was being taken very seriously. But criticism from opponents of nuclear power was equally strenuous, as Anders made clear his commitment to continued development of nuclear power. In March, 1976, the White House announced that Anders would be leaving the NRC to accept the post of Ambassador to Norway. The NRC issued a statement that Anders had not planned to remain beyond the expiration of his current term in June, and in fact had asked at the outset that his term as chairman be limited to a single year.

Anders was succeeded as chairman by Marcus A. Rowden, a forty-eight-year-old lawyer, an NRC commissioner who had served for seventeen years as staff attorney and then General Counsel to the Atomic Energy Commission. The trade publication *Nuclear Industry*, in announcing his appointment, described him as a "conservative."

CHAPTER TWENTY

Corporations Are Persons

REVOLUTIONS aren't made by judges, and the Granger cases of 1877 were soon reduced to much less than they first appeared. Chief Justice Morrison R. Waite had said a good deal more than he meant. Perhaps some concessions were needed to curb the worst abuses of the railroads, but Waite and his colleagues were not prepared to go very far. In case after case, Waite and the other justices warned that the Granger decisions were not what they seemed to be. For one thing, the Court retained its right to review the regulations imposed by the states. And, "it is not to be inferred that this power of limitation or regulation is itself without limit. This power to regulate is not a power to destroy, and limitation is not the equivalent of confiscation."

But the difficulty was, of course, that regulation and confiscation were degrees on the same scale; to reduce a rate meant to deprive the railroad of profits, just as surely as if its bank accounts were taken by the state. Waite had become trapped in efforts to draw distinctions which he could not articulate. If the states had the power to regulate, how could it be limited this side of socialism?

Roscoe Conkling helped out. Senator Conkling, who had turned down Waite's post as chief justice, was lucratively retained by a group of railroads to argue an important case in California. These railroads had hired not one, but two United States senators in a case which was meant to, and did, restore the protection of the Constitution to corporate rights.

Senator Conkling had been a member of the congressional drafting committee which produced what is now the Fourteenth Amendment to the Constitution of the United States. The amendment was drafted and ratified in the aftermath of the Civil War and was meant to give constitutional support to civil-rights laws passed by Congress to combat Southern efforts to restore the equivalent of slavery. The Fourteenth Amendment was passed to make good the promise of freedom, and to give the federal government the power to insure the civil rights of all citizens. There was no real question about the amendment's purpose, and each time a question concerning it came before the Supreme Court, the Court decided that this purpose was the narrow one discernible on its face. The amendment stated that: "No state shall make or enforce any law which shall abridge the privileges or immunities of citizens of the United States; nor shall any state deprive any person of life, liberty, or property, without due process of law; nor deny to any person within its jurisdiction the equal protection of the laws." This seems to be clear in its intention, if somewhat vague in terminology. The Supreme Court, when first asked to construe this language, found that it was meant to restore black citizens to the status of white citizens, and to keep the states from reducing any other group of people below that level.

So narrowly was this language read that Chief Justice Waite could complacently hold that it did not apply to women; the Fourteenth Amendment did not guarantee them the right to vote (indeed, in a remarkable display of judicial obtuseness, he held that the Constitution did not guarantee *anyone* the right to vote). In effect, Waite had held that women were not "persons" in the meaning of the Fourteenth Amendment; the amendment extended its protections only, it seemed, to black men. And then, without discussion, Waite announced that the word "persons" in this amendment included railroads.

This was the argument which Roscoe Conkling had been hired to make. California had levied taxes on the railroads which they resented. The taxes were aimed in part at the huge pyramids of bonds erected on each mile of railroad track. In assessing property taxes, the state had decided, railroads would no longer be allowed to deduct the amount of mortgage bonds from the value of the property as a whole; but other corporations, and all individuals, were permitted to deduct mortgages from the assessed valuation. Railroads were heavily mortgaged and planned to become more so. In 1882 and 1884, the taxes were challenged in the California state courts and in the federal courts.

Roscoe Conkling made a simple argument on the railroads' behalf. He, as a senator, had been one of the Republican members of the drafting committee which produced what became the Fourteenth Amendment. Senator Conkling said he had hand-written minutes of those meetings which showed that the drafters intended to include corporations within the term "persons" as used in the Fourteenth Amendment; this was done, he said, to extend the protection of the federal government to business enterprises, large and small. None of this appears to have been true; the notes didn't exist.

Justice Stephen Field, the justice of the Supreme Court responsible for the ninth circuit, which included California, sat on the appeals court in which these arguments were made; he approved of them. They echoed opinions he had been writing for twenty-five years in defense of free enterprise. Justice Field, who with a flowing white beard looked like Michelangelo's Moses, quoted Adam Smith in arguing that the rights of property were "sacred"; the quotation was a well-worn one, for it was taken from the unsuccessful argument by John A. Campbell in the famous Slaughterhouse cases, thirteen years before, in which the Court had first held squarely that the Fourteenth Amendment did *not* apply to business enterprises. Field had dissented in that case, citing the Adam Smith quotation for his own. He repeated the quotation and parts of Campbell's arguments in a later opinion which triumphantly reversed the effect, if not the principle, of the Slaughterhouse cases; and then he found much of the same language useful again when he decided for the appeals court that railroads and other corporations were persons and so were entitled to constitutional protection. Field apparently believed that the state governments were powerless to treat a railroad corporation any differently from a human being.

The state of California appealed Field's decision in favor of the railroads and the case was argued before the Supreme Court during four days in January, 1886. But Roscoe Conkling was not called upon to repeat his bogus argument. The court reporter recorded the following at the beginning of his account of the case: "Announcement by Chief Justice Waite: The Court does not wish to hear argument on the question whether the provision in the Fourteenth Amendment to the Constitution, which forbids a state to deny to any person within its jurisdiction the equal protection of the laws, applies to those corporations. We are all of the opinion that it does." In his customarily awkward language, Waite had announced that corporations were persons entitled to all the protections of the Fourteenth Amendment. This amendment restrained the exercise of state power; corporations were therefore protected by the federal government with those safeguards erected for the benefit of the emancipated slaves. Corporations were entitled to due process of law and all the legal safeguards which had been extracted bit by bit from the English kings to protect the helpless individual against the overwhelming power of the state. Corporations were guaranteed equal protection of the law, which meant they were protected from discrimination at the hands of state legislatures. Federal courts could thenceforth review any state action which deprived a corporation of property; and, as later cases were to show, the federal courts would strike down, as unconstitutional deprivations of property, any state laws which regulated prices or wages.

This revolutionary doctrine was adopted without argument by Waite's court. Like Marshall, Waite enunciated the conclusion without providing any argument in its support. The case in which it was announced did not, in fact, hinge on this issue and was decided on other grounds; but the Court from this time forward cited this case as if it had determined the question and, within a few years, the Court would say that the question was well settled. By this announcement, therefore, Waite was able to read into the Constitution of the United States a new provision of great importance, without having to expose the question to argument or having to offer any justification in its behalf.

While it had seemed for a time that corporations might be subject to the tyranny of the legislature, they were thenceforward clothed in all the protections from despotism that the Constitution afforded each man (but not yet each woman). But the corporation had many more resources than a lone man to insure its well-being. It began existence with a charter of privileges and immunities that granted governmental powers: utility corporations, including railroads, had the right of eminent domain and could condemn the property of others for their own use. But most of all, railroads had money, enough money to hire senators in their defense, to make extensive use of all the legal machinery provided by the Constitution for the protection of individual rights. The Granger cases were thus left as an anomaly; the states could regulate the prices charged by public utilities like railroads, gas companies and grain elevators. But states must extend to these corporations, themselves armed with the power of governments, all the protections against state oppression guaranteed to the most impoverished ex-slave.

The court decisions which judged corporations persons "substantially reversed the position of *all* corporations in their relation to government. That is to say, once corporations were declared persons they were able to challenge the validity of any action adversely affecting their interests; from agencies traditionally presumed subject to strict legislative control, they were transformed overnight into agencies able to compel government to justify its regulations affecting them. . . . Practically speaking, it was quite as if all laws and regulations were *presumed* to be hostile or discriminating until the judicial branch ruled to the contrary." The attorney from whom this passage is quoted goes on to say, correctly, that this view, attributed to "historians" is faulty in that it attributes to a single judicial action what was, in fact, a situation dictated by the economic and social realities of the nineteenth century. This passage nonetheless accurately describes the relation between corporations and government, as embodied in the legal machinery built on the Supreme Court decision of 1886.

One surprising result of the doctrine that corporations were persons appeared in 1896, when the Southern Pacific Railroad challenged Nebraska's rate regulations, imposed under the authority of the Granger decisions. William Jennings Bryan, fresh from defeat as the Democratic candidate for president, argued the case for the state; the railroad deployed the customary platoons of Eastern legal talent.

Southern Pacific claimed that it was entitled to rates which would allow it to meet operating expenses, as well as to pay dividends on its stock and interest on the great volume of bonds it had issued. Like other railroads of the time, the Southern Pacific had issued a great many bonds, or mortgages against its future income, to stockholders and government officials for their own profit; the farmers and other customers of the railroad were to supply the cash. It would be an unconstitutional deprivation of the corporation's property, the railroad lawyers maintained, to keep Southern Pacific from meeting all of these dividend and interest payments.

The state of Nebraska wished to restrict the railroad's rates to what a well-managed enterprise would have charged; enough for operating expenses and a reasonable return on an assessed value of its physical assets; the historical cost of building the railroad, inflated by corruption and incompetence, was to be ignored, thus treating great masses of bonds, accurately enough, as profits rather than investments.

The Supreme Court took a middle course and completed the process of compromise begun in the years following the Dartmouth decision. Like Solomon it decreed that each should have half of what was asked; the railroad would not be limited to what a well-run railroad would charge; it was beyond the power of the state to interfere with the management of the railroad. But neither could the corporation exceed what would be reasonable rates. The standard by which the rates were to be measured was a compromise based on the constitutional rights of the corporation; it would be limited to obtaining "a fair return upon the value of that which it employs for the public convenience."

The state could control a public utility only to insure rates which provided a fair return on the utility's physical assets—the actual plant and facilities used to provide services to the public. The state could interfere no further than to limit rates to this return; it could not judge the reasonableness of the investment or set rates on the basis of what it thought a well-run railroad would charge. And the federal courts would offer their protection to utility corporations to assure that state governments attempted to do no more than was permitted to them.

The Supreme Court could hardly have been expected to foresee that this decision, in time, would make it possible and desirable for electric power companies to build nuclear powerplants.

CHAPTER TWENTY-ONE

The Environment

SECURITY IS TIGHTER at the Environmental Protection Agency than at the former headquarters of the Atomic Energy Commission. The old AEC Washington office, dingy and decayed, is still guarded by one of the women who sit at the portals of all seats of power, but the signs warning visitors against carrying cameras or firearms into the once-secret chambers, now occupied by the much-diminished Nuclear Regulatory Commission, are tattered reminders of an earlier day. The Environmental Protection Agency, by contrast, is well guarded. A visitor travelling by car or taxi is halted at a guardhouse and barrier, where a uniformed guard inspects occupants of the car but asks for no identification; travelers on foot are scrutinized but not halted. The office building itself can be reached only through a single entrance guarded by a uniformed woman who *does* ask for identification; a visitor is then permitted to travel to the top floor but one, where there is a second, more comfortable waiting room, where another female guardian, in civilian clothes this time, asks for a name and the visitor's purpose; after a wait of a few moments, the visitor passes by a separate shuttle elevator to the top floor of the building, where Russell Train, administrator of the Environmental Protection Agency, has his office and where there is yet another anteroom and receptionist.

The oddities of this arrangement are partly explained by the fact that the Environmental Protection Agency is housed in what was to have been a high-rise apartment building, part of a complex of apartments and shops of the kind many cities are now building in an effort to lure a middle class back from the suburbs; only two of the three apartment towers were rented, however, and the federal government, as it so often does, came to the rescue by renting the remaining empty tower for the offices of the EPA. The security measures are in part relics of what was to have been reassuring protections for hesitant apartment dwellers (presumably white) in a sea of black poverty. Russell Train's office is what would have been the penthouse of the luxurious apartment building. Although the effort to lure back apartment dwellers has been abandoned, the security measures have been stiffened, because teenagers have been wandering into the EPA headquarters, jostling federal employees and perhaps pilfering from unguarded desks. The repeated checkpoints are intended to intimidate prospective vandals.

Russell Train is a genial, very likeable, sandy-haired man, a former tax-court judge, who for several years headed the Conservation Foundation, a Washington-based group supported in part by the Rockefeller family charities. The Conservation Foundation was one of the older, more conservative groups in the field, with strong interests in the preservation of wildlife and a faintly imperial concern for the environments of developing nations.

Train moved on to the Council on Environmental Quality in 1970, one of the three initial members appointed by Richard Nixon. The council's work was largely advisory and perhaps cosmetic; in 1973, however, when Watergate devastated the Nixon administration, Train was caught up in the swirl of changes. William Ruckelshaus, who had taken a strong independent stance as head of the

EPA, was shifted to the Justice Department and resigned shortly thereafter rather than fire special prosecutor Archibald Cox. Nixon hastily shifted Train to the EPA, an appointment much praised as time passes. Train, as both a judge and an environmentalist, seems extremely well suited by background and disposition to his new role, and he has so far managed to retain the support of most environmental groups, although his agency has followed the dictates of the Ford Administration—which lacks enthusiasm for environmental protection—so far as the Congress has permitted him to. Train, in fact, heads an anomalous agency. While called upon to carry out the judicial role of the independent regulator, the agency is firmly a part of the administration, and Train is responsible to the president. Unlike the heads of the Nuclear Regulatory Commission and other independent agencies, Train serves no fixed term and can be removed, with or without cause, at the whim of the president. It is difficult to imagine a judge able to make truly independent judgments in such circumstances, but Train insists that with regard to enforcement decisions, the quasi-judicial side of the agency's work, no interference from any other agency or from the president is permitted, and that none has been offered. In issuing rules and regulations, however, the EPA works closely with a wide variety of other agencies, with private industry as well as the general public, seeking to obtain consensus or at least an expression of differing views.

In its combination of managerial and regulatory functions, the EPA resembles the old Atomic Energy Commission; the combination lent unusual executive power to that agency, and it was much criticized on that score. Train doesn't think EPA is subject to the same criticism; while EPA makes rules and also enforces them, combining legislative, executive and judicial functions, Train says, "But I would contend that the ax we are grinding is a public-interest ax—that's a somewhat self-serving statement, and I don't believe we necessarily have a monopoly on public interest here. But I think it's a different situation than, I'm sure, with the AEC, really quite different."

Train is not much interested in these theoretical questions, however. If on paper it seems that the EPA is extraordinarily powerful, the reality of daily work shows that it is not. Environmental problems span the entire range of industrial and governmental activity; there are few organized efforts of any kind which do not affect the environment or do not come within the purview of the EPA in one way or another. For the agency to have the kind of executive function that William Anders thinks the nuclear industry needs, it would have to become a kind of supergovernment, an executive agency for the entire nation's industrial enterprise, balancing foreign-policy considerations, domestic economic demands, the conflicting requirements of air- and water-pollution control, the conservation of resources and the planning of land use. The EPA has extensive theoretical authority to manage the US economy, but it is very much like William Anders's theoretical authority to halt the nuclear-power program—neither man would think of exercising such authority, and it clearly would be withdrawn by Congress if he should make the attempt. Train, like any judge, takes each case as it comes before him and balances conflicting interests as well as he can. The Congress provides by statute that in setting air-pollution regulations, the administrator is to consider only the health of the general public; the Ford administration insists that sulphur-control limits be eased to permit the substitution of coal for Arab oil; private power companies insist the technology to comply with the EPA's regulations simply won't work or is too expensive. Train makes decisions as well as he can and in response to a question confesses that he does not think of himself as regulating the electric-power industry—although, of course, studies are performed to show the impact of his agency's actions on the industry as a whole. For Russell Train's decisions affect the power industry, and nuclear power, as profoundly as any one person's.

Train is not particularly conscious of nuclear power; he is not familiar with the other federal agencies which regulate radiation releases and exposures, although he "had lunch once with Bill Anders" and the Nuclear Regulatory Commission, to agree on boundaries between their activities. Train's agency has responsibility for setting over-all radiation standards for the environment (as opposed to the specific regulation of particular bits of machinery, which is the NRC's job). Train, however, has difficulty with the very terminology of this field; he draws a blank when a reporter asks him about the NRC's value of a thousand dollars per man-rem, a standard measure of radiation exposure. Train, in announcing his own agency's standard (a value of five thousand dollars for the radiation exposure assumed to cause one case of cancer) avoided the use of technical terms, and he does not know whether his agency's standard is in conflict with that set by NRC.

How do you feel about nuclear power, personally, and quite aside from (laughter). . . . *Well, do you want to go off the record?*

TRAIN: No, no, no, no, I do this all the time. In the first place, I find it, as a lawyer, just goddamn hard to understand *(laughter)* the technicalities of the issues and the problems of the whole nuclear science. It's a tough one, I find, to deal with. I don't tend to be a wild enthusiast for nuclear power, I don't think very many environmentalists are. I'm not out calling for a moratorium because I think we do have to have energy, and I think some of the alternatives to nuclear power may be just as bad as nuclear, represent just as many problems as nuclear power. I'm not at all sure that coal may not be in that ilk. I am, like many, very concerned about some aspects of nuclear power . . . the safety issues, the waste-disposal problem, the plutonium-safeguard problems, I don't think we have satisfactorily solved any of these as yet, and I think that we really should find solutions. . . .

I think from our standpoint, just on a practical [basis], energy conservation can be exceedingly important in helping buy time for doing the research, finding the answers to some of the energy problems, and in permitting us, where we do feel that we can go ahead in something like the [oil drilling on the] outer continental shelf, to do a decent job and take the time to do a decent job. . . . [W]ith an effective energy-conservation program, I think we can avoid a crash, panic-driven supply program, I hope. I must say, I'm becoming increasingly concerned over failure to do really anything. We're really not doing a goddamn thing. And I think the public attitude is very troublesome, I don't think the public is really much concerned over the energy problem any more, and I don't know what the answer is. Maybe another embargo.

Do you think that the executive should be taking the leadership in that, or is it something that requires whole new programs in legislation?

TRAIN: Oh, I think both. I think there's a very real limit in what the executive can do without the legislation, and there seem to be very real limits on what the Congress is willing to do. And I must say I'm depressed at the moment about this picture. I don't know where the fault lies, probably shared all around.

CHAPTER TWENTY-TWO

The World Capital

THE NEW POST OFFICE BUILDING, as it is known in Washington, is a massive gray stone structure on 12th Street and Pennsylvania Avenue, one of the great antheaps of federal workers scattered between the Capitol and the White House. A visitor entering from 12th Street comes through a grubby branch post office on the ground floor, where tired customers shuffle slowly across scuffed tile floors toward barred windows. A small lobby is carpeted in some aged red fabric; a dropped ceiling conceals the air-conditioning ducts and artifical lights, all on full blast although the lobby has ample windows. Dominating the small area is a larger-than-life bronze statue of Benjamin Franklin, presumed founder of the US Postal Service. A chiselled inscription on the wall informs the passersby that this building was "Erected under the Act of Congress Approved May 25, 1926, and July 3, 1930. Completed under the Administration of Franklin D. Roosevelt, President of the United States, James A. Farley of New York, Postmaster General." The inscription seems calculated to distribute credit or blame between Republican and Democratic administrations.

Past the lobby, a uniformed woman guard sits at a rickety wooden table before the elevators ("Can I see some ID?"). Ramshackle elevators carry the visitor upward to the third floor, where he wanders for some time along curved, Kafkaesque corridors lined with featureless doors. A passerby gives directions, and the visitor sets off again along the beige corridors. Turning a corner, he comes upon a circular gallery that seems wildly out of place. A graceful circular metal staircase passes through the terrazzo floor; large, lightly curtained windows illuminate the stairwell. In a wooden archway facing the stairwell is an incised wooden sign which identifies this as the Office of the Administrator of the Federal Energy Administration. The visitor is seeking Frank Zarb, head of this newly established agency and chief energy official of the Ford administration; the agency is only a few months old, and the visitor has been warned to expect only temporary quarters in the New Post Office Building.

Surprised, the visitor passes through the carved archway into a comfortably furnished reception area with high dark-wood wainscotting. Behind a low wooden railing, at a very large blond wood desk, is a strikingly beautiful young woman, with masses of curly black hair and a straight Greek nose. She greets the visitor with a broad smile of obvious warmth and sincerity—she is evidently *pleased* to see him; won't he take a seat. The visitor, now entirely off balance, remarks inanely that this is a very nice office—as indeed it is. The receptionist seems even more pleased and rises from her seat to hold open the gate in the low railing which separates them.

"Would you like to see the room in which we hold our press conferences?" she asks eagerly.

Certainly, the visitor would. He is taken behind the railing, and through tall oak double doors. He steps forward as the receptionist throws open the double doors, and then he stops abruptly.

The room is like nothing he has seen in the United States. The ceiling of carved dark oak is thirty feet above him. To his right, the wall is an unbroken expanse of carved oak, rising in columns and arches to the ceiling; ahead, the room stretches what seems an impossible length; the absence of anything to give it ordinary scale makes distance hard to estimate, but the far wall seems to be a hundred feet away; in the intervening space, three gigantic cut-glass chandeliers, illuminated by thousands of dim lights, hang suspended by massive chains from the ceiling; the left wall is windows, from floor to ceiling, lightly covered with translucent white drapes which admit a dim illumination but conceal whatever is to be seen outside. A few scattered rows of folding chairs are dwarfed by the dimensions of the room; at the far wall is a slightly raised dais; the oak panelling somehow suggests by its design that a throne was once on that dais. Indeed, the whole room gives an overwhelming impact of European royalty; it is an audience chamber, a throne room. The visitor is literally open-mouthed.

The receptionist was waiting for this; once the full impact of the room has registered, she says gleefully: "This was Jim Farley's office."

The sign in the lobby; the New Post Office Building; James Farley, postmaster-general, patronage chief for the Roosevelt Administration, dispenser of jobs and favors for the governor of New York and then for the president of the United States; the king of patronage for the whole New Deal era; this was his office. The receptionist watches all this sinking in, and she laughs delightedly when the visitor, too, begins to laugh.

Frank Zarb couldn't bring himself to use Jim Farley's office, so the Audience Room is now used only for press conferences; the receptionist says that in Farley's time job seekers entered through the reception area and then had to traverse the whole awful length of the room to Farley's desk on the dais at the far end; one imagines the supplicants making this passage on their knees. Zarb has moved into the only slightly more modest private apartments that Farley had constructed for himself, which are separated from the Audience Room by a carpeted and panelled hallway. Zarb's office, which was probably Farley's bedroom, is no more than forty feet on a side, with a twelve-foot ceiling and carved oak panelling over the walls, floor-to-ceiling windows and an immense oak desk behind which Zarb reclines, in shirt sleeves, in a high-backed leather swivel chair. The exaggerated proportions of the office make Zarb seem smaller than he is; a slight, forty-year-old, black-haired man with startling blue eyes which he turns widely on a visitor to emphasize a point or engage his empathy. Zarb seeks to be engaging, but he is tired; he has spent the morning testifying before a congressional committee on oil price controls; he will spend much of the afternoon at a televised press conference and in further meetings. The visitor seems to be taking his lunch hour, but nothing is said of this; it is evidently important to Zarb to cultivate the press. A public-relations man listens patiently to the conversation and afterward asks how well Zarb did—an odd, unprofessional expression of anxiety about the reporter's opinion.

Frank Zarb himself, however, does not seem ill at ease. He has been in government since Richard Nixon appointed him assistant secretary of labor; he had been a senior officer of a large Wall Street bond brokerage. In the second Nixon administration, he was promoted to a role in the Office of Management and Budget, where he had considerable authority over the planning and budgets of the Atomic Energy Commission and the Department of Interior. It was presumably this experience which led Gerald Ford to appoint Zarb as administrator of the Federal Energy Administration. Now, like one of Napoleon's marshals given a Bourbon palace and a kingdom, Zarb sits in James Farley's office, an office that Zarb says was once the "patronage capital of the world." Zarb has been to see Farley and reports that the old man says of the earlier days only that "we found jobs for a lot of good Democrats in that office." Zarb tells the story of their meeting with much relish.

Zarb is knowledgeable about nuclear energy in general, but he does not condescend to specifics. His realm is that of broad policy, international questions of great import:

"In the next ten years, I suppose, three things ought to be happening; we ought to be using the light-water reactor [the type of nuclear powerplant now in use] to generate more electricity. In a way that we can raise it from roughly five percent [actually two percent] of total energy now, to approximately twenty percent. We ought to be thinking in those terms if we are going back to the needed [low levels of] imported oil, and bring consumer prices back to some semblance of reason, with respect to electricity. The second thing that should be happening in that period is a satisfactory solution to the fuel-cycle question [supplies of fuel and disposal of waste] and a satisfactory solution to the safety—excuse me, [plutonium] safeguards—question. The third thing that should be happening is that we should make some meaningful advances, both with the breeder reactor and with fusion. Now those are the three general things that we should achieve over the next five to ten years."

These appear to Zarb to be modestly difficult technical issues, by no means decisive for the power program. He concedes the theoretical possibility of accidents in powerplants which would do widespread damage, but says the probability of such accidents is too low to be of concern. His discourse is on a very grand plane: the United States must maintain its participation in nuclear-power programs so that it can provide leadership elsewhere in the world to the nations who will move in this direction with or without us; plutonium must be safeguarded through international agreements.

Until asked specifically about it, Zarb fails to mention energy conservation or the development of energy sources which do not rely on fuel, such as solar power. When asked, he says that energy conservation is, in fact, the highest priority of his agency.

It is difficult to get any sense of the reality underlying these pronouncements. The national nuclear-power program is to be greatly expanded; Zarb agrees that his agency should be the focus of this great expansion, which is needed for foreign policy and domestic reasons. But there seems to be little or nothing that the FEA is called upon to do. Press reports later indicated that Zarb had created a fifteen-man office to carry out whatever role FEA will have in the planned expansion of nuclear output over the next ten years, at a rate that seems impossible without heroic national efforts. Because the FEA is not within the jurisdiction of the powerful Joint Congressional Committee on Atomic Energy, this effort—however minuscule—to assume leadership on nuclear matters was strongly resisted in Congress; the nascent nuclear executive was shifted to an unobtrusive place in FEA's planning office.

The movement toward energy conservation, which purportedly is as important to the administration as the development of fuel substitutes for imported oil, is almost indiscernible: the only tool which Zarb thinks is needed is the higher price of fuel occurring independently of any government action—higher prices will encourage business people and homeowners to conserve fuel; and tax breaks and other federal benefits for industry, which Zarb calls "incentivising." The incentivising process, in his view, makes use of the free market and does not involve government telling people what to do.

Zarb also accepts the idea that his agency, or he himself as secretary of the cabinet-level Energy Council, is in the position of managing over-all federal energy programs. But here again Zarb steps back from asserting any true managerial role. His function with regard to other federal agencies he describes as "catalyzing."

With regard to state agencies, the FEA proposes modest legislation intended to grant rate increases to gas and electric-power companies when state agencies are too slow in granting such increases. The present financial difficulties of the power companies, Zarb claims, can be traced to the states' failure to provide rate increases with sufficient promptness.

The only positive program to secure energy conservation that Zarb names is his "Energy Ant" conservation campaign: he personally chose the ant as a symbol. (The public-relations office later supplied the reporter with copies of art work and complete booklets apparently intended for grade-school children, in which the principal figure was a cartoon ant dressed in warm clothing, from which an abdomen (the ant stood erect on two legs) projected grotesquely. Newspapers later report that the Energy Ant comic books are the work of Frank Zarb's wife.)

At the end of the allotted hour, Zarb wearily gathers himself for a press conference. He has been treated well by the press, for he gives them a great deal of his time and flatters them. At the close of the interview, Zarb grips the reporter's hand, fixes him with wide blue eyes and earnestly asks for a copy of the book the reporter is writing. He repeats the request, to make sure the reporter understands it is sincerely meant.

CHAPTER TWENTY-THREE

Public Service

IN THE NATIONAL mobilization of the Cold War, the federal government was able to, or was impelled to, launch a civilian nuclear-power industry. The initiating impulse is spent, the Atomic Energy Commission has been dissolved, and the government now finds itself helpless to foster or direct the continued growth of the civilian industry.

There is no agency of the federal government whose purpose is to implement a national energy policy. There is no such policy to be implemented, in any case, except the vague generalities which Frank Zarb expresses; still less is there a national policy regarding nuclear power. National policy emerges piecemeal from thousands of case-by-case determinations and individual conflicts, like the one in

Missouri. The state public-service commission, so reluctant to assume the responsibility for deciding on a single nuclear powerplant, is the agency effectively charged with establishing a policy to balance fuel-supply, economic and environmental considerations against the public health and safety; such decisions are beyond a state agency's competence or resources, but by deferring to federal agencies, the state commission defers its decision to a vacuity. There is no federal agency, nor any federal policy, which will resolve the questions put before it. The Nuclear Regulatory Commission will not consider the risk of financial loss to the company's stockholders or the effect of price rises on Missouri customers; it will not review the desirability of expanding electric-power production; it will not weigh the preferences of Missourians concerning the statistical risks inherent in nuclear power. It will simply decide whether the Union Electric Company has complied with NRC regulations.

The Missouri Public Service Commission has no intention of remedying these omissions. The basic contracts between the Union Electric Company and the Westinghouse Corporation were withheld from the Commission. They are private property and may not be divulged; Westinghouse stipulates in the contracts that Union Electric may not reveal their terms, and so Union Electric will allow the state to view them only if it, too, swears to maintain the secrecy. The citizens' groups which oppose the company may only examine these contracts under the same restraints. Neither the citizens nor the Commission show much interest in acquiring information which they cannot use, however, and the contracts for the supply of a nuclear-power system and for the fuel to run it remain secret. The entire proceedings before the Public Service Commission therefore have an unreal quality; one witness is permitted to testify that he has examined the contracts for fuel supply and that they are very favorable to the utility; but within a year after the close of the hearings, Westinghouse announced that they will not comply with the terms of the contract because uranium prices have risen so high as to make compliance "commercially impracticable." Union Electric promptly brought suit. Alberta Slavin's witnesses had tried to testify about the difficulty of obtaining uranium fuel, but were not permitted to make this point because of the secret contract which purportedly guaranteed the reactors' fuel for twenty years.

In the midst of the hearings, Commissioner William Clark resigned in the face of press allegations that he had accepted numerous favors from the Southwestern Bell Telephone Company, including free hunting trips to a lodge which the company allegedly maintains. In the wake of his resignation, the Commission began an investigation into allegations of widespread bribery by the telephone company, which it also is responsible for regulating. The report of its investigation exonerated Bell, and found no purpose to be served by prosecuting Clark.

Within a few weeks of the close of its hearings on the nuclear powerplant, the Commission was confronted with further allegations of wrongdoing. A nuclear engineer from the University of Missouri, who had submitted prepared testimony, was stricken from the list of witnesses and his testimony was never made part of the proceedings. The testimony was damaging to the utility company and would have questioned the adequacy of industry standards intended to prevent the bursting of a nuclear plant's boiler; it seemed odd that the testimony had been suppressed by the Commission or its staff. Colleagues of the nuclear engineer were permitted to testify and said under oath that they had no conflicts of interest, but it was later alleged that the professors all worked for a university department which was receiving, and which expected to receive, substantial research support from the utility company.

Utility officials announced shortly after the nuclear hearings had closed that a large coal-burning plant which had been planned would be cancelled. The announcement made no mention of the officers' sworn testimony that the coal plant as well as the nuclear plant would be required to meet expanding demand for electricity. If the announcement of cancellation had been made during the hearings, the Commission might have had to consider whether to retain the coal plants which were already planned and postpone or cancel the nuclear plants which were less certain.

Despite the difficulties it labored under and its lack of relevant information, the Public Service Commission decided to grant to the Union Electric Company authorization to build a nuclear powerplant. The Commission found that a new plant was desirable and that it had no authority to determine whether or not a nuclear plant would be safe. A state court upheld its findings.

The Commission had not so much made a decision as evaded one; and as had happened so many times before, the conflict moved toward the federal courts, where ultimately it will have its resolution. The Union Electric Company, protected by the constitutional guarantees of due process and equal protection of the laws, will have its day in court. The federal government, unable and unwilling to intervene directly in the affairs of a private company, will again act as adjudicator of the conflict between the corporation and the public, a conflict in which the corporation will have overwhelmingly greater legal resources than its opponents.

III. POWERPLANTS

W. Donham Crawford, president, Edison Electric Institute (Chapter 25)

Gregory Vassell, vice-president, American Electric Power
Service Company (Chapter 27)

Dayton Clewell, senior vice-president for research and development, Mobil Oil
Corporation, interviewed in the office of *Environment* magazine (Chapter 32)

CHAPTER TWENTY-FOUR

An Invention

A T THE AGE OF TWENTY-ONE, Samuel Insull was a slight blond with muttonchop whiskers and a contemptuous curl to his upper lip. His eyes were prominent and noticeable. In 1881, the year of his majority, Insull became private secretary to Thomas Edison, then just past thirty. When they later recalled the meeting, each remembered thinking how very young the other had seemed.

Insull was English, from an earnest, teetotalling family of modest means. He had learned shorthand and was advancing nicely in a London firm of auctioneers when, in 1879, at the age of nineteen, through one of those improbable coincidences which were common in the rags-to-riches stories of which young Samuel was fond, he answered a newspaper ad and was hired by a banker who served as Thomas Edison's London agent. Edison, the fabulous inventor—or mountebank, as many still called him in England—and Phineas T. Barnum were Insull's flamboyant heroes; within two years, the young man was able to obtain a favorable recommendation to Edison himself, who, finding himself in need of a secretary, summoned Insull across the Atlantic.

Edison needed a secretary because he had undertaken an ambitious enterprise. For several years, rumors had flown concerning Edison's electrical inventions; in 1879, the year that Insull joined his London agent, a rumor that Edison had made a practical electric light touched off a panic in gas-illuminating company stocks. But Edison had not invented an electric light; this had already been done, by Sir Joseph Swan in England, and perhaps others; Swan, in fact, patented the incandescent electric light (a carbon filament in an evacuated glass bulb) which American school children are taught was invented by Edison. Edison was at work on something much more grandiose: he was trying to create an entire industry. In 1881, just as Insull arrived, Edison was shutting down his famous Menlo Park laboratory in New Jersey and moving into 65 Fifth Avenue, a narrow four-story brownstone building just north of Washington Square in Manhattan. His object was to create the first system for the sale of electricity.

All of the elements of a commercial electrical power system had already been invented. The dynamo, the first practical generator of electricity, had been developed by Zenobe Gramme, a Belgian mechanic working in Paris. Electric lights, brilliant arc lamps used for street lighting, were already widespread in Europe and had been installed in several municipal lighting systems in the US; the first such was installed in San Francisco in 1879. The incandescent light, needed for interior lighting, had been invented, as had the electric motor to harness electricity for industry. What was lacking, however, was a means of putting these elements together into a single workable system.

Arc lights, for instance, were then always arranged along a single circuit (wired in a "series") and powered by a generator devoted to that single purpose. If one of the lights burned out, the entire series would be extinguished, leaving the generator idle. This awkward arrangement seemed to be necessary, and the early development of incandescent lamps followed the same pattern: Joseph Swan's lamps were designed for use in a single circuit. As a result, each lamp was made to have low resistance to electricity: the voltage drop across each lamp was small, but the flow of current through each was heavy. Early incandescent lamps thus suffered from the difficulties of arc lights—if one failed, all would fail, and the heavy flow of current seemed to assure that each lamp would be short-lived. If motors were to be driven by electric power, they could not be part of this same system but needed generators of their own.

The difficulties of these early systems were widely discussed; the popular press lumped them together as the problem of "subdividing the electric light" and there were prominent engineers who held that it could not be done. Edison, like some others at about the same time, saw that there was a solution in what we now call parallel wiring (which was then referred to as "multiple-arc" wiring). The first to suggest this solution was Professor Moses G. Farmer of Newport, Rhode Island. Each lamp or motor was to have a separate circuit to and from a common central generator. Each lamp would then have a very large electrical resistance, unlike those so far developed, which would insure only a small flow of current through each lamp. A large number of lamps and motors, each with different demands for current, could be put on separate circuits powered by the same generator. One bulb burning out (and lamps had lives of only a few hours at first) would not disturb any other part of the system; each lamp would be, in effect, on its own circuit and could be replaced independently of the others.

This was a basic principle which made possible a new kind of industry—the manufacture and sale of electric power. What Edison saw and developed was this new industry.

This was not a concept which had not occurred to, or attracted, the financiers who had until then supplied the money for Edison's research. Rather, this group of bankers, including the railroad organizers J.P. Morgan and Henry Villard, thought of making and selling lighting equipment: small generators with a loop or "arc" of incandescent lights; other generators to power single large motors for elevators and pumps. The company they wanted to organize would sell power equipment, with which each customer could supply electricity to his own office building or factory.

Edison had a much grander scheme; he wanted to sell *electricity*. This was the scheme which was to be hatched from 65 Fifth Avenue and for which Samuel Insull was required. What Edison proposed to do was to establish the first central-station powerplants—the first powerplants which would sell electricity to customers over a wide area. To do so would be a forbidding task. While the main elements of the system already existed—the generators, lamps and motors—an extraordinary number of details needed to be resolved. If electricity was to be sold in the densely populated cities, distribution wires would have to be laid underground. This had not been done before: proper insulation therefore had to be developed. Every detail from the burning of coal to the flipping of the customer's switch had to be thought through, planned and designed; every switch, connector, insulator, conductor, socket and lamp had to be designed to mesh with the over-all system. The generating plant had to be built, and the wires laid in city streets; private homes and office buildings had to be wired for power, and motors installed or adapted to mesh with the system. A sales force had to be assembled to persuade potential customers to permit the wiring of their homes and offices. And the whole elaborate scheme had to be financed and managed.

The plan was at first a success. The world's first central-station electric pow-erplant for the sale of electricity went into operation in Appleton, Wisconsin, in 1882; the second went into operation a few weeks later on Pearl Street, in the heart of New York's financial district.

Appleton, a rural Wisconsin town with a population of fifty-five thousand today, has a reproduction of that first central-station plant, the Vulcan Street Power Station. In the spring of 1882, it was powered by a three-foot wooden paddle wheel under a ten-foot fall of water in a small stream and it lighted sixteen incandescent lamps of about fifty watts each. The site of this plant, which was also the nation's first hydroelectric plant, is marked by a modest sign that fails to note it was the first of the Edison system. Few noted its advent in 1882, but soon afterward, in the heart of New York's financial district, the Pearl Street station began operating with extraordinary publicity. A new industry was born, and Mr. Edison had conceived of it *in toto;* it sprang fully armed, as it were, from his brow. The system as a whole, and in all its details, was patented by Edison, and if it had been successful he would have controlled the entire industry for many years.

But the effort was overambitious. Not everything could be foreseen and in-vented, and Edison's effort to create a new industry over which he would have complete patent control failed. He had made a fundamental error by adopting direct current.

Direct current is distributed from a powerplant at the same voltage or "pres-sure" at which it is produced. This is ordinarily a fairly low voltage—the Apple-ton plant produced power at 110 volts. But because of this low voltage, the power cannot be distributed very far—the maximum practical distance is about a mile. A city like New York would require hundreds of separate power companies to serve it, which Edison assumed would be the case. Dozens of companies were, in fact, formed in New York City alone (their eventual merger producing today's "Consolidated Edison Company").

The Edison system quickly gave way to George Westinghouse's plan. Westing-house reproduced Edison's feat of creating or adapting all the components of a complete power system, but Westinghouse's was based on alternating current. Generators in an alternating-current system could produce power at high volt-ages, or the voltage could be stepped up for transmission by transformers; other transformers at distant points would then reduce the voltage to the low levels needed for home or commercial use.

For a few years, both direct and alternating current systems competed. The Edison systems had the great advantage of an early start and of their inventor's renown and his genius for publicity. Edison had also assembled an impressive sales staff, and Samuel Insull, his secretary, assisted in this intense rivalry. The diminutive Insull went on the road, touring small towns throughout the country, inducing their municipal governments to buy the Edison system. The Edison construction division would then put up the plants; exclusive contracts with the Edison companies would bind the cities to purchase lamps and replacement parts from Edison. Municipal ownership of power companies, which Insull would spend a good part of his life battling, began in these early commercial efforts of the Edison Electric and its rivals.

When Westinghouse entered the field he took a similar tack in trying to market his own system, and the rivalry for municipal franchises was occasionally intense. There were a number of city governments for sale in those days, and where persuasion didn't work bribery often did. The distribution of inducements seems to have been one of Insull's responsibilities for Edison.

The alternating-current system was superior, but Edison, unwilling to give up the chance for complete control of a national industry, fought it bitterly. As the advantage swung to other companies, he began calling alternating current the "killing current" because of its high voltage, hoping to mobilize public opposition to this new technology. There were warnings of what would happen if power lines should fall on city streets. Undeterred, the Thomson-Huston Company, an alternating-current company which had some patents of its own—and which pirated others from Westinghouse—secured heavy financial backing, which Edison was unable to obtain, and began taking over an increasing share of the sales of power systems.

The battle was pretty well over by 1889. Under J.P. Morgan's aegis, the Thomson-Huston Company absorbed Edison's; Thomas Edison surrendered his patent rights for a lump sum payment of $1,750,000 in cash and stock, of which about $1,000,000 went to his associates, all stockholders in the original Edison companies (Insull got $75,000). The great adventure of creating a new industry was over for Edison. Under J.P. Morgan's control, the former competitors emerged as a new and dominant firm—the General Electric Company, whose sole rival was the Westinghouse Company. Today, these remain the only two manufacturers of heavy electrical generating equipment in the United States—and the predominant makers of nuclear powerplants for military and peacetime uses.

One curious relic of this battle of the currents remains: the electric chair. In the desperate days of the struggle to secure control of the industry, Edison had not only called alternating current the "killing current" because of its high voltage, he had persuaded the New York legislature to publicize his view by adopting alternating current as that state's mode of execution. Despite a formal protest from the National Electric Light Association, the governor of New York in 1888 signed into law a bill providing for the execution of criminals in an electric chair, perhaps the most gruesome publicity stunt ever conceived.

CHAPTER TWENTY-FIVE

Management

T HE EDISON ELECTRIC INSTITUTE is housed in a forty-story skyscraper at Park Avenue and 40th Street in New York City, just two blocks
from Grand Central Station, that extraordinary architectural monument to the
buccaneers of railroading. The Institute's skyscraper reflects a more recent industrial era—it is smooth and featureless; sealed glass windows, artificial lighting
and air conditioning frown down upon the Grand Central's baroque ornamenta-
tion and its enormous interior spaces which make air conditioning impossible.
The Institute's own offices are impeccably anonymous and modern, but there are
some odd touches in the waiting room. A life-size bronze bust of Thomas A.
Edison (1847–1931) stands at almost shoulder height on a pedestal set into a niche,
giving the appearance of a household shrine. The usual female receptionist is
customarily polite; there is the usual modern uncomfortable sofa, and a coffee
table. But here the reading matter seems to consist largely of comic books glorify-
ing the private power industry, featuring Reddy Kilowatt and other trademarks
of the power business. A number of the comic books are devoted to nuclear
power, and all are evidently intended for free distribution in public schools; there
is a blank space on each for a local power company to imprint its name.

W. Donham Crawford is president of the Edison Electric Institute, the private
power industry's trade association. He is tall, slim, with short light hair, at
fifty-two a very young man among power-company executives. He attended the
Naval Academy and worked in the Atomic Energy Commission's military reactor
facilities from 1951 to 1954, at the plants built to further our H-bomb program at
Aiken, South Carolina. Those four years provided his entry into the power
business; from the AEC he stepped over to Middle South Utilities and then to
New York's Consolidated Edison Company, where as a vice-president he supervised that company's move into the nuclear-power business. When the chairman-
ship of Con Ed changed hands, Crawford continued on to the Edison Electric
Institute, which he now heads. Like many utility-company executives he appears
to believe that his industry has been unfairly attacked; he agreed to be interviewed by a reporter who has written critically about his industry in the past, but
he took the precaution of asking two public-relations assistants to sit in on the
interview. There was an odd pause at one point in the conversation; Crawford
had become more relaxed and said that critics of the nuclear-power program have
been led "down the garden path"; when asked who has done the leading, he
altered his manner and named a single scientist who can hardly be blamed for all
of the criticism of nuclear power. Moments later, Crawford's slip of the tongue
seemed to give a hint of what was on his mind. Referring to a scientist who had
made well-publicized attacks on radiation releases, Crawford called him
"Greenglass"—and then laughingly corrected himself. Greenglass, he recalled,
was the figure in the Rosenberg spy trials; Sternglass was the one he meant.

Over the years the Edison Electric Institute, through various subsidiaries such
as Reddy Kilowatt, Inc., has conducted the extensive public-information cam-

paigns on behalf of the private power industry that are represented by the comic books in the anteroom.

I wonder if you could very briefly summarize what you view the Edison Institute's role and purpose to be?

CRAWFORD: Well, very succinctly, the EEI's role is to do those things which can be better be done by the companies acting together than by themselves individually. For example, taking positions on legislation; it doesn't make much sense for two hundred electric-power companies each to take a position. Those positions are taken by the EEI before the Congress. The funding of research, national marketing programs, the repository of information and statistics on the industry as a whole, these are the things we do here at the Institute.

Is lobbying part of your function?

CRAWFORD: No, it is not. We are somewhat unusual in this industry in that regard—in that the legislative activities that are done in lobbying, so to speak, are done by a different organization. EEI is the principal organization of the power companies, but we do not lobby. We do the testifying, we adopt the policies, EEI, in terms of the policies of the industry. But we draw the line when it comes to lobbying, another organization does that.

What organization does that, sir?

CRAWFORD: It is called the National Association of Electric Companies.

The Edison Electric Institute was formed in the 1930s, at a time of great turmoil for the industry. Do you feel there has been any residue of the problems of those years, in terms of difficulties in public relations, or in any way?

CRAWFORD: No, I really don't, in the way that you pose the problem. I guess what you are saying is, Are there problems that existed then still with us today? I think that there will always be some element of one of the problems, namely, the question of public ownership. That, of course, in those days was a very serious problem, but today, I think it is not a serious problem. There is no particular move in that direction. But the problems today are so different than they were then, that I would not say that there is any substantial residue from forty years ago.

There has been, at least judging from press reports, there have been very recently some new efforts concerning public ownership for the first time in some years. Do you think that represents a trend of any kind?

CRAWFORD: No, I really don't. I can understand why these matters are looked into, because people are upset about the fact that power bills have gone up and they are casting about looking wherever they might, to see some way to reduce the power bills. And, they figure if they don't have to pay taxes on the power bills, that would be good. But, as we always say, the question of public ownership is not an answer to the problem of the power bill. What you do is to transfer the responsibility of part of the power-bill payment from one sector of society, namely, people who use electricity, to the general tax payer. The cost of producing electricity is the same regardless of who produces it. We have the same kind of machines and people as TVA.

Now, it is true what you say, there have been some signs of interest and movement towards public power. I think they are really quite minimal. There is a matter up in Massachusetts that is developing. That is a referendum, as I understand it, [which] will be on the ballot next year having to do with setting up a state power agency. Whether this will come about or not is anybody's guess. I understand, also, that the governor of Massachusetts is endorsing a bill which is quite different from what would come out of the referendum, and how that all is going to eventuate is very difficult to say at the moment. Now, there was a case in New York State where the town voted to take over the facilities of Niagara Mohawk. But this was a very unusual case. This particular town, you can look out and see one of the dams of the New York Power Authority from the town and there was power available for that particular town. But this is not a decided matter, either. Whether that eventually will change ownership of Niagara Mohawk to the town itself remains to be seen. There will be quite a dispute, I am sure, over the value of the properties involved. I had an occasion to testify up in Erie County last year. They asked me to come up to talk to a task force, or committee, that had been put together, to consider whether Erie County ought to think in terms of public power, and their conclusion was that the county should not become involved in that, that there was no reason for them to do so. So, there have been some moves in the direction of municipal ownership. On the other hand, there have been recent votes against this sort of thing, and so over-all, I don't think there is any particular momentum towards government ownership.

Do you think there is any difference in management between, say, the Los Angeles power company [which is municipally owned] and an investor-owned power company in similar circumstances?

CRAWFORD: Well, it is very difficult for me to be specific about this because I don't know these people of the city of Los Angeles, but I would put it this way: there is certainly nothing to be said in the direction that the managers of public power organizations are any more motivated, are any more expert, than managers of investor-owned companies. We are at least as efficient as government power people.

Some of the executives in your industry are in favor of greater consolidation of power companies, that some of the smaller companies could be more efficiently drawn into larger entities. Do you think that is a trend in the power business?

CRAWFORD: Well, I don't know that it is a trend, and the reason I say that is because of the following: I think it could be a trend, but the federal government so discourages it, that the trend which I thought was developing a few years ago has been stopped. The SEC [Securities Exchange Commission] resists mergers of that kind. If you have a combination company, and there are many combination electric and gas companies, when two of them want to merge, they say, All right, get rid of your gas property. Well, this is a big business in some of these companies and they don't want to divest their gas properties. This, plus the specter of bigness itself when it's before the Justice Department, has tended to essentially put a damper on any moves in that direction. If that were to change, I think you would begin to see more mergers.

Does EEI feel that it is important to consider whether or not to influence the rate of growth of power demand? Should there be sort of negative marketing programs, to discourage sales, the way some years ago there were the positive marketing programs to increase sales?

CRAWFORD: Yes, from the standpoint of doing whatever can be done to use energy more efficiently. It is not in the utility companies' interest, certainly, to have to continue to build and build and build, to carry smaller and smaller peaks. What we try to do is to level that out. And that would have the effect of decreasing, certainly, the peaks and perhaps the kilowatt-hour growth.

Is that a major issue for the industry, do you think, the load factor [the use which a company makes of its full generating capacity], is that a big problem, or is that something that's really just an ordinary sort of thing?

CRAWFORD: That's the name of the game. Load factor is the name of the game. You have to build, you have to install the capacity to carry the peaks. The question is how much use, how many kilowatt-hours, and thus how much revenue can we get out of it. And if you are Con Edison with a forty-seven percent load factor and nobody uses energy at night, then you have a hell of a problem. If you're an American Electric Power, or Gulf States Utilities and your load factor is sixty-four percent, you have a lot less of a problem. So, that is one of the keys, it is just not a number, it is a very key consideration in the whole structure of the business.

CHAPTER TWENTY-SIX

Empire

IT WAS SAMUEL INSULL, the slender blond secretary with intent protruding eyes, rather than Thomas Edison, who came to dominate the electric-power business. Insull invented nothing mechanical, but he was as great an inventor as Edison. He invented the electric-power monopoly.

Well into the twentieth century, the electric-power industry was a minor and scattered affair. The sale of electricity itself was not a large business: more important, by far, was the sale of generating plants to single-purpose customers, as J.P. Morgan and his associates had foreseen. This pattern had been established early, with the street-lighting systems first installed by cities. The financiers who supported Edison, as well as his rivals, realized that the sale of equipment would be the quickest and easiest way of making electricity pay: owners of large office buildings and retail stores were persuaded to install their own generators and lighting systems, often simply for electricity's novelty or publicity value. The sale of electricity itself, energy which could not be seen or handled or stored, was not yet taken seriously.

Even equipment sales was a marginal and insecure business at best. The electric lights were not particularly good; with the appearance of gas mantles at about this time, gas lights were better and cheaper for almost every application (except for ships, which could not carry their own gas supplies). Electric lights were used largely for advertising and display purposes, where their novelty counterbalanced their high price. The cost of electric lighting did not drop to that of gas lighting for many years, nor did the quality of the electric light compete with gas mantles until the introduction of the tungsten-filament lamp in 1909 (the earlier carbon filaments gave off a dull red light and were short-lived and expensive).

The electric-power business, in fact, stagnated after Edison's dramatic demonstration at Pearl Street in 1882. Central-station power was possible, but no one seemed to be interested in making the huge investment in generating plants and transmission systems which were necessary before any power could be sold. For many years, the central stations were merely curiosities, despite the high hopes of those battling for control of their future.

The most important electrical industry of those times was the electric streetcar, or "traction," company. The first streetcars were installed in 1885 in New Orleans, South Bend and Minneapolis by an inventor named Charles J. Van Doepele, and they were an immediate and dramatic success. In a few years most of the cities of the nation had one or several competing traction companies, and the popular magazines were heralding the imminent demise of the horse as a means of transportation. These traction companies ran on electricity and consumed it in amounts which completely overshadowed the output of the tiny electric-lighting plants which had so far been built. But streetcar companies didn't buy electricity from the Edison central stations; they bought powerplants and made their own electricity, as did owners of office buildings and department stores. Very few found any reason to buy power from the power companies, and when they did, the price was high: in the late 1880s, electricity for lighting cost twenty or twenty-five cents per kilowatt-hour, a very large sum in those days, far higher than the price of gas lighting.

Samuel Insull had emerged from the collapse of the Edison Electric Company with some cash, some stock, and the offer of a job as president of the Chicago Edison Company, a small electric-lighting company which had contracted to buy Edison equipment. This was a considerable step down for Insull, who had but a short time before been president or secretary of a network of Edison companies which the two men had hoped would control a vast national industry.

Insull was twenty-nine years old when he began his second career. When he took control of the Chicago Edison company, most of the electric power generated in Chicago was still being made by streetcar companies, which bought generators from both General Electric and Westinghouse. Large business establishments had their own plants for lighting and for equipment. Chicago Edison was a minor competitor of its own potential customers. But Insull had plans. With the assistance of Chicago bankers, his Chicago Edison company began buying up the numerous small power companies which had been created in the rivalry between Edison and the alternating-current companies; most of these now had exclusive sales contracts with the merged General Electric Company, and Insull had no difficulty in acquiring them. They were small but, in the aggregate, justified the purchase of some expanded generating plants. From this base, Insull was able to assault and eventually capture the streetcar companies, which used more power than all his previous customers combined.

Insull's experience in dealing with corrupt municipal governments probably was useful to him here; in 1907, the city of Chicago passed ordinances designating Insull's company as supplier of power to the traction companies, doubling his sales by a single stroke. With the traction companies secure, Insull was able to build a virtual monopoly of the power business in Chicago and the surrounding suburban areas. By acquisitions, mergers and the purchase of stock through holding companies, Insull's domain gradually spread over Illinois and the Midwest, and then over much of the country. By 1932, when this vast empire collapsed, Insull was president of eleven power companies, chairman of sixty-five, and the director of eighty-five. The actual extent of his control was never quite clear, but it seemed to at least some observers that he and J.P. Morgan, his greatest rival, controlled almost all of the nation's electric power at the time of the Great Depression.

Insull, who grew portly in these years, but whose prominent, hyperthyroid eyes and boundless energy remained with him, built this empire on a central idea: the electric-power monopoly.

CHAPTER TWENTY-SEVEN

A Profitable Concern

THE AMERICAN ELECTRIC POWER SYSTEM is the country's biggest network of privately owned power companies; it is a holding company that controls the power supply of a vast region spanning seven states of the upper Midwest; its headquarters are at 2 Broadway, in New York City's financial district. It is the largest remnant of the great holding companies formed during the 1920s, which survived the trust-busting of the Depression era only by justifying its existence with superior efficiency and superior profits. Its load factor, the efficiency with which it uses its generating capacity, is among the highest in the industry, as W. Donham Crawford noted, in part because of the heavy industrial demand of the area which it serves, a steel-making, heavy industrial region. The American Electric Power Service Company is the management core of the system, and Gregory S. Vassell is one of its vice-presidents. Vassell is pleased with himself, as well he might be: a senior officer of one of the largest and most successful industrial firms in the world. It is a singular firm in many ways, not the least being its striving after efficiency. AEP is also unusual in being the only major utility to have decided not to undertake any major nuclear powerplants. John Tillinghast, senior executive vice-president, engineering and construction, explains:

"As you are perhaps aware, American Electric Power Company's commitment to nuclear-power technology has been limited to the partial completion of a single powerplant. The motivation for this limited nuclear commitment has been our proximity to the abundant coal resources of the Appalachian region, which makes coal a preferable economic alternative [like other large utilities, AEP owns its own extensive coal reserves]. Furthermore, we are dissatisfied with the lengthening lead times [time between the decision to build and the completion of facilities] and uncertainties associated with the scheduling of nuclear generation.

"This should not, however, be interpreted as a criticism of the decisions of others."

AEP has maintained this position for some time; its former chairman, Philip Sporn, for many years was an advisor to the Joint Congressional Committee on Atomic Energy and was well known as a critic of the early economic claims made for nuclear power. His equally outspoken successor, Donald Cook, has refrained from criticizing fellow utility executives' decisions to buy nuclear powerplants; instead, he has conducted a very vigorous and extensive advertising campaign in favor of the use of coal, attacking all federal efforts to control the emissions of sulphur dioxide.

Gregory Vassell seems to be far removed from this vast clash. He is a pleasant, gray-haired man with a round face and pudgy hands, who seems faintly irritated by his visitor's ignorance of electrical engineering. Vassell only becomes animated when describing the technology of his industry; in talking about the way powerplants work, Vassell, an electrical engineer with an additional master's in business administration to assist in the climb through the corporate ranks, becomes enlivened and eager.

VASSELL: The electric-power industry is the most capital-intensive industry. In electric power we need close to five dollars investment in assets, fixed assets, plant, in order to obtain one dollar of revenue. By comparison, a heavy industry such as steel may require a dollar, a dollar and a half of assets for one dollar of revenue. If we talk in terms of light industry, it goes down to as low as thirty cents of fixed investment in facilities for one dollar of revenue; and by the time you get into merchandising, it may be an entirely different situation and certainly as far as publishing, you mentioned publishing, it's also a different situation. So [with] electric power, the very substantial investment goes, to begin with, in the facilities. Then the other element is, the lead times for those facilities is very long. To build an automobile assembly plant may take a couple of years; to build a shopping center maybe a year; to build a house maybe a year; to install an air conditioner, maybe a week to make up your mind and then one day to put it in. As far as electric-power facilities, it takes six, seven, eight, ten years to build from the moment that you actually make a decision to build it and commit resources to proceed with it. It takes that long. So that you have the situation here where-
. . . the need for power, [the] use for power, is operating from a much shorter cycle, while establishing the basis for the availability of power has a much longer cycle. You have the problem that you have to commit resources a long time before you can be sure exactly who is going to take it; when he is going to take it; how he is going to take it. You have to make commitments now for 1980, '82, '83, '84, '85.

Nuclear plants have a longer lead time than most other plants.

VASSELL: Yes.

So to that extent they exacerbate this problem of capital intensity. . . .

VASSELL: They first have longer lead time, and also they cost more, in terms of capital cost. Of course energy [fuel] cost is less.

You have to build powerplants to meet demands and these plants very often stand idle much of the time. Does this not make your problem worse in terms of the large investments that are required for any company?

VASSELL: I believe there is a certain degree of misconception in the way you phrase it and that misconception carries over quite a bit in the public debate now. It's true, that you need to have capacity available to meet your peak demand, whatever it is, or as an alternative you have to reduce your peak demand to meet whatever capacity is available. It's true also that at times when the demand is less, some of the capacity may then not be used at the full output. But most of the capacity that is not being used to the fullest is still being used effectively because the entire design of the power system—design of the composition of your types of generating capacity that you have—is predicated on the knowledge that the demand does vary, for example, seasonally, from the summer months when the demand is high into the fall when the demand is lower and so on. And, for example, scheduled maintenance of capacity would have to take equipment out of service for regular maintenance. Generating plants are very complicated assemblies of equipment that require regular maintenance in order to operate properly and to maintain reliability of performance over long periods of time; we keep them for thirty, thirty-five, forty years. So that the fact that demand is, for example, seasonally, at times less than at peak is taken into account in the type of generation that is being installed. We could not, for example, suddenly today find ourselves having continuous demands at the same value and be able to meet it, simply because the entire system development is predicated on the kind of load shape that we now have.

Now are you referring in part to the different kinds of generating plants?

VASSELL: Right, right.

Could you say a couple of words about them?

VASSELL: Well, for example, peaking capacity which is designed specifically to meet peak demands is predicated from the assumption that that capacity will operate only a relatively few hours during the year, and its entire design is based on that. It costs less to install, it costs a little more to operate. And we have a hell of a lot of that capacity now installed.

These are, particularly, gas turbines now?

VASSELL: Mostly gas turbines. The other type, I might say, is pumped-storage hydro, which is another type of capacity that, in effect, is an energy-storage device, if you will. [In pumped-storage plants, water is pumped uphill to an elevated reservoir when generators are otherwise idle. In times of peak demand the water is allowed to run back down through hydroelectric turbines, generating added power.]

Again, returning to other kinds of businesses where there are big variations in demand, for instance, Thanksgiving or Christmas trade, people who sell things for Christmas trade make them all year, and they sell them at Christmas. They tend to use their plant very steadily during the year, even though the sales are very seasonal. Why has it not been possible for the electric-power business to arrange itself in that fashion?

VASSELL: Because the basic, intrinsic, physical characteristic of electricity is that it has to be used the moment it is produced. Now, when it is produced it has to be used at that very instant and, of course, you know electricity moves with the speed of light, essentially. So that it's instantaneous. What it means is that all your generators that are connected to the network, that at each instant they generate exactly the amount of energy that is being consumed at the time. Now, when somebody turns on a switch on an air conditioner, then that additional energy requirement has to be produced at the moment when it is called for. Now the way it happens, actually, is that, of course, the generators have certain inertia, so that at the moment additional electric load is injected, then there is a slight reduction in the speed of the generators, which, in effect, means that they give up some of [their] kinetic energy—[turn their] mechanical energy into electric energy—and that expresses itself in a slight change of frequency. And then, through various control mechanisms in the electric-power network, then the change in frequency causes the generators all over the network to burn more fuel, if you will, and make up this loss in kinetic energy and the frequency is restored again. The fact is that energy has to be used the moment it's produced. If you try to produce more energy than you use, then you are in trouble.

Why has it been so difficult to reduce the capital intensity of the industry?

VASSELL: Well, generally you could say that maybe forty-five [or] fifty percent of the total capital investment in the utility industry is in the form of generating facilities. And maybe twenty percent roughly in transmission and then thirty-five or so percent in distribution facilities. And then maybe the remaining five [for] everything else, for the buildings, offices, and everything else, [which] is just a very small part of it. I might say that capital intensive as it is—the production of electric energy on the large scale as it is done, through the large powerplants, large distribution networks that we have now—is still by far much more economical than if we tried to do it in some other way. You know, you read a lot about windmills and so on, and except in specific situations . . . well, of course it very well may have its place. When I go through the Western part of the country you will see a few windmills here and there, you know, on the farms. But it's a special situation for special purposes.

CHAPTER TWENTY-EIGHT

Monopoly

I N THE POWER BUSINESS, ELECTRICITY is not stored; no one can understand the power business, or nuclear power, without grasping this central idea.

Electricity is manufactured, delivered and used in the same instant. When you snap your wall switch, a central generating plant miles away produces an iota of additional current which is fed into the circuit you have opened; the current passes through transformers, over miles of high-voltage transmission lines, through other transformers, through a buried distribution system, through your house wiring, to heat your lamp to incandescence. All of this takes place at the speed of light.

The generator must be there and must have enough capacity to light lamps when the switches are thrown. Electricity is not stored. The generator cannot produce power during slack times and then deliver it later, when wanted: the generator must operate whenever power is demanded.

Until Lord Kelvin travelled from Paris to Glasgow with a "box of electricity," the first storage battery, in 1881, there existed no method of storing electricity. And despite the drama of Lord Kelvin's demonstration of the battery (invented by Fauré), interest in storage batteries was meager for many years, and few were produced. In the early days of the power business, therefore, it was a simple fact that electricity could not be stored. And this had profound implications, as Samuel Insull was able to see.

In 1908, a bill was pending in the Chicago city council to set rates under the new contracts Insull had obtained to supply power to the city's street railway companies. In his successful battle to prevent municipal regulation of these rates, Insull addressed a group of Chicago businessmen and tried to explain the method by which rates for electric power were to be set. He explained why his company could offer such suspiciously low rates to the traction companies, effectively closing off competition.

"That is the main cause, that question of the maximum load, of all the trouble in fixing electric-lighting rates. The gas company would be subject to precisely the same condition if it were not for the cheapness of storage. Storage with us is a practical impossibility."

Yet it was this inability to store electricity which paradoxically allowed Insull's company to provide power more cheaply than the street railroads. Insull used a favorite illustration, a block of northside Chicago houses, to make this point. There were 193 apartments on that block, and 189 of them were customers of the Chicago Edison Company. There were no appliances or motors or other electrical devices to speak of in that block of dingy apartments—just the dim electric lamps of the day. The power demanded by all the separate apartments on the block, if totalled, was 68.5 kilowatts. But, as Insull pointed out, the different lamps would be lighted at different times, and the actual maximum demand for power from that block of apartments was only 20 kilowatts.

To supply all of these customers from a single source would therefore require generating power of only 20 kilowatts. But if each household were to be equipped with a separate generating plant to meet its own needs, an aggregate of 68.5 kilowatts would be needed—more than three times as much.

This situation was a result of the fact that electricity could not be stored. If each home's powerplant, like a hot-water heater, could have stored its output in a tank during periods of low use, there would have been adequate stored electricity for the moment when all the lights were on (as there was enough hot water when a bath was drawn). But electricity could not be stored; there was no equivalent of the hot-water tank. Generating power had to be available at the moment it was needed. A single generating plant could serve many customers, if each were to make his demands at a different time. But if each customer were to have a separate generating plant, there would be enormous, and expensive, duplication of facilities.

Gas and water companies had long been recognized as "natural" monopolies because of the enormous difficulty of laying down several competing systems of distribution; it made no sense for a city to have more than one set of gas pipes or water pipes. But this reasoning did not then apply to electric power, since most people still made their own electricity: distribution systems were a *result*, not a *cause* of the power company's monopoly. It was as if Samuel Insull had announced a monopoly of hot-water production and had proceeded to dig up the streets to lay hot-water distribution pipes.

Insull argued that the same diversity in demands which made it more economical to supply all the apartments on a block from a single powerplant would make it more economical to supply all electricity, whatever the purpose, from a single source.

The reasoning clearly was correct when applied to streetcar companies. People get up in the morning and turn on their lights. They turn off their lights before leaving home (with lighting at ten cents per kilowatt-hour, people were careful about turning off lights in 1908) and then take the streetcar to work. The maximum demand for power from the streetcars would be at a different time from maximum demand for electric lighting. And when people were neither at home nor on their way to or from work, they would be *at* work, using electrically driven machinery—which once again would demand power at a different time. The same generating equipment could supply these multiple demands, so long as they came at different times, more cheaply than any one customer could meet his own needs for power.

The savings were not as dramatic as Insull's apartment-block illustration suggested, but they were real enough. In his sales pitch to Chicago businessmen in 1908, Insull concluded that he would need to add about 17 percent less generating plants to produce the power the streetcars needed than the traction companies would have had to build to supply their own power.

This difference was important because powerplants were expensive to build. In the years in which Insull was creating his monopoly, coal was burned in a boiler that generated steam, which in turn ran a piston-driven steam engine, which in turn cranked a dynamo, which produced electricity. The power had then to be carried through massive transformers and converters, over transmission lines and through buried distribution lines to the customer. The capital investment involved in all of this was tremendous. Even modest proportional savings in the amount of generating plant could be important.

For this reason, and also because larger plants were simply more efficient than small plants, Insull argued, a single supplier could provide power more cheaply than scattered generating plants. The argument was both persuasive and correct, so long as there were no means of storing electricity.

A means of storing electricity would have completely changed the situation. For even with a very diverse group of customers, Insull's powerplants were only producing a third of the electricity they could make if operated at full capacity. Two-thirds of the plant was therefore, in a sense, idle capacity, except for the brief times of maximum demand (when most lights were on in December or January evenings). If power could be stored, a generating system one-third the size, but running continuously, could have manufactured enough electricity to supply the same customers. Economies of scale were still very limited; the piston-driven generators of the day could not be built in large sizes, and most potential electric-power customers could build the largest and most efficient plants then available. Steam turbines, which do permit very large plants to be built, were introduced by Insull only after his monopoly had already been firmly established; no efforts were made to develop more efficient small generators.

After the turn of the century, storage batteries did become more widespread, although they were still expensive. Insull's company, on the verge of absorbing the street railways, installed storage batteries worth $1.5 million. But Insull and his company had long since been launched on a different course, toward monopoly of power production; batteries were of little interest to him.

The storage battery has not been much improved since 1900, and there has been little development of other means of storing electricity. From Insull's day forward, therefore, the huge size and complexity of central power stations has been an asset, rather than a drawback, for the electric-power companies. It is the very expense of power-generating equipment which makes electric-power companies possible at all, and which has allowed them to justify their "natural" monopoly of electric power—a monopoly otherwise no more natural than one on the heating of hot water.

CHAPTER TWENTY-NINE

Ecology

WHEN SAMUEL INSULL set out openly to make a monopoly of electric power, he was seeking to achieve what financiers J.P. Morgan, John D. Rockefeller and their brethren were battling to create in other industries. But Insull was not a wealthy man; he was president of a diminutive Midwestern power company. Monopoly was just a step above the Antichrist in public esteem; the monopolists were among the most vilified (if also the most admired) people of their day. To advocate monopoly openly was more than even the brazen Robber Barons would dare.

In this climate, Samuel Insull, without the financial resources to impose monopoly rule on an industry in the fashion of J.P. Morgan, undertook to sell the *desirability* of monopoly to the public.

It was Insull's audacity, rather than his vision of monopoly, which was singular, of course. The wild turmoil which had afflicted the railroad and other industries was then being replaced by rapid consolidation. The new dominant corporations were not called "monopolies"—this was so hated a word that a euphemism was needed. Just as in modern times the newspapers, reluctant to call the president a liar, invented the "credibility gap," the great new conglomerations of 1900 were called "trusts," and every politician gave at least vocal opposition to their power. But the end of the nineteenth century and the beginning of the twentieth was the time of trust formation more than of trust-busting. Historians now commonly refer to the "merger movement" of 1895–1905, when many of the great modern corporations were formed. In those years more than three thousand industrial firms disappeared into mergers. Most of this conglomeration occurred in the brief period from 1898 to 1902: 1028 firms were absorbed into mergers in the year 1899 alone. Among the modern companies organized to absorb these smaller firms were US Steel, American Tobacco, International Harvester, Du Pont, Corn Products, Anaconda Copper, and American Smelting and Refining (ASARCO).

The creation of these combinations yielded enormous profits to the financiers who organized them—J.P. Morgan's fee for floating US Steel Corporation's initial stock offering is estimated at sixty-five million dollars in 1901—but they also stabilized prices and ended much of the more violent competition of the time. Many of the firms organized then are still in existence and still dominant in their industries.

Not all of these firms were created by mergers. Some, like the Aluminum Company of America, achieved monopoly status in their industries by control of patents. One economist tabulated the industries still dominated in 1971 by firms created by mergers in the period 1895–1905. In the copper industry, the first and third largest producers, Anaconda and American Smelting and Refining, were created by mergers during that time, as were the first and third largest makers of farm machinery, International Harvester and Allis-Chalmers, and the two largest sellers of tobacco products, American Tobacco and R.J. Reynolds. The single largest firm in each of the following industries was also a result of the mergers during this brief period: chemicals (Du Pont), electrical machinery (General Electric), paper (International Paper), steel (US Steel), metal cans (American Can), photographic equipment (Eastman Kodak), meat packing (Swift), nickel (International Nickel), shoe machines (United Shoe Machinery), corn products (Corn Products Refining), and silver (International Silver). The second largest firms in glass (Pittsburgh Plate Glass) and dairy products (Borden) were also put together during this decade.

If we do not restrict ourselves to companies formed by merger, we can see that this was a period of even greater importance. Also formed before 1910 were the Standard Oil Companies, (Exxon, Mobil, Standard of Ohio, of Indiana, of California), General Motors, Westinghouse, Ford, American Telephone and Telegraph, the largest coal miners and railroads, and IBM.

Seven of the ten largest corporations in the United States in 1974 (and eleven of the twenty largest industrial corporations) were formed before 1910. Most of these began as monopolies or near monopolies through the absorption of large numbers of competitors or through the control of patents. Of the remainder of the twenty largest industrials, most are oil companies, formed later in an industry already dominated by Standard Oil.

A very large proportion of our national industry is clearly dominated by firms which appeared in American life before 1910 and which were intended to exercise monopoly control over their respective industries. Samuel Insull's effort to create an electric power monopoly was therefore quite in accord with the character of his times.

While it is usual to talk about "industries," and even individual firms, as if they were autonomous entities, this is somewhat artificial. Language and habits of thought make it easier to talk about autos or General Motors than about the nation's industrial enterprise as a whole, for which we do not even have a convenient term. But every industrial company depends so intimately on its fellows that it would be convenient to have a word similar to the biologists' "ecosystem" to describe a community of interdependent enterprises. Each industrial plant is served by raw materials provided by others and, in turn, sends its products to other plants for further elaboration or to other companies for transportation and sale. As in a biological community, the waste products of one plant become the raw materials of another, and the viability of any company is determined, in part, by its ability to fit within the larger community of exchange.

The "ecosystem" of Insull's day was one of national monopolies, in which the railroads were the dominant genus. By the turn of the century, the concatenation of industries mentioned above appeared, a kind of industrial climax community, which depended on and required a basic compatibility among the technological processes in use and a complementarity of products and wastes. In a competitive system, whether biological or economic, only those species which fit themselves into the preexisting pattern—into a "niche" in the environment, to continue the analogy—will survive. Only those companies which fit into the preexisting pattern are able to take advantage of the technical and financial support provided by the structure as a whole.

It is this unity of the industrial process which makes technological change so difficult except where it follows the predetermined course; where, as in the electric-power industry, existing technology was compatible with large-scale centralized industry and with the formation of monopolies, there was no inducement for change. Thousands of competing technologies, like the fuel cells and storage batteries of the nineteenth century, failed to flourish.

In the setting of the rapid consolidations of 1900, we should not be surprised to find the electric-power industry undergoing concentration and monopolization; to see it oriented toward mass production for industrial customers. The electric-power companies found their place in the national scheme; it is not surprising that there were only modest changes of technology after the basis of monopoly had been laid, by 1903, when Insull's company installed the first steam turbine.

Even the introduction of nuclear power half a century later did not create any new patterns. In nuclear powerplants, uranium instead of coal was used to generate steam. The only change was in the fuel; the rest of the plant—boiler, turbine and dynamo—remained as elaborate and expensive as before. In fact, nuclear powerplants were even more elaborate and expensive to build; the savings in fuel cost were hardly dramatic, since the cost of capital was the largest element in power prices. By 1913, the cost of fuel had dropped to 0.6 cents per kilowatt-hour out of a total cost of 2.85 cents in the Commonwealth Edison system of Chicago; industrial customers paid less than a cent per kilowatt-hour, and residential customers paid 10 cents or more. Even a dramatic reduction in fuel costs would hardly have affected the general cost of electricity very much (although it would have made possible even more extraordinarily cheap rates for industrial customers). Nuclear power was not expected to work any great changes in the electric-power business; it was just one more variation of a long-standing pattern —reduced fuel costs, to allow monopoly-securing rates to large customers, but undisturbed high capital costs, which made the monopoly desirable. As one perceptive writer put the prospects for nuclear power in 1949:

> "Atomic energy opens a vast vista of new power potentialities. As yet, however, its practical scope of applicability is far from clear. It may prove to be a direct substitute for electricty, but this seems doubtful. Its chief power application will probably be to replace coal or other fuels in the generation of electricity. Its apparent function will be to release heat and to produce steam for the turbines. . . . Presumably atomic energy will require special plant installations to assure protection and safe utilization. The total costs, including interest on investment, depreciation, and operating expenses will have to be balanced against corresponding costs of commonly produced steam. The prospective plant outlays are likely to exceed substantially those of present available large steam installations. The relative over-all economy of atomic energy will thus be probably limited to the fuel item, and this for a large modern boiler plant comes only to about 1.5 mills (0.15 cents) per kilowatt-hour . . . If this view is correct, then at best the margin of total advantage for atomic energy will be small." (John Baner, *Public Organization of Electric Power.*)

The interest which electric-power companies took in nuclear energy was for many years correspondingly minor. But nuclear power was fundamentally compatible with the form that electric power had taken. Low fuel costs allowed low promotional rates; and state regulators would always allow a utility to recover expanded profits on an expanded capital investment.

Samuel Insull no more created this pattern than he could have opposed it. What Insull did was to sell the *idea* of monopolization of electric power as an economically desirable thing; and, by the effectiveness of his salesmanship rather than by the weight of finance, to take personal control of one of the vast conglomerations which emerged at the turn of the century. In other countries and at other times, the same technology and the same reasoning seemed to dictate the creation of government-owned power systems. In both cases, the axioms of electric power—that electricity could not be stored; that powerplants must be large and expensive—quite naturally went unchallenged. They allowed, even dictated, a power system that could only be run by large regional monopolies. If small, diverse, and widely dispersed powerplants were possible, they were not of interest to either power companies or governments.

CHAPTER THIRTY

The Crisis of the Power Business

NOW THAT THE SYSTEM devised by Insull is beginning to collapse, and the electric-power companies face a financial crisis, we can see the importance of the principles on which they are founded.

Generating plants are expensive to build. Because electricity cannot be cheaply stored, the central generating plant with a diversity of customers has an economic advantage over a plant serving only one customer's needs. The central monopoly can enhance its advantage by keeping operating costs—largely fuel costs—low. There is no particular disadvantage for generating plants that are expensive to build—the evolution of law in this field has guaranteed utility profits on their investment, no matter how large—but low labor and fuel costs are desirable. Nuclear powerplants match the requirements of the central monopoly precisely: they are expensive to build, which puts them out of reach of all but the very largest consumers of electricity (only Dow Chemical has so far built a nuclear plant of its own). The enormous quantities of energy stored in the atom seemed to assure that fuel costs would be low; and labor costs in a nuclear plant, which is highly automated, are also low. Nuclear power therefore seems to be a natural if modest extension of the fundamental principles of the power business. But no principle can be infinitely extended; by 1975 it began to seem that the limits of expansion of the power business had been reached.

Because its historical origin as a supplier of electricity for home and office lighting, every power company was obliged to supply thousands of small customers and, of course, had to have sufficient generating plants to supply the aggregate of these tiny demands for power. During most hours of the day and night, this plant would sit idle. The plant was expensive, but electricity could not be stored. During these idle hours, Samuel Insull found, the power company could sell electricity to industrial customers for little more than the cost of the coal burned. One had only to find industries like the streetcar companies, capable of buying such large blocks of power. The great capital costs of the system would be charged to the thousands of small customers who determined its size; its expansion into a monopoly would thus be financed by the thousands of homes and small businesses wired for electric lighting.

Insull did not, in fact, invent this system—he found it already underway in Bristol's government-owned lighting system, where the newly invented "demand meter" allowed the optimum allocation of capital costs among customers. Insull's contribution lay in seeing the application of this principle to free enterprise. Rates based on anything but numbers of kilowatt-hours were surprisingly difficult for many power companies to grasp, but Insull preached his new discovery with a religious fervor. Through preferential rates he obtained the traction power business and secured industrial customers who otherwise would have built their own generating plants. This was what Insull meant when he delivered speeches with the audacious title, "Sell Your Product at a Price Which Will Enable You to Get a Monopoly." The largest market for electric power was among industrial customers: the traction companies, the factories, the deparment stores with day-long lighting demand. But these customers could build their own powerplants—and often their plants would be bigger than those of the central-station company. What Insull could supply was off-peak power at extremely low rates.

Residential customers, of course—the households with a few lamps—were charged the same high rates for small quantities of current, no matter when they were used. There was no competition for the household electric-lighting business; powerplants were too expensive to be built to supply single houses or even apartment buildings. But industrial customers could, and did, build their own plants, and to secure this business, industrial and commercial rates had to be very low indeed: in 1912, Insull was selling off-peak industrial power at a half a cent per kilowatt-hour, while home-lighting rates for small customers were up to twelve cents per kilowatt-hour.

In speech after speech—to the electric-power trade associations, to sales conventions, to business-club meetings, to his employees and to every audience he could reach—Insull sold the idea of low industrial power rates as the basis of a monopoly in electric power. The success of the technique was its best recommendation.

Everywhere that power was used, preferential low rates captured the industrial and commercial customers—the factories, office buildings and department stores. These large potential competitors were induced to abandon their own generating plants and purchase power from the central utility.

There are two ways of looking at this process. Industrial customers, who use power steadily, or at different times from residential customers, allow powerplants to run more nearly continuously. This means more efficient operation—greater revenues for each dollar of investment in generating and distribution systems. Unless the industrial customers are given low rates and so reap the rewards of this increase in efficiency, they will build their own powerplants and destroy the whole basis of the central-station system. (The economies of scale available to power companies are only slightly more advantageous than those of large industrial customers, and are counterbalanced by the need for expensive transmission lines.) It is in this sense that power companies, both public and private, argue the lower rates given industrial customers enhance efficiency and result in savings for all customers.

But it is also clear that the industrial customers must receive the lion's share of these savings to induce them to buy electricity. These low rates are promotional—sales devices to increase business in areas where competition is effective. There is no competition for the single household's lighting and power needs; the central utility has a legally protected monopoly in this area. So the householder pays higher rates—often very much higher—than the industrial customer. He is, in effect, subsidizing the promotional rates offered to industry (and to the larger home-heating customers). This subsidy makes possible the massive capital investment on which the system depends.

A fundamental principle of the power industry is that capital investment is profitable. State regulatory commissions can limit power prices, but they are restricted by the legal doctrines laid down in the Supreme Court decisions concerning railroads a hundred years ago. Public-utility commissions are compelled to allow power companies a "fair return" on their investment.

A fair return on investment seems like a harmless sort of guideline. In practice, the effect has been to permit power companies to charge rates sufficient to pay for the physical plants that they build, regardless of the motive or prudence of their construction, and to encourage new investments in still larger plants.

This legal doctrine has meshed in striking fashion with Insull's program of expansion to secure a monopoly. Large powerplants can be built, even though they will be only partially used; for a time, existing customers pay for the expansion, while the huge new surplus capacity is used to solicit new customers for large blocks of power at very low prices. Once the powerplants are producing as much as can easily be sold, the cycle is repeated. State regulation permits the company to charge enough to *existing* customers to bring itself a fair return on investment. Expansion carries no financial risk and a promise of substantial gain; there is no reward for increased efficiency or innovation.

The state regulation system therefore has fuelled and almost guaranteed rapid expansion of the power business—which in Samuel Insull's time seemed an unmitigated benefit. The state would allow a fair return on investment, no matter how large; the company could charge whatever the state would permit. The more expensive the company's plants, therefore, the more it could charge. Expensive generating plants would expand the profits allowed to the company in absolute terms (although, of course, it would *not* increase the *fraction* of gross revenues which were profits); the company would make more money, but it would not necessarily grow more efficient. Residential consumers, of course, would pay the major part of the increased rates, so that the expanded production of an expanded plant could be sold to industrial customers at low rates.

Profits of expansion to shareholders grow even more rapidly than total returns, through the workings of leverage in an expanding company. Suppose, for instance, that a utility wants to double the size of its powerplants, and needs five hundred million dollars to do so. It issues bonds for this amount, which are most likely purchased by insurance companies. The insurance company wants a stable, long-term rate of return, which is provided in the form of comparatively low interest payments on the bonds. The utility's capital has been doubled, and it is allowed to charge its customers enough to retire the bonds and maintain its previous rate of profit, but on a larger base. The number of shares of stock have not increased; each year, after the fixed return on the bonds is paid, the increased total profits are divided among the same small group of stockholders. In modern parlance, the earnings per share have increased.

Utility companies have made extensive use of this device to enrich stockholders; more than half of all utility financing is in the form of bonds, and power companies are the most highly "leveraged" industry in America.

Earnings can also be inflated by deferring income-tax payments by the use of accelerated depreciation and accounting techniques like the "allowance for funds during construction"; these techniques, too, depend on continual growth of investments. Expansion is now the only route by which utility earnings can be steadily increased—except, of course, for regular rate rises—because there are no longer any great technological advances being made in the industry. The efficiency of utility power production has in fact declined in recent years and seems likely to decline further. In coal-burning plants, the measure of efficiency is the "heat rate," the number of BTUs or the heat energy needed to produce one kilowatt-hour of electricity. In 1962, the power industry's best plant, which represented the leading edge of technology in the industry, used 8588 BTUs of fuel to produce one kilowatt-hour. By 1971, ten years later, there had been a slight decline in efficiency, so that the industry's best plant in that year required

8695 BTUs to produce one kilowatt-hour. Industry averages showed an even worse trend because of the increasing reliance on older plants as sales efforts outstripped powerplant construction.

Without increases in efficiency and with limits on rate increases set by the size of the existing capital investment, the way in which utilities increase their earnings is by expansion.

There is a self-reinforcing aspect to this process of growth. In order to expand, a company must raise money; but to attract investment, it must be able to show increasing future earnings. The growth of earnings is therefore needed for expansion, to allow a company to compete in capital markets for funds; but expansion is the source of increased earnings. This cycle, in which growth feeds on itself, is not restricted to the utility industry, of course, but it is most clear and marked there.

It is now easy to see why electric-power companies have developed their sales techniques so highly, for expansion is synonymous with survival. The company which ceases to expand quickly sees its stocks' value decline, finds the interest rate it must pay on bonds rising dramatically, further reducing earnings, and sees nothing in the future but a rapid downward spiral. In the words of James Coatsworth, retiring after twenty-two years as commercial director of the Edison Electric Institute, "The first forty years [of the electric-power industry from 1879 to 1919] were devoted to the technical side of electricity—developing new ways to generate, distribute and utilize electricity—but the second forty years were devoted to expanding the market. . . . " The estimate of forty years devoted to technical development may be generous; the subsequent devotion to marketing has always been frank and obvious. Power could be sold to industrial customers as a cheap substitute for labor or other energy supplies. To support the base on which all power production rested, sales to homeowners had to be expanded as well. But low prices could not be offered to the residential consumer, or the calculus of the system would be defeated; other inducements to consume power at home would be required.

In 1904, at the twenty-seventh annual convention of the National Electric Light Association, predecessor of the Edison Electric Institute, a large report on advertising, with sample advertisements, was read. In previous years, the report pointed out, electricity's spectacular novelty had served to advertise itself, but this period was ending: "Whereas hitherto the chief question has been to increase the capacity for making the supply meet the demand, the ridge is about passed, and from now on more energy must be devoted to marketing the product." Two years before, a previous president of NELA had explained the simple logic behind this position: "Our earnings through the economy of operation have well-defined limits, but the possibilities of increasing our earnings by developing our market have a much wider range."

This has been the touchstone of the industry for seventy years, during which it has succeeded admirably in its first object, to increase sales. Aside from a brief dip in the early years of the Depression, electric-power sales have grown with remarkable steadiness, doubling every ten years over nearly this whole period. Some of this growth was needed and valuable, but as the electric-power industry continued to grow by seeking new markets, the usefulness of growth became more and more doubtful. And the residential customer bore and still bears the principal burden of this growth.

By the 1930s, the electric lighting market was close to saturation. Labor-saving appliances like vacuum cleaners and washing machines do not consume very much power, and radios consume even less. The big new areas for growth were in the use of electricity for cooling and for heating, particularly in stoves, space heaters and water heaters. These are purposes more cheaply and cleanly served by gas or oil; further expansion of the power industry in the residential area therefore required increasingly wasteful and dubious applications. (The great exception was the refrigerator, an invaluable appliance most efficiently run on electricity.) In the lighting field, emphasis was on increasing the intensity of existing illumination rather than installing new lighting (this campaign, under the slogan "Better Light—Better Sight," was heavily waged in the schools). In 1935, C.F. Greenwood, commercial director of the Edison Electric Institute, summarized the new strategy: "The services that help the most in load building are refrigeration, cooking and water heating . . . consumer room cooling will also help materially. It is the long hour 'automatic' electric services that roll up the kilowatt-hours in volume sufficient to make possible lower average rates . . . but there is no magic wand that can be waved over the household to increase electric consumption. . . . To attain that goal electric cooking and water heating must be vigorously pushed."

Heating and cooling operations consume much more energy than electronic devices (radios and televisions) or than mechanical devices run by electricity (motor-driven fans or washing machines); but electricity is not well suited to these more energy-intensive applications. Space heating can serve as an example. When fuel is burned at an electric powerplant, only about 30 percent of its energy is converted to electricity—the rest is lost as waste heat. There is a further loss in transmitting electricity to the customer, who then converts the electricity back into heat to warm his home. A well-designed home furnace, however, can convert as much as seventy-five percent of its fuel energy to useful heat when burning oil or gas directly at the home. All other things being equal, two or three times as much fuel must be burned to provide electric space heating than would be needed for gas or oil heat. This means, roughly, three times as much pollution and three times as rapid a depletion of natural resources.

As far as the consumer is concerned, there is no perceptible difference between heat provided by electricity and heat provided directly by burning fuel. In the case of hot-water heaters, many home owners would not know which form they had without consulting their monthly bills. Electric stoves and ovens seem to be at a positive disadvantage with respect to natural gas, since they take so long to heat and are so difficult to regulate quickly; but the mystique of electricity and the very able and extensive promotion campaigns put on by the utilities and appliance manufacturers have had their effect. Electricity is supposed to be "clean heat," and new owners of electric stoves seem sometimes to believe that their kitchens become cleaner. This is simple fantasy. The only products of the combustion of natural gas, burning with an ample supply of air as it does in a stove or furnace, are carbon dioxide and water vapor, neither of which is ordinarily perceptible to the senses. Oil heating may produce some soot and other pollutants, but the total production of pollutants is less than that arising from the same energy provided through the production of electricity.

In short, much of the new market for electricity sold to the public over the past forty years consists in satisfying fantasies about the supposed superiority of electricity in many applications where it is actually less desirable than older energy sources.

The latest manifestation of this long campaign is the push toward the all-electric home, which differs from other homes in that it has electric heating devices—stove, water heater and space heating. The thirst for energy which these heating devices have can be judged from the fact that an all-electric home consumes twenty thousand kilowatt-hours per year, triple what the average home of today uses. The projected tripling of residential demand with the installation of all-electric homes represents little or no advantage to the consumer. And, in fact, it is not the consumer to whom electric heating is sold. It is sold to the building contractor, for whom it has two attractions. First, installation costs are lower than for gas or oil heat, and the construction savings can be realized as profits. Second, the electric company may give him a direct cash bonus to install electric heating.

Through techniques like these, sales of power to homes and small businesses expanded fast enough to support the far larger structure of industrial sales built on them. Until very recently, in a rapidly-growing power business, homes bought about one fourth of the electricity and paid about half the bill. Industry bought half the power and paid about one fourth of the bill.

In 1973, this entire system began to collapse. The oil-producing countries of the Middle East announced an embargo on oil to the United States. Oil supplies did not, in fact, decline appreciably, but oil prices went up very rapidly, followed shortly by the prices of coal and uranium. All of the fuels used by power companies began to grow rapidly more expensive.

In accordance with the principles discovered by Insull, the prices charged to large industrial customers for electricity were barely above the cost of fuel needed to produce the power; the bulk of capital costs was carried by the millions of smaller customers.

But power could not be sold to industrial customers at a loss. When fuel prices went up, industrial rates began to rise with equal speed. In many states, power companies sought and obtained new regulations which allowed them to pass on increased fuel costs directly to consumers, without undertaking the time-consuming ritual of seeking state permission for rate increases. In 1974 and 1975, industrial power rates rose rapidly and it was clear they would continue to rise as fuel prices soared. Industry responded, predictably, by cutting back on its consumption of electric power and by generating more of its own heat and process energy. In the year ending April, 1975, sales of electricity to industrial customers declined 2.2 percent. The decline was not simply a result of the recession: commercial customers, as severely affected by the recession as industry but with less flexibility in energy requirements, used 2.1 percent more electricity than the year before; residential customers, with the least flexibility of all, increased their consumption by 2.4 percent. The change in some areas was even more dramatic: in South Carolina, industrial demand dropped 10 or 12 percent; and some New Jersey and Ohio utilities sold 12 percent less power to industry in the first eight months of 1975 than they had in the same period of 1974.

Power companies tried to maintain their earnings by massive rate increases for residential and commercial consumers, and rates rose more rapidly than inflation or the massive interest payments on utility bonds, but the traditional system had begun to break down. Home users of power began to refuse payment on the higher bills; in early 1974, fully a third of Consolidated Edison's customers in New York were in default of their electric bills. And state agencies, swept by political reforms in the wake of Watergate, were refusing some increases and delaying others. The over-all picture showed that the gap between industrial and residential rates was beginning to close. Traditionally, residential consumers paid about half the bill, while consuming only about a fourth the output of the nation's power companies; industrial customers paid a quarter of the bill and consumed half the output. By 1975, rising fuel prices and consumer resistance had almost put the two classes on an equal footing. Homes consumed one-third of all electricity and paid one-third of all power company revenues. Industry consumed forty percent of the power and paid just under a third of the bill.

There were still enormous disparities between the prices paid by the smallest residence and the largest industrial consumer; in each class, rates declined as consumption increased. But the wide difference between the classes, on which the growth of the electric power monopoly was founded, had disappeared. Industrial consumers no longer received rates low enough to encourage them to expand their consumption; residential consumers were no longer willing to pay the cost of such expansion. By 1975, growth in the power business had all but come to an end, and the industry faced a financial crisis. Growth in electric-power sales could no longer be predicted or controlled, nor could new generating plants be financed. Utilities across the country began to cancel their orders for powerplants and to halt the construction of plants already being built. Con Edison sold two plants which it could not afford to complete to the state of New York; other companies absorbed their losses or passed them on to customers, further dampening demand. Hundreds of powerplants were removed from the order books, and most of them were nuclear powerplants. By the fall of 1975, half the nuclear powerplants on order had been cancelled or delayed for years.

The system built by Samuel Insull, on which nuclear power had rested, seemed to be collapsing.

CHAPTER THIRTY-ONE

Solar Power

POWERPLANTS are very expensive to build because they use indirect means for producing electricity. The usual system is not abstruse or very new—electricity has been made in about the same way since the turn of the century—but it is complicated. First, one usually has a boiler full of water. A fire is built under the boiler (using coal or oil). The water in the boiler is heated and steam is produced. This steam is carried off under pressure by large pipes which lead to a turbine. The hot steam under pressure strikes the blades of the turbine, which spins like a pinwheel held in a stream of air. The turbine, turning, rotates a coil of copper wire in a magnetic field—a dynamo—and electricity flows in the wires, electricity which can now be drawn off for use.

There are several wasteful conversions of energy here. Coal is burned to produce heat; the heat is transmitted to water and then carried in the form of steam. At the turbine, heat is converted to mechanical energy, the spinning of the blades; and this mechanical energy is finally converted to electricity. For each of the conversions, elaborate equipment is required, equipment which is expensive to build, difficult to maintain and inherently inefficient.

Nuclear power did not alter this basic scheme. Nuclear powerplants employ all of these steps but the first; instead of burning coal, there is uranium in long rods, inserted into the water of the boiler. These rods become hot and cause the water to boil. From this point on, the process is generally the same.

Powerplants are therefore expensive to build, no matter what their fuel. Modern powerplants, especially nuclear powerplants, are even more expensive than their predecessors; the extra expense is justified by greater efficiency in the use of fuel and human labor. Without changing in any fundamental way, powerplants have become larger, more expensive and, in some ways, more efficient.

Because the power business is founded on the expense of powerplants, and because the legal system allows power companies to base their charges on their investment and hence to expand their business simply by buying more expensive powerplants, alternatives to this cumbersome system have not been explored.

The alternatives are simple and inexpensive, in theory. Preeminent among them is solar power, which need not require the several stages of inefficient energy conversion that create the high cost of existing powerplants. The fuel cell, which converts chemical energy directly to electricity without intermediate steps, is another potential alternative. Both these alternatives have been understood in theory for some years, but they are fundamentally incompatible with the electric-power business as it has grown up over the past century.

Sunlight, of course, has been used for centuries to boil and distill water, to heat homes and more recently to cool them. There is no doubt that solar energy could be used immediately—at least so far as the technology is concerned—to heat homes and drive air conditioners in all parts of the United States. Dozens of buildings using solar energy for these purposes are under construction, and in the next few years it is likely that solar energy will begin to displace some fuels, and electricity, for space heating and hot-water heating. The well-warranted enthusiasm for these applications of solar energy has led to some confusion and the impression that sunlight can quickly replace all other energy sources. There is a widespread but incorrect impression that solar energy is on the verge of being used to produce electricity.

There are two ways in which this might be done. Sunlight can be converted directly into electricity in systems which are called "photovoltaic." And sunlight can be used to generate heat to produce electricity in systems generally called "solar-thermal." Photovoltaic systems are also called "direct conversion" setups, since they turn sunlight into electricity without intermediate steps. Thermal systems use sunlight to boil water or some other fluid, which is then used to drive a turbine and generator in the time-honored fashion.

Direct conversion certainly sounds like the best bet. Sunlight falls on solar cells made of silicon, the second most abundant material in the earth's crust. Electricity comes out of the cells. There are no moving parts, no pollutants, and there is no consumption of any scarce resource. The only obvious problem is in making the silicon cells cheaply. And even this engineering problem, although difficult, seems soluble.

But what does one do when the sun doesn't shine? Spacecraft have been using direct solar conversion for years, but in space the sun does not set and there are no clouds. Here on earth the problem is a vexatious one.

At night, on cloudy days, in storms, the output from a direct solar-conversion plant declines to zero. Because electricity is not stored, the only way of avoiding this problem is to provide additional power-generating capacity to take over in sunless periods; ordinary coal-burning or oil-burning plants must be built to provide back-up. Such back-up plants must be large enough to meet the largest demand placed on the solar generators.

Building coal-burning or nuclear plants of about equal size to replace solar plants during the night removes most of the original attraction of the solar plants. To be sure, much fuel would be saved if solar plants operated during the days, but the costs would be very high—fuel accounts for only about one-sixth of the cost of electricity to the consumer. And there would still be substantial environmental damage from nighttime fuel-burning service.

Batteries, our only present means of dealing with this difficulty, are bulky, inefficient and expensive. Lead-acid batteries, for instance, of the type used in cars, are the cheapest available. They store about ten watt-hours per pound of battery. A family in the US uses about twenty thousand watt-hours on an average day; on a hot day with air conditioners going, the same family may use twice as much energy. To store its electricity needs for a single peak day would therefore require about two tons of batteries costing several thousand dollars. To store just one day's output of a modern million-kilowatt powerplant would require about a million tons of lead-acid batteries. Batteries to supply a city during a week of overcast weather are clearly out of the question. Batteries which can store more power per unit of weight are known, but they are also much more expensive.

A battery does not really store electricity, of course. Electricity forces the dissociation of chemicals in the storage battery, and the recombination of those chemicals generates an electric current. It is really chemical energy, and not electricity, which is being stored. A number of other systems for converting electricity to different forms of energy for storage have been investigated. Flywheels are one such. Electric motors would spin heavy flywheels operating in a vacuum on low-friction bearings. Once spun up to high speed, the flywheel continues spinning for some time, thus storing the energy used to bring it up to speed. To retrieve this energy, the system operates in reverse: the flywheel is used to drive a generator to produce electricity. Flywheel systems developed so far have not improved on the cost and efficiency of batteries, but there are no theoretical obstacles to this approach. Better materials from which to build the flywheels are needed.

Another possibility is to store the electricity in the form of chemical energy, by using solar energy in the form of electricity to break up water molecules into hydrogen and oxygen. When the sun is not shining, the oxygen and hydrogen can be recombined, releasing energy; these chemicals might be burned in a conventional powerplant, or they might be used in a fuel cell.

The trouble with storing energy in the form of hydrogen and oxygen to be used as fuel in an ordinary powerplant is the low efficiency with which energy is recovered. Electrolysis of water to produce hydrogen is only about one-third efficient. Steam-driven powerplants are also about one-third efficient, which means that the solar plant used to generate the hydrogen fuel would be nine times as large as needed for direct power output—a considerable penalty, far overshadowing any possible fuel savings.

Fuel cells are much more efficient than steam plants, however, and theoretically could recover almost all the chemical energy of recombining hydrogen and oxygen in the form of electricity. Fuel cells are quite close to commercial development. Pratt and Whitney Aircraft Corporation has for several years been testing household-sized fuel-cell powerplants which run on natural gas. A Pratt and Whitney official in an interview said that these are not considered to be economically attractive, although they are technically feasible, because of the large capacity and low utilization of a powerplant which serves only one household. Fuel cells are apparently still expensive enough to be vulnerable to Samuel Insull's monopoly theory.

More recently, Pratt and Whitney announced substantial progress in developing power-station-sized fuel cells. According to press announcements made in January, 1974, and confirmed in 1975, the company hopes to have a demonstration plant operating in 1976 and to begin filling orders for fuel-cell plants in 1978. These would be twenty-six-million-kilowatt units which would use oil for fuel. The oil would be first converted to hydrogen and carbon dioxide, and the hydrogen would be the actual fuel of the cell. These fuel cells would be highly suitable for direct operation on hydrogen. Using oil for fuel, their efficiency is expected to be about forty percent, or equal to the design efficiency of the largest powerplants now made. Operating directly on hydrogen, their efficiency would probably be even greater—perhaps fifty percent. Fuel cells would therefore greatly reduce the inefficiency of hydrogen storage; the inefficiency with which hydrogen is first produced remains a problem, however. There are no theoretical obstacles to highly efficient electrolysis. Assuming this can be accomplished, and if the fuel cells themselves are sufficiently inexpensive, one can imagine a system combining solar energy, hydrogen and fuel cells for power production.

The fuel cell is very attractive in its own right, of course. Fuel cells can operate essentially without pollutants; a hydrogen-burning fuel cell produces only water as a waste product. Because of the low temperature of operation there is no thermal pollution problem. The process of converting natural gas or oil (or synthetic fuel from coal) to hydrogen should produce only carbon dioxide and some small quantity of sludge for disposal. The environmental impact of fuel cells, assuming they are developed for power generation, might therefore be nearly as low as that of solar plants.

Solar plants, even without energy storage, could be used to generate hydrogen or other fuels for use in transportation, industry and space heating, rather than for power generation. But electric power would still be needed. And there are formidable technical problems to be resolved—such as the embrittlement of steel it causes—before hydrogen can safely and cheaply be transmitted long distances and used as a substitute for existing fuels.

Direct conversion of solar energy to electricity, which seems so attractive on its face, is therefore blocked by the lack of suitable energy storage. Large generating plants based on this technique do not seem likely for many years. A nearer possibility is house-top direct conversion units, which would provide a single-family residence or a larger building with some power during most of the year; the ordinary power system would supply electricity during periods of cloudiness and at night. This would not reduce the need for large central generating plants of other types, but would reduce the use of fuel and the demand for power at peak times on summer afternoons. Small house-top electric generators may become available in the next few years, but their cost will be high, particularly as they will be in use only part of the time.

The problems of storage are much less severe in the case of solar-thermal plants—which can be built immediately at a cost which will be within the range of other forms of power in the next few years. Perhaps the most attractive proposal to appear is one made by the Lawrence Livermore Laboratory in a study called "The Shallow Solar Pond Energy Conversion System," by A.F. Clark, J.A. Day, W.C. Dickerson and L.F. Wouters. Like many other recent proposals, it would store the energy of sunlight in the form of heat.

Heat can be stored easily. Sunlight can be converted to heat quite simply by putting almost any object in the sun. And converting heat to electricity is precisely what is done in most existing powerplants. Solar-thermal electricity requires no radical developments; in fact, the authors of the Livermore study note that a scheme similar to theirs was proposed in 1909 by H.E. Wilsie.

The basic idea is a simple one. Shallow ponds, filled with water and insulated to prevent loss of heat or water, would be warmed in the sun. The warm water would be pumped into a large reservoir, which would serve to store the collected heat. Heat from the reservoir would be used to boil some liquid (like ammonia) that boils at low temperatures, and the resulting vapor would turn conventional turbines and generators. While the over-all efficiency of the system is low, the cost of the components is also low. The authors of this study calculated costs from a small, ten-thousand-kilowatt plant at just under three cents per kilowatt-hour, or about three times the present cost (at the generating plant) of electricity in the Southwest. This is a plant built immediately from existing materials and components. With modest development work, the authors feel, not unreasonably, that construction costs could be cut in half over a few years, during which time the price of fuel and so of electricity from other sources would be rising. Solar electricity might therefore be as cheap in dollar terms as electricity from coal or uranium in as little as ten years.

Because the Livermore system would operate at low temperatures, it would require no special heat-storage systems, and the water used to collect solar energy could easily double as the storage medium. Low temperatures mean low efficiencies, however, and a correspondingly large amount of collecting surface must be used. The scheme would work, therefore, only where land costs are relatively low and sunshine is abundant. These are conditions which obtain in the deserts of the US Southwest. They also exist in many other less developed countries. Since the Livermore scheme is a relatively simple one, it might be well suited for tropical countries which are just developing and which are short of conventional fuels.

The Livermore shallow-pond scheme would store enough heat to equal the power output of the plant for a period of days, quite enough to cover nighttime and cloudy-day interruptions in the sunny climate of the Southwest. The authors propose to build their plant in Arizona. In other parts of the country, cloudy conditions persist for much more than a few days, particularly in the winter—and so even solar-thermal powerplants are likely to be limited to the Southwest by energy-storage problems.

Although they are the oldest and closest to completion, low-temperature solar-thermal systems are not often noticed. Discussion far more often centers on high-temperature solar-thermal systems. These are systems in which sunlight is concentrated by mirrors or lenses so that the originally diffuse energy is collected or focused in a small area. Anyone who has used a burning glass knows that very high temperatures can be created in this way. The federal government is now supporting study of a variety of schemes to use sunlight in this fashion. Lenses or mirrors would focus sunlight on water, which would then boil. The steam would be further heated to perhaps a thousand degrees, or to temperatures typical of powerplants which burn coal or oil. The steam could be used directly for power production, but here again we run into storage problems. And storage of heat at high temperatures is a much more complex problem than storage at low temperatures. Water, for instance, is not a practical storage medium if temperatures are above the boiling point at ordinary pressures. One common suggestion is the use of a large reservoir of molten salt and, while this is technically feasible, there are a large number of engineering problems to solve. These problems may be soluble, but they put high-temperature solar systems some years further into the future than low-temperature systems. They also add greatly to the capital costs of the system. High-temperature systems are even more capital-intensive than existing generators. Despite these difficulties (or because of them, the mistrustful observer may suspect) the federal government is pouring most of its meager development funds into high-temperature systems which are many years from practicality.

Even with the problem of short-term energy storage solved in low-temperature solar-thermal systems, a problem of long-term storage asserts itself.

The Livermore study, which deals with Arizona, notes that variations of sunlight from winter to summer present a serious difficulty. The authors calculate that a plant of their design which would generate eighteen thousand kilowatts in the sunniest month of summer would produce only three thousand kilowatts in January. This is a result of the variation in sunlight intensity over the year. And while heat energy can perhaps be stored cheaply by some existing methods for short periods, it cannot easily be stored for a period of months. Solar-thermal plants, therefore, are simply not suitable for a dominant role in power production. Even in the sunniest areas of the Southwest, winter sunlight levels reduce power output sixfold. In other parts of the country, solar-thermal plants would fare still worse, with storms, snowfall and persistent cloudiness further reducing the sunlight available. Where frequent storms are not a problem, solar plants could be made large enough to provide sufficient power in the winter. The large excess capacity could be used in summer months to generate surplus power for sale. A few such plants might be built to meet year-round demand in small communities, with excess power shipped to other areas during the summer—but transmission systems to carry power from Arizona to Vermont might cost far more than the solar plants themselves.

Solar power, we are forced to conclude, will not play an important role in electric-power production until efficient and economical means of storing electricity are developed. Not only is there very little research in this direction, but there is very little awareness that it is even required.

There is no lack of interesting ideas to explore. Since the nineteenth century science has uncovered information about the natural world which should assist us in devising a means of storing electricity more efficiently than the lead-acid storage battery. For instance, we now know of the phenomenon called "superconductivity," found in metals at very low temperatures. As the temperature drops toward absolute zero, all resistance to the flow of electricity ceases; an electric current in a loop of wire would continue flowing forever. In theory, therefore, we can now store electricity without converting it to another form of energy—the "hot-water tank" storage of electric power is a possibility.

There are no present means, however, of doing this in a practical fashion. Machinery to produce the extremely low temperatures required is both bulky and expensive, far too expensive to warrant commercial use. Some physicists believe that it may be possible to exhibit the phenomenon of superconductivity at room temperatures, but this possibility is not yet clear even in the realms of theory.

Flywheels and hydrogen fuel-cell systems have been mentioned; there are other possibilities. They have in common only a lack of attention from public and private research organizations. It is difficult to generate enthusiasm for energy storage, and almost impossible to persuade anyone that our national interest requires a Manhattan Project or an Apollo Program to devise a means of storing electricity cheaply. But the use of solar energy for power production depends on energy storage—and wouldn't it be nice to have an unlimited, safe and nonpolluting form of energy that falls from the skies? Every house could then gather its own electricity, the monopoly of the big power companies would be broken, and we would have at least the technological basis for that decentralized, pastoral existence so many of us claim to want.

CHAPTER THIRTY-TWO

Research

THE ELECTRIC-POWER INDUSTRY, public and private, had revenues in 1975 of more than forty billion dollars, an increase of more than thirty percent over the previous year (expenses rose even more rapidly, however, primarily because of rising fuel costs). Research is not a sufficiently large item among utility expenditures to be included in most published reports. The industry as a whole plans to spend two hundred fifty million dollars, over a period of years, on the development of the breeder reactor; almost all other research is supported through the Electric Power Research Institute, in Stanford, California, the research subsidiary of the Edison Electric Institute which disburses somewhat less money than the utilities contribute to breeder development—about one hundred million dollars per year, or one-quarter of one percent of the industry's revenues. And of this meager total, only three million dollars will be spent in 1976 to do research on energy storage. About an equal amount is spent by the manufacturing corporations—General Electric, Westinghouse and others— according to Dr. Fritz Kalhammer of the EPRI. Almost all of these tiny budgets are devoted to refining existing technology—compressed-air and water-storage devices, primarily. The federal government is spending slightly more, roughly ten million dollars per year, on developing new batteries, but most of this research is carried out by the auto makers and is directed toward developing a battery suitable for electric cars, not to energy storage for power companies. As Gregory Vassell of American Electric Power points out, the existing technology of the power industry is very well suited to the way the industry is organized; there does not seem to be any present need for radical departures, from the industry's own point of view. While energy storage would reduce the industry's need for capital to build new plants, as W. Donham Crawford of Edison Electric Institute concedes, demand for new capacity is diminishing; the great urgency is for financing to support construction projects which already have been under-taken with short-term funds. If energy storage happens to be necessary for solar energy to become a reality, that is perhaps not the utilities' concern.

The Mobil Oil Corporation (née Standard Oil of New York) is among the firms which have recently been advertising their desire to see new energy sources like solar power relieve our dependency on imported oil. A reporter therefore asked Dr. Dayton Clewell what his firm is doing in this field. Clewell is a physicist, a senior vice-president of Mobil Oil and also president of the subsidiary which conducts most of the company's research, Mobil Research and Development Corporation. Clewell is a gray-haired, middle-sized person in his sixties, who lives in Darien, Connecticut, and is a director of the Darien Historical Society.

Now you say that you have known, and the oil industry has been telling us for years, that we're going to run out of oil in the United States. . . .

CLEWELL: We never said that about the Middle East, except for . . . the year 2000. . . . Sure we will.

Now why is it that all these big companies, with very sophisticated scientific personnel, who were in the best position of anyone to see what was coming, and, in fact, warned everybody about what was coming, basically didn't do a damn thing on the research side to develop alternative energy supplies?

CLEWELL: Well, you mean that we should be walking away from cheap oil and gas right now and go into the expensive coal liquefaction?

I don't mean you should have gone into that business but why—

CLEWELL: We've done it . . . we've done the technology.

Okay, well, how about, for instance, solar energy? I think people are pretty clear that, if given development, it will be ultimately cheap and in the long run everything's going to be solar energy. . . .

CLEWELL: In the long run. In the long run it will, and that's why we're into it.

Yeah, but you only spent how much?

CLEWELL: Thirty million dollars over a period of five or six years.

Okay. Now what were the gross revenues of the Mobil Oil Company last year?

CLEWELL: Eighteen billion dollars.

And they devoted thirty million dollars to research in solar energy.

CLEWELL: Right, absolutely. And I know the next question: why don't you throw more money at it? Because more money wouldn't do a damn bit of good.

We're in the early stages of this sort of thing, and if you don't have some ideas, just throwing money at it, like the federal government does, is nothing but waste.

What about solar electric power? That's a large enough market to be attractive to your company, isn't it?

CLEWELL: Yeah, but we're sure not gonna do it until we really see we can get the price [of solar collectors] down. And we're not gonna move until the price is down, because otherwise you're just risking way too much money.

Do you want to say what kind of collectors you're working on?

CLEWELL: Sure, silicon.

Silicon. Are you working on a ribbon?

CLEWELL: Ribbon . . . the ribbon technology [for producing continuous strips of silicon cells]. We formed a joint venture to develop some ideas. . . .

Now isn't there another component that you need? Don't you need a storage technology?

CLEWELL: Yes.

Now what kind of work are you doing on that?

CLEWELL: Well, we're not . . . we're not doing any ourselves. We know they're a lot of people [who] are, but one company can't do everything. We try to pick out a piece where we have some expertise and think we can go ahead and I think that we find that the government agencies, other companies and so on, will do the other.

IV. NUCLEAR POWER

E. P. Swisher, vice-president of the Oil, Chemical and Atomic Workers
International Union (Chapter 35)

David Brower, founder and head of Friends of the Earth (Chapter 39)

CHAPTER THIRTY-THREE

Uranium

G RANTS, NEW MEXICO, was a coaling station on the old Santa Fe Rail-road, one of the transcontinental lines built by the railroad buccaneers of the 1860s. In 1863, Colonel Kit Carson rounded up the Navajo, with whom the railroads and the Utes were at war, and deposited them on a reservation far from the huge tracts of northwestern New Mexico, Utah and Colorado which they then considered their own. The Navajo were devastated by disease and crop failures on the reservation chosen for them, Bosque Redondo, at a place named Fort Sumner in the New Mexico Territory, where the soil was too alkaline for successful farming. Thousands died; others left the reservation for the only other life permitted the Navajo, slavery in the Mexican communities of the area.

In 1868, General William Tecumseh Sherman met with Barboncito, chief of the remaining Navajo, then no more than seventy-three hundred, and other chiefs of the Navajo families, to negotiate a new treaty. Barboncito eloquently refused the proposal that the Navajo be relocated to the south of the Arkansas river, where the government had placed the Cherokee. Barboncito said,

> "I hope to God you will not ask me to go to any other country except my own. It might turn out another Bosque Redondo. They told us this was a good place when we came but it is not. . . . After we get back to our own country it will brighten up again and the Navajos will be as happy as the land, black clouds will rise and there will be plenty of rain. Corn will grow in abundance and everything look happy."

General Sherman finally agreed, and the Navajo were permitted to return to a reservation on the San Juan river, where they still remain in a small corner of the land they once claimed. It was never solely theirs: the pueblo builders had been there long before, but, like the builders of Central America, had inexplicably vanished shortly before the European invaders appeared.

For at least a thousand years, the Pueblo Indians have been here; perhaps they are descendants of the great city builders. While the Ute and Comanche battled the Navajo for dominion, the Acoma Pueblo Indians have lived on and near the four-hundred-foot-high mesa on which their pueblo still stands; somehow, over the centuries of warfare, as nomads, railroads, Spaniards, Mexicans, North American soldiers, miners and road builders have swept in successive waves over their lands, the peaceful Acoma have maintained their centuries-long way of life, their own language and their peace.

New Mexico has only been a state since 1912, and the modern world has put a strange veneer over its thousand-year-old culture and its more recent history of violence. A visitor to the state is quickly reminded of the modern world's far greater potentiality for destruction. At Albuquerque's airport, arriving passengers, expecting to see the usual displays of local industry and the chamber of commerce, are greeted with a large sign and illuminated exhibit:

Sandia Laboratories is an Energy Research and Development Administration installation engaged primarily in design of nuclear weapons.

The nation's first atom bombs were assembled in New Mexico, and the first nuclear explosion on earth took place at Alamagordo to the north. Not far away is the White Sands missile range; at Sandia, near Albuquerque, nuclear weapons are still designed and assembled; rumor has it that the weapons are stored deep beneath the Sandia Mountains.

To the west of Albuquerque is a bizarre volcanic flatland, scattered with twisted shapes of lava that have not yet been weathered to softness. The immense tableland is ringed with mountains, but the mountains are very far away, no more than a rim around the flatlands, over which an oppressive sky arches. One can see extraordinary distances in the thin air; storms can be watched forming, dozens of miles away. From a single point one may be able to watch both a snow storm high in the distant mountains and a thunderstorm approaching from the opposite direction, which, in an hour or two, will scatter one of the area's frequent showers that have too little water to wet the soil; and over it all there will still be the oppressive dome of blue sky, crushing in its immensity.

Driving westward from Albuquerque in this flat brown desert, one encounters Grants, New Mexico, seemingly crushed and scattered along the highway by the weight of the sky pressing down on it. The town is stretched along two miles of old US 66, now bypassed by the modern Interstate 40. Along the road one finds only motels, trailer courts, a dusty shopping center, some bars and luncheonettes, the chamber of commerce; the Uranium Cafe; the Lariat Motel, The Sands, The It'll Do Inn; Acoma Pottery For Sale; The Roaring '20s. There are no pedestrians in Grants during the sunbleached day; houses are low, scattered, pale. A white and red banner slung across the thoroughfare proclaims prayer week, to be celebrated in a local high school gymnasium.

At the east end of town is the Holiday Inn, which keeps a fastidious distance between itself and the sleazy motels of Grants. This Holiday Inn is identical to hundreds of others across the United States, except that the tall welcoming sign, which in other parts of the country gives the name of whatever convention or wedding party is in residence, here bears the legend:

SMILE, SUPPORT LOCAL AND STATE POLICE

Scattered into the desert for a little way along the whole length of the two miles of Route 66 which make up Grants's spine are low, pastel-painted one-story houses. Some are of concrete-block and plaster construction, but most are trailers, mobile homes anchored against the steady winds. There do not seem to be many permanent structures in Grants, and most of its shops and facilities seem to be designed for visitors. For Grants is a boom town again, as it was once twenty years ago; and the chamber-of-commerce slogan for Grants is "Uranium Capital of the World."

The upper-echelon mining-company people do not live in Grants. Anaconda, one of the larger mine operators in the area, built a town—Anaconda, New Mexico—for its supervisory personnel, who pay nominal rents and are provided with a swimming pool, rifle range and gymnasium. The trailers and shacks of Grants belong to the Spanish-speaking miners who have come here from elsewhere in the Southwest. The local Indians do not go to the mines; they remain at their pueblo communities, scratching out a bare living at farming, struggling for a share of water rights from the impounded streams. Bluewater is a nearby prosperous farming community, a relic of yet another of the migrations that scarred the Southwest; it is a Mormon community, and the Mormons have first rights to the water from Bluewater reservoir in an area where water is life. A few Navajo have returned to the area from their reservation to the west, but most of the miners are Spanish-speaking descendants of the first European conquerors, now themselves conquered.

The second uranium boom in Grants is now just beginning. Gulf Oil is opening a new mine in Mt. Taylor, which holds at least a hundred million pounds of uranium, but little exploration has as yet been done. Uranium mining is skilled work, and there is a shortage of miners in Grants. The shortage is made worse by the unwillingness of miners to remain very long at their work; turnover is high and strikes are frequent. Kerr-McGee, United Nuclear, Anaconda, Homestake and other giants of the uranium business are competing for the few available skilled uranium miners and importing new miners from the depressed copper mines of the region. There is a frontier, lawless tone to Grants. When, after exposing faulty safety practices at a Kerr-McGee plant, a young woman was killed in Oklahoma in a car accident that a union charged was murder, people in Grants were not much surprised or interested. They don't think the company murdered her, but they wouldn't be concerned if it had.

Uranium has suddenly become valuable again, after twenty years of depression. After a boom in the 1950s, which uncovered far more ore than the nation could possibly use, the uranium industry of Grants was supported by a kind of federal welfare program of scheduled purchases for a weapons stockpile. The federal stockpile of uranium still hangs over the market like one of the region's towering thunderclouds, but federal surpluses have not dampened a new private uranium rush. The nuclear powerplants which have been ordered will require a lot of uranium, and most of it will come from the Colorado Plateau on which Grants sits. After years of hovering below the eight dollars per pound at which federal purchases were pegged, uranium prices have shot up to forty dollars per pound. New mines are being opened and exploratory drilling is beginning again. Companies trapped in long-term contracts at the old prices are fighting to free themselves; in October, 1975, the Westinghouse Corporation simply announced that it would not honor its long-term contracts to supply uranium fuel for reactors it had supplied. While the fuel contracts it signs are kept secret, Westinghouse must have agreed to supply at eight dollars a pound uranium which now brings five times that amount—and Westinghouse claims it did not purchase enough uranium when prices were low to meet its commitments. The scramble is on; power companies and reactor manufacturers must have uranium, from which eight percent of the country's electricity is already derived. And Grants, New Mexico, has become as important to the energy planners as Teheran and Riyadh.

CHAPTER THIRTY-FOUR

Blow-Up

IN 1903, every powerplant had a boiler in which water was heated; since then, the basic design has not changed, but there have been refinements. The boiler, once a large steel pot, is now an array of small tubes through which water is pumped. Coal is now pulverized to a fine powder and blown flaming around the boiler tubes. The result is a very much hotter steam, which can accordingly be used economically with more efficient turbines. So high have these temperatures become, in fact, that the physical limits of efficiency are being reached in the most modern coal-burning electric powerplants.

Nuclear powerplants were a step backward in this development, as those who studied it in the 1950s realized it would be. In nuclear powerplants, we returned to the old potlike boilers and lower temperatures of the turn of the century.

In old-fashioned powerplants, one burned coal under a large boiler, but in nuclear powerplants heating elements are inserted into the boiler, somewhat in the fashion of the heating elements one can use to make a cup of tea. Long, thin nuclear fuel rods heat the water, steam is produced, and from that point onward, little has changed. The steam spins a turbine, the turbine turns a generator, and power is produced. In the type of reactor sold by General Electric, the "boiling water" reactor, this scheme remains unchanged. In Westinghouse's "pressurized water" reactors, the boiler is kept under higher pressure; the water does not boil, but is carried to a heat-exchanger—a second boiler—where steam is produced.

All of the problems of nuclear power stem from events which happen inside the heating elements inserted into the nuclear powerplant's boiler, the long, thin rods of uranium oxide, wrapped in thin metal sheaths. A single fuel rod may be twelve feet long and only one-half inch thick. Like the heating elements intended for a cup of tea, these fuel rods become hot and the water surrounding them begins to boil. The fuel, unlike the devices for cups of tea, is not heated by electricity, but by the splitting of uranium atoms. The fuel rods become hot and continue to generate heat until the uranium 235 they contain has been depleted. This can take a period of time ranging from a few months to several years.

Thousands of such fuel rods, gathered into bundles, are used in the large powerplants now in operation; each of these thousands of finger-thin rods is a separate and intense source of heat. Interspersed among the fuel rods are safety and control rods, which absorb neutrons and which are more than sufficient to halt all nuclear reactions in the plant.

However, these safety rods and control rods are not the principal safety devices in the plant. Their function is really to start and stop the plant or alter its power level. The principal safety measure in the plant is its fundamental nuclear design.

The water which a powerplant's boiler holds is an essential part of the nuclear reactions going on in the fuel. The nuclear reaction depends on slow neutrons and neutrons must collide with the water molecules before they can be slowed sufficiently to split uranium 235 atoms. If the water is lost, the nuclear reaction will stop. Further, if the temperature of the fuel rises greatly, the reaction also will stop; usually, if anything goes seriously wrong with the operation of the plant, it will simply shut down. It is to this extent "fail safe"—major failures will leave it in a safe condition. At least so far as these water-boiler reactors are concerned, there is no fear of a nuclear explosion. In the early military days of nuclear development, this appeared a strong advantage over reactor types which could, in theory, blow up like atom bombs.

But there is another set of quite different problems which have plagued the water-boiler powerplants and which have a common source. Within the long rods of uranium which heat the water in these powerplants, uranium atoms are splitting. The result of this splitting is a good deal of heat, which is useful, and the fragments of the uranium atom, which are not. These fragments are tiny bits of matter, unstable fragments with no place in nature. They disintegrate further, in abrupt bursts, until they reach a stable place in the table of the elements. These further disintegrations of the fragments of uranium produce what we know as radioactivity. Each abrupt disintegration may produce particles or rays which are still smaller bits of these fragments of atoms. Such emissions have varying effects, depending on their nature and circumstances, but collectively they are measured as radiation.

It is this radioactivity, produced as a by-product in nuclear powerplants, that creates the safety and pollution problems which have plagued that industry in the years since its inception.

Very large quantities of radioactive material accumulate in a powerplant during operation. This is a necessary result of the fissioning of large quantities of uranium—much more uranium than would be used in nuclear weapons. As a consequence, nuclear powerplants ordinarily contain far more radioactive material than is released in even the very largest military explosion. Most of us have no experience with radioactive materials or the units in which radioactivity is measured, and so it is difficult to get an accurate sense of this accumulation of material. A million-kilowatt nuclear powerplant, after a year of operation, contains roughly ten billion curies of radioactive material. To most people, this number means nothing at all.

A curie is the radioactivity of a single gram of radium—or about one-thirtieth of an ounce. Ten billion curies of radioactivity represents the equivalent, in energy release, of many tons of radium and is certainly enough, if properly distributed, to kill everyone in the United States. But then, there is enough water in Lake Superior—if properly distributed—to drown everyone in the United States.

There does not seem to be any easy way for someone who has not worked regularly with radiation to get a good feeling for the meaning of the units of measurement which are used in this field. The best one can say is that the radioactivity contained in a nuclear powerplant is equivalent in effect to tons of a very powerful poison, and that there is quite enough of this material in a powerplant to kill millions of people. To do so, however, it would have to be released and dispersed. There may very well be equally large collections of equally fatal material in the chemical industry; there certainly are in biological- and chemical-warfare programs.

The presence of all this hazardous material is not in itself a matter of unusually great concern; it is at least no worse in this regard than other horrors of modern life. We would become worried only if there were also present means for dispersal of the radiation. But while nuclear plants cannot explode like bombs—at least those now in operation cannot—there is such a mechanism of dispersal present in each powerplant.

The radioactive material which accumulates in the fuel is not only poisonous—it is hot, in the ordinary sense: if one could safely touch it, it would be very hot to the touch. During normal operation, in fact, about seven percent of the heat output of a powerplant comes from the radioactivity created in the fuel. So long as nothing goes seriously wrong, this is a beneficial effect—this heat, like that produced in nuclear reactions, can be converted to electricity.

Unfortunately, the heat produced by the disintegration of radioactive materials is not under the power company's control. Radioactive fragments of uranium which accumulate in the fuel of the plant are hot by virtue of their radioactivity, which is a process of decay that cannot be easily affected by outside forces. Once created, these radioactive materials will continue spewing out radiation and heat until they have a stable weight and place among the elements, a process which may take centuries.

Seven percent of the power produced in the fuel of a nuclear plant, therefore, is self-generating and cannot be shut off. The fuel can be removed from the plant, but it will still be hot; for centuries, the radioactive decay will continue and heat will still be produced, although at steadily declining rates. Once started up, a nuclear plant cannot ever really be shut down completely. And this is the source of the most serious problems to afflict nuclear power.

Seven percent of the output of a million-kilowatt plant is substantial—seventy thousand kilowatts, enough power to light up a small town. It is also enough, if concentrated in one place, as it is in the fuel of a powerplant, to produce extremely high temperatures. It is quite enough to melt the fuel of a powerplant, the steel supporting structures, the boiler itself and the concrete structure beneath it. And if the water is lost from the plant's boiler, that is what will happen.

Because nuclear fuel cannot be completely shut off, the power it continually produces is a potential means of dispersing the poison it contains. This, rather than any uniquely poisonous character of radioactive materials, is the source of the unusual safety hazards of nuclear powerplants. If it were not for this continual power production, which can never be completely halted once it is put underway, the poisonous qualities of radioactive material would be a matter of much more limited concern. The powerplant yonder would be full of poison, indeed, but it would be *there* and I would be *here*. It is only the specter of poisons which can carry themselves from there to here that is frightening.

The specifics of how radioactive materials might become dispersed from a powerplant are a good deal more complex, at least in detail, but the principles are simple. If cooling water is lost or fails to flow, the fuel or the boiler as a whole will heat up. Either the fuel will melt, or the boiler will burst; in either case, the physical integrity of the containers will be lost. The fuel will then collect at some place in molten form. If it drops into a pool of water, there may be a steam explosion. There may be various chemical reactions and even chemical explosions as a result of the high temperatures generated by the fuel. Simply pouring water on the fuel at this stage will make matters worse.

Once molten, the fuel might collect at the bottom of the plant's boiler, through which it would slowly melt; the molten mass of fuel—perhaps two hundred tons of liquid metal and uranium oxide—would melt downwards, slowly, through the concrete floor of the building and into the earth or rock below. It would then continue melting its way downward for some time—nuclear engineers, with gallows humor, long ago named this sequence of events the "China Syndrome." After a year or two, the molten fuel would come to rest at the center of a large sphere of molten rock, perhaps fifty to a hundred feet below the surface of the ground, and there it would remain for some centuries, gradually cooling as radioactivity spent itself, or until someone came to dig it up.

At some point during this process it is likely that radioactive gases would be released to the outside air and that radioactive materials would be carried into ground water. The amounts of radiation released would depend greatly on the circumstances, which cannot be predicted; but a portion of the total radioactivity contained in the plant would be in gaseous form at the temperatures which would be found in a fuel-melting accident. One can imagine the release to the outside air of many millions of curies of radioactivity—only one-tenth of one percent of the total, perhaps, but as much as all the long-lived radioactivity released in all the United States' weapons tests. These gases would drift slowly with whatever wind was blowing; they would be warm and so would also rise slowly away from the ground, perhaps decreasing the hazard somewhat.

For this sequence of events to occur, the powerplant's fuel must be deprived of its usual flow of cooling water. This might happen in a number of ways. The plant's boiler might burst. One of the large pipes carrying water into or out of the boiler might break—or simply develop a large leak.

To forestall such occurences, the manufacturers of powerplants are required to take considerable care in the design and fabrication of boilers, pipes, valves and other components. But metals fracture, and people make mistakes or cut corners, despite all efforts to see that they don't. Accordingly, nuclear powerplants are fitted with a number of devices to supply cooling water to the fuel whenever the normal flow of water to the boiler is cut off. There are a large variety of such systems even within a single plant, and collectively they are known as emergency core-cooling systems. The debate over their adequacy has been at the center of the controversy over nuclear powerplants for several years, ever since Henry Kendall, Dan Ford, and several colleagues at MIT cast a critical look at the systems as they were then designed. Criticisms published by Kendall, Ford and others forced extensive and exhausting public hearings on the adequacy of these systems. At the conclusion of the hearings in 1973, the Atomic Energy Commission slightly tightened its requirements for emergency cooling, but Kendall and others still doubt that the systems will work under emergency conditions.

The safety systems are not particularly complex in design. In the powerplants sold by General Electric there are, for instance, what amount to sprinkler systems above the fuel. If for some reason the water is lost from the boiler, these systems are to spray the fuel to keep it below melting temperatures. In Westinghouse plants, there are large tanks under pressure to flood the boiler with water.

Serious questions about the adequacy of such devices remain. While their principles are very simple, the factors affecting their successful operation are not. The bursting of a pipe in an operating nuclear plant would be a dramatic event— tons of water would suddenly pour through the breach under pressure; much or all of the water would quickly flash into steam. Temperatures would quickly rise from the hundreds into the thousands of degrees; metal structures would begin reacting chemically with the steam, generating potentially explosive hydrogen. Within a minute or two, the fuel could be so badly buckled and molten that further functioning of the cooling systems would be impossible.

In the first minute or two after a pipe ruptures, therefore, while enormous pressures are at work, the emergency cooling systems must function properly. Such systems have never been tested under realistic circumstances; to do so would require the destruction of very large and expensive powerplants. Instead, their functioning has been predicted by the use of complicated calculations conducted on computers. The results of these predictions have been mixed and are, given the nature of the task, somewhat uncertain. One of the mathematicians who helped develop such calculation techniques, Carl Hocevar, has joined Kendall and Ford in their argument that emergency cooling systems are unproven. In October, 1975, Hocevar published an extensive report on the failure of calculations to predict even simple experimental results.

Unfortunately, the question of emergency cooling has become more, rather than less, acute as the industry has developed. In its early years, it was still possible to argue that the systems would not be needed; pipes would not break; boilers would not burst. But there has now been enough experience to show that pipes will crack; perhaps even boilers may burst. The most ambitious study of reactor safety undertaken by the Atomic Energy Commission showed that fuel melting was more likely than not, unless emergency core-cooling systems functioned properly. This study, directed by Norman Rasmussen, was released in draft form in September, 1974. (The revised final version of this study was released in 1975. References throughout this volume are to the earlier draft. Please see Appendix A for further information on the final version.)

Newspapers dutifully reported the study's conclusion that the risk of anyone being killed by a nuclear powerplant accident was smaller than that of being struck by a meteor. Industry spokesmen and the AEC itself thereupon hailed the report as a vindication of nuclear power. But the report and its twelve volumes of appendices are far more interesting and disturbing than the press handouts of the AEC.

The report expressly or implicitly confirms the huge damage estimates made in earlier and less detailed studies of reactor accidents. In the worst case it examined, the Rasmussen study estimated twenty-three hundred immediate deaths, fifty-six hundred cases of radiation sickness, sixty-four hundred cases of induced cancer and genetic effects, and $6.2 billion in property damage. The numbers of deaths and injuries are somewhat lower, proportionately, than earlier studies because of the Rasmussen report's assumptions—notably that evacuation would remove most people from the path of radioactivity—but still they are hardly reassuring. More pessimistic assumptions about evacuation and weather conditions, which in earlier studies produced estimates of tens of thousands of casualties, would have produced similar results in the Rasmussen study.

Still worse accidents may be possible. In most discussions of safety, it is assumed the molten fuel melts its way downward and that, as a result, radiation is released some hours after the accident is initiated, after some of the radioactivity has decayed and there has been time to evacuate some people from the expected path of release. This need not be the case. The plant's boiler may burst, piercing the surrounding concrete containment building and releasing radiation directly and quickly to the outside air, seconds after the reactor ceases operating at full power.

Furthermore, under some special conditions of equipment failure, a reactor could run out of control. Such a nuclear "excursion," as it is called, might result only in melting of the fuel, or it might itself release enough energy to explode the boiler, piercing the surrounding structures and releasing radiation to the air while the reactor is operating far in excess of its normal power. Such incidents, called "transients without scram" in the opaque language of the nuclear initiate, are a matter of serious concern. Richard Webb, holder of a doctorate in nuclear engineering, has energetically tried to bring this possibility to official and public attention and has forced the disclosure of some worrisome research reports. It does not seem possible with present information to say how likely or how severe such excursions may be.

Overshadowing all other possibilities in magnitude of damage is the steam explosion. This curious phenomenon results when a hot liquid is dropped into a cold liquid. If the hot liquid is hot enough and other conditions as yet not understood are fulfilled, the cold liquid does not simply boil—it explodes.

Such an explosion seems to have occurred unexpectedly in 1950, during a nuclear safety experiment in which a small reactor was purposely destroyed; an energy release much larger than expected was later attributed to a steam explosion. The phenomenon has been observed in metal foundries and other industrial settings in which molten metal may come into contact with water. It has also been observed when liquid natural gas is dropped into water—there water is the "hot" liquid and it is the liquid natural gas which produces the vapor.

When a reactor's fuel melts, it may come into contact with water. Even if the melting has occurred because of the loss of cooling water, there is likely to be some water remaining in the bottom of the boiler or in the reservoirs used in some reactors as a safety device. If the melting occurs because of an excursion, the boiler will be full of water when the fuel melts. Will a steam explosion result?

No one seems to know. The Nuclear Regulatory Commission is investigating the problem at Sandia Laboratories, but research has not yet established when or why such explosions occur or how severe they may be. The research is not being pursued vigorously. A long-planned experiment, to be conducted at the reactor testing station in Idaho, involving the melting of a reactor's fuel, has been delayed year after year.

One theory proposed by Stirling Colgate is being actively investigated by NRC; if it proves correct, the potential damage of a nuclear accident will have to be revised severely upward. Colgate, dean of the New Mexico Institute of Mining and Technology, at first did not develop his theory with an eye to nuclear powerplants. He had been retained by the Icelandic Government to deal with a very different sort of problem. The eruption of a volcano on Heimay had produced a flow of lava underwater that threatened to close the island's harbor. Colgate and a colleague from Reykjavik, Thorbjörn Sigurgeirsson, were to set off an underwater explosive, on the theory this would force the lava to mix with seawater and so cool and harden itself, thus impeding the flow that threatened the harbor. On August 31, 1973, the account of that effort appeared in the prestigious journal *Nature*. Colgate and Sigurgeirsson wrote:

> We planned such an explosion with the help of the Icelandic government, the Icelandic Coast Guard and the US Navy, but the day before the detonation we realized the awesome possibility that once mixing was initiated [between lava and sea water] it might be self-sustaining in that the high pressure steam produced might cause further mixing until all the lava had exchanged its heat with the water above it. The energy released might have come to between 2 and 4 megatons [equivalent to the explosion of 2 to 4 million tons of TNT]. Naturally the experiment was called off.

Colgate and Sigurgeirsson theorized that the explosion of Krakatoa, with a force of hundreds of megatons, might have been produced by the same kind of steam explosion they feared would occur in Iceland.

On his return to the United States, Colgate gave more thought to the possibility of steam explosions. He was aware that two liquids of very different temperatures would result from the melting of a reactor's fuel or from spilling the contents of a liquid-natural-gas tanker. On August 7, 1974, Colgate prepared a paper which calculated in more detail what might result from the kind of hot-cold mixing he and Sigurgeirsson had feared at Iceland. By now Colgate called the phenomenon "explosive self-mixing." Steam produced by a slight mixing would cause a further mixing, releasing further steam, etc. The process would happen very quickly.

The calculations were disturbing, although Colgate used only very rough approximations: "If the reactor core has reached the melting point, the explosion could be the equivalent of several tons of TNT, enough to shred the containment vessel." Colgate had calculated on the assumption that the molten fuel would be five tons of core material; the larger modern reactors now contain more than a hundred tons of such fuel. At the temperatures he felt would be reached, Colgate concluded the explosion might occur even if there were no free water remaining in the reactor. Once the fuel melted its way into the soil beneath the plant, water of hydration in concrete, rock or soil "becomes an equally hazardous potential source of 'cold fluid'" for explosive self-mixing.

With regard to liquid natural-gas tankers, Colgate estimated a spill of the tanker's contents in a harbor, even if very limited at first, would propagate into an explosion equivalent to ten thousand tons of TNT; the resulting natural-gas and air mixture might then be set off by any random spark, with a chemical explosion in the megaton range—large enough to level any city on earth.

Colgate submitted portions of the paper with these calculations to scientific journals, and the responses persuaded him to withdraw the paper from circulation. He said he became concerned that the presentation was not balanced; that there might be a relatively simple means of forestalling the explosions his theory predicted; and that in any case the theory itself needed to be developed completely before such extreme consequences could be predicted.

Colgate expresses entire confidence that the theory is correct in its essentials, however. Reviews of the theory by independent engineers obtained by *Environment* magazine confirmed the theory's apparent validity, and NRC concedes it is sufficiently reasonable to warrant experimental tests—but experimentation is being carried on at a lackadaisical pace at Sandia, where Colgate has been retained as a consultant.

Steam explosions do, of course, occur, and Colgate's theory may predict the circumstances under which they happen. If he is correct, they can be far more severe than any official report has yet recognized. The entire heat content of the molten fuel in a reactor accident would be released in a steam explosion by explosive self-mixing; the result would be an explosion equivalent to that of about 20 tons of TNT that would burst all protective structures and disperse the fuel of the reactor into the air; it would be far worse than the mere venting of some radioactive gases. The results might be on the scale of the very worst cases ever considered, theoretical accidents investigated by the Atomic Energy Commission in 1965 which a then-secret government report concluded might result in forty-five thousand deaths, hundreds of thousands of casualties and land contamination over an area the size of Pennsylvania.

Although more than fifty reactors are already operating in the US, we do not know whether such an accident is possible.

The above is, of course, a "worst case" analysis, an estimate of the combination of all the adverse factors considered possible. No one disputes that such an accident is unlikely. The point of the exercise is not to predict in detail the consequences of a powerplant accident, but to establish reasonable outer bounds of the risk involved. In qualitative terms, all efforts to establish these outer bounds have agreed that thousands of deaths, tens of thousands of injuries, and billions of dollars of property damage could, in theory, be caused by a power-plant accident. If one assumes the release of a cloud of radiation and a steady wind blowing toward a city, the results are predictable.

This raises a number of questions. Insurance companies, as former Pennsylvania insurance commissioner Herbert Denenberg has repeatedly pointed out, will not accept such risks, no matter how small their calculated probability, when the upper bound is so high. The federal government, therefore, is required to and does limit the liability of power companies and insurance companies to an acceptable (to them) $560 million in case of accident. Congress presumably would act to provide indemnity for damages beyond this amount.

Should the public be called upon to assume the risks rejected by insurance companies? This issue has plagued nuclear power from the outset, since the magnitude of potential risks has always been clear. It is not at all obvious how the question should or could be decided. Certainly, much opposition to nuclear power stems from an unwillingness to accept such risks at any level of probability, without much stronger inducement than the claimed benefits of nuclear power, which generally can be reduced to savings in fuel and pollution-control costs.

The Rasmussen report, the federal government's most recent review of nuclear safety, does not address itself to this question. What the report does, or attempts to do, is to identify every equipment failure and human error which might be a link in the chain leading to release of radiation to the environment. By multiplying together the observed or estimated probabilities of these separate events happening, it is hoped the probability of a large accident can be estimated. The analysis is limited to large nuclear powerplants of the type now being built, and it excludes other nuclear facilities which are part of the power industry's fuel supply. It explicitly excludes intentional events—sabotage.

Despite the exclusion of a large class of the problems of the nuclear-power industry—sabotage of power plants, theft of plutonium, accidents in fuel-reprocessing plants, and so on—the report's central findings are not necessarily reassuring. The report concludes, for instance, that the chance of a powerplant's fuel melting is much greater than critics of the program have been alleging. The report puts the chances of this accident occurring as one in seventeen thousand per reactor per year. For the one hundred reactors considered by the study, over the thirty years they can be expected to operate, the chance of a fuel meltdown is therefore about one in six. Considering that two hundred and fifty reactors are already operating or have been planned, a core meltdown begins to be roughly as likely to occur as not.

A core meltdown is a necessary condition for the release of large quantities of radiation to the outside air; it is a step toward the worst-case accident already discussed. It would be prudent to assume, on the basis of the Rasmussen report, that such a meltdown will, in fact, occur in one of the plants now operating or under way. At the very least, such an accident would result in the complete loss of the powerplant in which it occurred; the loss to the company and its customers would certainly be measured in the billions of dollars, counting outright losses and the costs of replacing the power the plant was to produce. That such an accident is likely to occur is disturbing, quite aside from the potential for public injury it would create.

The consequences for the power industry as a whole are more difficult to predict. In the summer of 1974, the AEC ordered twenty nuclear powerplants shut down for inspection because of cracks which had been detected in the piping of four of the plants. Following a core meltdown, would the entire nuclear-power capacity of the nation be shut down? For how long? To what extent will we be dependent on this capacity when an accident happens?

There remains the risk of public injury. A fuel meltdown can be expected to happen. The events which would follow such an accident are not well understood, but can be estimated. One possible consequence is the worst-case accident in which there are tens of thousands of casualties: one imagines a meltdown at Indian Point, with a steady wind under an inversion blowing toward New York City, twenty-five miles away. Are we willing to take even a very small chance that such an accident would occur?

The report does review the limited experience the industry has had with nuclear power so far, but if this experience were sufficient to settle the question, there would be no debate and no Rasmussen report. We are all aware that there have not yet been catastrophic nuclear powerplant accidents. The question is whether such accidents will happen in the future in circumstances which have not yet appeared in our still-limited experience. Not every bridge falls; not every dam fails. These are unusual events, and it may be many years before one occurs, and accordingly many years before one knows the probability of its happening. In the interim, if we are to proceed with construction, we can only rely on the judgment of the engineers. This sort of judgment is presented by the Rasmussen report. Whether we are willing to rely on this judgment, given the stakes involved, is the question which remains unsettled.

Such uncertainties are not inherent in nuclear power, but they are inherent in the water-cooled reactors chosen for us by the military some twenty-five years ago. The enriched uranium which makes them possible is also the source of their difficulties.

This simple design was available only because the nation was already building massive gaseous-diffusion plants to make uranium bomb material and fuel for submarines. These enrichment plants could extract the rare uranium 235 from the natural and more abundant form; this enrichment was a means of making nuclear explosives, but it was also a prerequisite for making water-cooled reactors. Ordinary water absorbs too many neutrons to be usable in reactors fuelled with natural uranium. Enrichment of the uranium, to enhance its uranium 235 content, is required if the simple water-cooled design is to be employed.

The enrichment plants were already under construction for the massive bomb effort; the Navy could therefore make use of the simplest, most easily developed reactor to power its submarines. Eventually, electric-power companies purchased scaled-up versions of these military reactors for civilian electric-power production.

Simplicity of development did not mean low costs, however. It meant only that no time was wasted in research. The enrichment plants represented an investment of billions of dollars; reactors using enriched uranium need massive boilers and elaborate and expensive protective devices. While easy to design and build, they were not necessarily safe or inexpensive to operate. But they could be ready quickly.

The water-cooled, enriched-uranium design developed for nuclear submarines in the 1950s is now the basis of the United States' nuclear-power industry, and of the industries of those nations like France and Japan which are dependent on US nuclear technology.

Other nations which have developed nuclear power have chosen different kinds of nuclear powerplants, of which there are a great variety. Reactors which use natural uranium, which contains only the seven-tenths of one percent of uranium 235 found in nature, are somewhat more complex to build. To keep them functioning despite the small proportion of uranium 235, heavy water or graphite must be used to slow the neutrons in the reaction (a process called "moderation"); these materials absorb fewer neutrons than does ordinary water. Gases like helium or carbon dioxide may be required for cooling; or ordinary water may be run in pipes through fuel immersed in heavy water. For ship or submarine nuclear propulsion, these designs have no obvious advantages and have many drawbacks. But for electric-power production, they have certain attractions.

The principal one is that, with a lower concentration of uranium 235 in their fuel, reactors which have not benefitted from enrichment can withstand a loss of cooling much more easily. The density of power generation in their fuel is much less, and ordinary flows of air or water, rather than the forced cooling needed in US reactors, may be quite sufficient to keep them from melting.

Britain has for many years relied on gas-cooled reactors; Canada has built and operated heavy-water moderated reactors. The Soviet Union, although exporting water reactors much like ours, is apparently making extensive use of a graphite-moderated, water-cooled design domestically. All of these reactor designs are largely free of the risk of meltdown—and hence of catastrophic accident—which so plagues the reactors the United States and its dependencies inherited from arms building for the Cold War.

CHAPTER THIRTY-FIVE

Radiation Workers

NO ONE SEEMS to know just how many radiation workers there are, where they work or what their working conditions are like. They are scattered among a number of different industries—mining, chemical work, military programs, government employment, electric power companies—and are represented, if at all, by a variety of industrial unions in which they form a small minority. The Teamsters, Steel Workers, International Brotherhood of Electrical Workers and many other very large unions include a few radiation workers among their members. The Oil, Chemical and Atomic Workers' International Union, whose headquarters are in Denver, possibly represents the largest number. No union official could say precisely how many radiation workers the union does represent. The vice-president with primary responsibility for this work thought there were fewer than ten thousand; Anthony Mazzocchi, the Washington-based legislative and citizenship director for the union, puts the figure at twelve to fifteen thousand. Mazzocchi thinks this figure may be only ten percent of the total of such workers in America, but he does not have any accurate information. Mazzocchi's union, the OCAW, represents perhaps two hundred thousand workers in the United States and Canada, most of them in the oil and petrochemical businesses. It represents some atomic workers largely because a chemical company, the Du Pont Corporation, built and operated the government's uranium-enrichment plants during World War II; its representation of workers in the enrichment plants, which continues to this day, is yet another artifact of the wartime military development program. The OCAW also represents some uranium miners in New Mexico. Mazzocchi thinks the average education of a radiation worker is a high-school diploma, if that; uranium miners are generally not even high-school graduates. They are ordinary factory-production workers and are not greatly concerned about their exposure to radiation. Mazzocchi attributes this lack of concern to inadequate educational programs, but in informal conversation he concedes that many workers are aware of risks derived from the material they work with. He compares radiation hazards to the risks faced by other chemical workers. In some chemical plants represented by his union, fully half the workers will die of lung cancer.

"What can you do about that?" Mazzocchi says. "How do these guys feel? It's like the guy who smokes cigarettes. Each one think's he's going to beat it. What can he do?"

In a hotel room in the Southwest, a young uranium miner, who likes his work and wants to make a career of it, talked about what it's like to work in a radioactive environment, a thousand feet underground. The mine he described was operated by the Kerr-McGee Company. Kerr-McGee officials declined to be interviewed or to answer written questions about conditions in their mines and mills; this miner's account was verified by other miners.

I'd like you to tell me what you did when you first began working in the mine.

MINER: Did a lot of shovelling. A lot. They have kind of a drainage ditch that runs alongside the tracks. It's constantly being filled up with the muck, the ore, so they constantly have someone shovelling it out. So I shovelled for about a month and a half, two months. Started working up in the stope [cavern], where they mine the uranium itself.

While you were still shovelling in the ditch, how much were you getting paid?

MINER: $4.19 an hour, and that was, that's a day's wage; you see they work on a day's wage and then a contract basis. But digging ditches was just a day's wages.

How do you get a promotion from shovelling muck?

MINER: Shovel a lot. (*Laughs.*) There's still people shovelling from when I started to work.

Uranium miners are individualist frontiersmen. They work as independent contractors and consider themselves self-employed. They are skilled, well-paid workers doing extremely difficult and dangerous work; in a part of the country that has been a frontier for so many races and nations, they fight for what they can get.

A uranium miner works alone, a thousand or four thousand feet underground. The mining company has located a body of ore—an "ore body"—and sunk a shaft to just below it; from this shaft, a horizontal tunnel—the haulage drift—is excavated by individual miners beneath the ore body. Then development miners, working alone, will each drive tunnels up to and through the ore body honeycombing it with narrow tunnels, leaving the ore in massive pillars; other specialized miners—pillar miners—each still working alone, will excavate the ore itself, sending it down the vertical shafts to the haulage drift, where electric trains carry it away. Each miner is paid for the length of tunnel he excavates or the tons of rock he digs from the ore body. The soft sandstone is broken up with explosives. The miner runs steel cable from the ore face and shovels ore into buckets strung on the cable.

You worked hard and were promoted. What did you do then?

MINER: Drove a train, for a while. Hauling supplies, ore.

These are electrified trains?

MINER: Electric and diesel.

Electric and diesel. And what are they like? What's that train underground like?

MINER: Pretty—fills up the drift [tunnel] pretty much. Underground, the track's really bad. Because of the water, you know. So you're always going off the tracks. And when you go off the track, you got to put it back on.

How big is the drift?

MINER: About eight feet, seven feet wide and usually about eight feet high.

How big is the train?

MINER: It's about four feet high and—let's see, you got about a foot on each side. But every fifty feet they got safety cutouts, and you've got trip lights in each drift that tell you if the train's in it and then they have a siren on the trains for when they're coming around corners. But, trains are one of the biggest dangers, supposedly.

Did anyone get hit by one?

MINER: No, not since I've been working there, no one's been hit by a train.

How long did you run the train?

MINER: Just for a couple of weeks. Three weeks.

Then what happened?

MINER: Started working in a pillar stope. And a pillar stope is a big room that's just blown out, and you have to get the muck out of there. The ore.
They go in first with the development miner. The track drift is down below and the ore bodies are up above. . . . The development miner cuts about thirty-foot squares, just little three- or four-foot-wide drifts. Sometimes two to four foot high. [These narrow drifts are spaced about thirty feet apart, leaving pillars of ore between them.] And when he gets all this developed out, the ore's still mostly right up above, and the pillar miner comes and blows out the [pillars]. And once you get, say, eighty or ninety feet square blown out, it starts caving in. And all the ore caves in from the top. Get it out that way. Saves a lot of blasting.

It's supposed to cave in in a controlled way?

MINER: No, there's no control. (*Laughs.*) No, there's no control. None whatsoever. Yeah. That's in the stope [the cavern left when the pillars are removed] itself. And actually, you're not supposed to go out there [because of the danger of cave-ins]. They'll fire you if they see you, if they catch you out in it.

Do people go out in it?

MINER: I haven't met *anybody* that doesn't. The safety book says, shoot a rope across, pull the cable over so you won't have to run through it.

That's to repair a cable, you mean?

MINER: Yeah. The cables [for hauling buckets of broken ore] are constantly breaking. You've got to move everything with cable. And then they've got a special gun, that you shoot across the stope and then pull it across [to repair it]. But evidently the company thinks the cables don't break very much, because I've never even seen a gun underground. (*Laughs.*)

Why is that, do you think?

MINER: Well, it's the safest way—the way the company says to do it. The company can't say well, take off. You know. It's the safest way to do it. But, if you did it, you wouldn't even make a day's wage and you've got to do so much work before you start getting the contract. There's just no way to do that, without running through there.

It doesn't sound like it would take that much time. Just shoot the rope and pull that thing back. Instead of running through.

MINER: It would, because it takes a lot of power. The cables are seven-eighths-inch [steel] cables, one hundred fifty, two hundred feet long, that's pretty heavy.

What are the cables doing out there in the first place?

The miner explains that he stands at the head of a vertical tunnel. The stope is a cavern cut into the ore itself, and as the roof of the cavern caves in, the miner breaks up the rock and hauls it in a bucket-and-cable device, called a "slusher," back to the head of the shaft where he stands. At the top of the shaft is a square grill called a "grizzly."

MINER: I was lucky. I got a miner, he's been underground for thirty-five years. I think. And he was really good, but he's been underground so long that he really—he's been out in an open stope so many times that he don't even think about it, you know. He'd just walk out there—I'd run—and look around.

The roof is about to come down, right?

MINER: Well, not really about to come down, because he's been underground so long that he can tell when it's gonna come down. And usually a few hours before it comes down, it'll start popping and grunting—I mean before the whole thing comes down. There's a lot of times, you know, just boulders fall down by themselves. But when the whole thing starts to come down, you can pretty well tell.

How big do the open stopes get?

MINER: About one hundred yards, and then things start caving in—before everything just caves in—[and] they just lose everything. And just move on down to ore bottom and start another one. It can cave in seven or eight hundred tons and you can still get it. That's a lot of rock.

That's a lot of rock. How is that done? What happens after it caves in?

MINER: Just have to run new cables, and new buckets and everything. Can't dig the old ones out . . . . It doesn't really matter how high the ceilings get. Stope I was working in the ceiling was about fifty feet, by the time we got out of there.

How deep are you working?

MINER: About a thousand feet. It's not all one big ore body. Sometimes they
have to drive a drift a mile to get a couple of ore bodies. It's over about fifty or
forty square miles. The whole area. It's quite a few mines there. A lot of them are
a lot deeper. A few of them are even shallower. I think they've got them around
four thousand feet. Which is still not really that deep. As far as mines go.

Is it hot?

MINER: No. As you get really deep—the copper mines are pretty hot because they're real deep, but it's pretty cool. In fact, in the winter it's cold.
Usually, you're moving so much that if you just wear a couple of pairs of long
underwear you keep pretty warm.
 You get your grizzly put in, get all your air hoses, water hoses. A lot of
equipment. Different kinds of equipment. And then just start blasting away. You
drill the holes and then, on a pillar stope you kind of slab it off, and blow it out.
When it goes—you've wired it up for when it blows—the muck [ore] will blow
away from the face toward the wall, so you can make your hookups and get it out.
That's a pretty important thing, how you drill your holes and how you load them
and how you wire them.
 The charges are stick, forty-five percent by volume. It depends on how hard
the ground is. If the ground's real soft, you don't have to put as much powder or
drill as many holes. But if you've got a lot of shale or something like that, you got
to load more. You can do it, you know, with a margin, but to do it real good you
got to be pretty much just right. To get it to go where you want it. Sometimes,
you know, you don't put enough and it might not pull all the way out, but if you
put too much it just might blow more of it. And you get paid by the time. So
actually you want to blow up as much of it as you can.
 It's all pretty much dangerous . . . you just got to watch what you're doing.
Pay attention to what's happening around you. But all of it's pretty [dangerous],
you could get hurt . . . .

Do people get hurt?

MINER: Yeah. From the charges themselves? Not really from the charges
themselves. I mean, maybe get hurt drilling them or loading them or whatever.
But they've got pretty tight control on when you blast. But then again, you
know, it varies. Depends on where you're working, where your air is going . . . .

 The miner explains another hazard: the shaft he stands next to and down which
he dumps the crushed rock.

MINER: [This shaft] can be anywhere from five feet, which is what they call a
Chinaman's chute, to about—the highest one I've seen is about two hundred feet.
The grizzly rails [at the top of the shaft] are only twelve inches apart, and you get
rocks in, you know, four, five feet, big rocks, and you got to bust them up to
where they'll fit in the grizzly. So you just have to stand on them with a chipping
hammer, there's no other way to—

On top of the grizzly?

MINER: Square steel beams.

Yeah, twelve inches apart. So you can't fall down the grizzly.

MINER: Yeah.

You can? Does that happen?

MINER: People falling all the way down—not very often. There's a safety rope that goes around, that if you start to fall you can usually grab. A lot of times people fall and catch theirselves with their arm or something like that; 'cause you fall, but still, it's twelve inches.

Getting close. And when the rocks are too big to go through you just stand there and break them up with a hammer on top of the grizzly?

MINER: Yeah. Chipping hammer.

There is a lot of manual labor in uranium mining—more than most kinds of mines.

MINER: A lot of different kinds of mining, they have machines that you can set on and drill, or set on and bolt, whatever. But in uranium mining you can't.

Now why is that?

MINER: Just the way that the ore is in there [in scattered bodies]. And to get all of it you have to go about it this way. I'm sure that there's probably machines that they could use, if they wanted to change their method of mining.

You were working for a thirty-five-year veteran. How old a man is he?

MINER: He's fifty years old.

He's been down there since he was fifteen. And he's been in uranium mining all that time?

MINER: Back then, each family had their own little mine. And the kids went to work, the old lady stayed outside and watched the compressors and stuff, and the kids went in and pushed the ore cars out and stuff. Most guys have been in the ground that long, it's how they started.

You work all by yourself now?

MINER: Yeah.

Yeah. Does that bother you?

MINER: No.

You're down a thousand feet underground. All alone in this little dark hole.

MINER: It's a solitary thing, I'll guarantee. I don't know, but sometimes I'll be standing there and be standing in mud, covered with mud and all this noise, you know, and manual labor and I'll think, What am I doing here? But, it's not bad working by yourself.

Does the darkness bother you?

MINER: No, it's kind of, I like it because it's not—it's not an everyday thing. There's—it's not really the risk but there's—it's not like going to work and looking at all the walls, four walls. It sounds nice right now. It's just that it's different. Like I'm, it seems like I like something different.

Is it exciting?

MINER: Yeah. It keeps your attention. Pretty well. But then sometimes it gets boring because it's the same thing, same thing.

How much does a pillar miner make?

MINER: The miner I worked for, he'd get fourteen, fifteen dollars an hour. Or more. He was making it pretty regularly. Every two weeks. I'm making less money now than when I was helping, but I figure it's gonna take a while.

Do you feel like you're working for yourself? Really self-employed?

MINER: Yeah. It's when you're underground, there's nothing to do anyway but work so when you really accomplish something you get a lot of work done. You're tired but it feels pretty good. To get a lot of it done.

Then you're pretty much your own boss?

MINER: Yeah, the foreman comes around and if you have a problem he'll tell you how to solve it and, when he leaves, if you want to try it, you can try it, and if you think he's full of shit you just don't try it . . . .
Sometimes you might have to walk a mile and climb a ladder one hundred feet down, just to get to the drift and walk a mile just to get to the lunchroom.

There must be safety inspectors come around, right?

MINER: State and federal both.

How often do they turn up?

MINER: Every thirty days. There's company safety inspectors, too. They show up about every two weeks.

You know when they're coming?

MINER: No. Usually they're on the surface when you're on the surface, so you know they're in the mine. But, no, they don't tell you when they're gonna come see you. But you know they're there. So if you see them there, you watch out what you do.

If you're in a stope, you don't do much travelling through the stope. A lot of times, when a safety inspector was coming we wouldn't do very much. A lot of barring down loose rocks and fixing things that have been broke, you know. A lot of that.

What is the rule about respirators? Are you supposed to be wearing it all the time?

MINER: We've got to carry it with us at all times, but the type respirators we carry, after fifteen minutes they start burning your throat and your lips off, so you don't use them unless you just gotta use them. We also have respirators for the dust. But very rarely wear them.

Are there regulations about those?

MINER: You're supposed to wear one when you're operating a chipping hammer or something like that, but very rarely. Very rarely. No one really wants to use one when you're working hard, because you gotta get a lot of air to do all that anyway. It's hard enough to get it without pulling it through those.

When you came to work were you given any training about safety?

MINER: Yeah, there was a two-day safety class that everybody has to go to. And then nine weeks later there's another safety test. The safety class itself is pretty good, because they teach you some first aid and what to do if whatever happens. And they teach you what to do in case of a fire or something like that. The company itself's got a pretty good safety program. Pretty good. But they have to. I mean, if they didn't it'd be a lot worse than it is.

Now, do you know, well, how do they keep track of accidents? The time-lost accidents . . . .

MINER: Lost-time accidents? You have to be pretty messed up to have lost-time accidents. Because they don't want them. I mean if you break an arm or something, like that, . . . they'd rather pay you just to sit up on the surface and do nothing rather than give you a lost time.

Does that happen very often? People sitting around?

MINER: Yeah. Broken foot or leg or . . . .

Is there usually somebody at the mine in that shape, or–

MINER: Uh huh. Nearly always. A couple.

A couple. How many at the most you've seen sitting around?

MINER: At one time, just two, three. It's usually a broke finger or broke arm or broken foot or something.

You can't work underground so you just . . . .

MINER: Yeah, so they just pay you to sit on the surface.

You guys are aware of the possibility of injury all the time, I guess. Because there's always somebody sitting around with a broken foot or something.

MINER: Yeah . . . . I, there's been a lot of people hurt real bad, but not killed. Really hurt, like broken backs and paralyzed, and things like that.

A lot?

MINER: Not a lot, permanently disabled. I've seen a few that were . . . just since I've been here. Even working in the area you see a lot of people that are kind of strange, and then you find out that five years ago they got hit with a big slab.

How many people can you think of, thinking of specific people, who were permanently disabled? That you know about?

MINER: I heard about two. This year. They've been like paralyzed.

That's in the whole area here, not just in your mine?

MINER: Yeah, well, no, not in my mine. In fact, I think the worst thing that's happened in where I'm working is probably some broken legs and stuff. People falling down a ladder, or something.

How does the ventilator system work?

MINER: They have a ventilator department and this is how they control the radon daughters [radioactive gases], with the ventilation. They come down and put in different fans and—

Okay. When you start working on a pillar stope, someone comes from the ventilating department and sets up ventilation in that . . . .

MINER: Yeah, figure out what size fan you need and everything. As you get into the stopes, the air slows down a lot. And the machines smoke sometimes.

They come and put a fan where you are?

MINER: Yeah.

Okay. You said something about "radon daughters." What's that all about?

MINER: That's just to control how much dust that you breathe. On your time card, every day, they have different numbers for different areas or different jobs that you'd be doing, and I guess they feed them to a computer, probably in Oklahoma City and figure out, you know, if you—say you've got fifteen years, they figure if you've been working in too hot a spot—which is not real hot underground, the radon—but if you've been breathing too much of it they'll change you to another job, supposedly.

How do they know how much you're breathing?

MINER: Ventilation checks. They come down and, like I said, in the haulage drifts [main tunnels] it's pretty fresh, but up in the stopes a lot of times it's not that fresh. And they come in and take samples and find out how much is in the air . . . .

How often do they do that?

MINER: About once a week. And the company keeps records of the workers and the measurements . . . .

Do you wear a film badge?

MINER: Yeah. It's in our helmet. We turn them in about every six months or every year. Well, they've never taken mine. I've had it six months. It's three months before I got it.

Do you ever look at it?

MINER: No. It's just tape. There's tape over it. It's not a dosimeter [exposure meter] thing. It's just film. I guess they have to develop it, or something.
I don't think you get that much radiation underground. Or not radiation, but radon daughters. Breathing them. Not that much of them to make you hot. I mean it might mess your lungs up, but it won't make you radioactive.

Radon daughters are what?

MINER: It's a dust. It looks kind of, you can shine your light through it, kind of a silvery particle . . . sandstone mixed in with it [they are actually gases and not visible].

There's just a lot of dust in the air that you're breathing.

MINER: Yeah, at certain times. Like, if you're using a chipping hammer, breaking rocks, or right after a blast, or something like that.

You think if there's a hazard it's mostly just from the dust. Not a real radiation problem.

MINER: Yeah. I know the dust isn't any good for your lungs.

Do you wear a respirator for the dust?

MINER: When I can. I wear one, but if you're hauling a 115-pound sledge hammer or standing over a 180-foot grizzly, you know, that goes straight down, breaking big rocks, it gets in the way. Sometimes.

Are you supposed to wear it?

MINER: Yeah. You're supposed to wear it. Very few people even have one. This is just a gas mask I have, it's not a respirator. I know when I'm setting, slushing [operating the bucket-and-cable machine] for a couple hours at a time, or something, I'll put it on to filter out asbestos smoke that comes up, plus the dust from moving the dirt around. And in about two days you have to change the filter because it's just clogged.

But most of the time you just don't wear it?

MINER: Yeah. Most of the time you just [don't], it's too much.

You're not very concerned about the radiation that you're exposed to, though?

MINER: No, I really, I don't think I'm exposed to that much. I'm just, the dust itself's like worse on me than the radiation.

You really know how much radiation you're exposed to?

MINER: No. Actually, no.

Does anybody ever talk to you about it?

MINER: Well, they told us that it wouldn't hurt us.

But they didn't tell you how much there was or anything?

MINER: No. Of course, the ore runs maybe fifteen to twenty hundredths of one percent pure ore [uranium]. So it's got a lot of rock there, too. All that ore's not just ore, there's a lot of rock in with it. But, then again, you know, you're right there in the middle of it, too.
   I think it's a lot better than the coal miners are getting. 'Cause, work in a coal mine ten years and you're eligible for black-lung benefits.
   I've heard a few of the miners saying that, like when you work around uranium, sometimes it'll take your energy away. Just kind of make you feel a little bit weak. At times. It's just like any other type mining, I guess. As far as the danger goes. Just one of the things that's present there. I never really thought about the radiation that much, but, strict as they are on the safety, I don't believe we're getting too much of it. You know, even at the mill where they process our ore, and make what they call the "yellow cake" out of it, those people are getting a lot more than we are, and they're still not even wearing any protective clothing. . . . They just put it in a fifty-gallon drum . . . .
   Asbestos and dynamite, smoke, too, is pretty bad. And the dynamite gives you a pretty bad headache. Like, when you're loading up your rounds I always get a headache. Like they say, after years you get used to it. But the dynamite will give you a headache, and then when you get in the smoke, the smoke will kill you because it displaces oxygen. If, you know, you get in it real thick. But I don't know, these guys have been working in the ground all this time, they can walk through it smoking a cigarette. (*Laughs.*) You know, not even feel it.

Are there rules about smoking?

MINER: Yeah, they'll fire you for smoking underground. But you're down there in a little bitty tunnel, you know, way up a ladder, so if you smoke . . . .

How many of the miners smoke?

MINER: I'd say about half of them. But, the reason they don't want you to smoke is the radon daughters and the cigarettes themselves work out some kind of combination that's really cancerous. That's the reason they don't want you to smoke underground. Which I can see their point. Cigarettes, you shouldn't smoke them anyway. It's one of those things.

In Denver, the headquarters of the Oil, Chemical and Atomic Workers are in a downtown building the union owns. An armed guard stops and questions visitors in the lobby, asks for identification and issues a visitor's pass; union officials say security measures have been tightened since bomb threats were received during a prolonged strike in the oil industry. E.P. Swisher is the vice-president of the OCAW with responsibility for atomic workers, among others. He cannot recall how many there are: he guesses seven thousand, about half Mazzocchi's estimate; after consulting his secretary, he guesses that there are more, but clearly has no idea how many. He is aware the uranium miners have been on strike in New Mexico for six months, but does not know the name of the president of the local which is on strike and does not support the strike. He expressed sympathy with the mine management, which is trying to do away with the contract system, the independence of the miners. Swisher thinks the contract system is just piecework and should be replaced by wages on an hourly basis; at first he asks that these sentiments be off the record, but then changes his mind. Swisher is a member of the Atomic Industrial Forum, the management-dominated trade organization and a strong advocate of nuclear power; he is almost completely unaware of the conditions under which it is produced.

The miners, for their part, are very little concerned with the nuclear power industry or the debate surrounding it. In Grants, New Mexico, a young former miner who calls himself "Nenyo" talked about nuclear power.

Most of the uranium that you guys mine goes for . . . ?

NENYO: Nuclear . . . it's also private contractors. And all of it goes for fuel . . . . I think it's a joke. I think they're crazy! I think nuclear power has to be . . . I think they're wasting their money, their time, and I was more than glad to take their money. (*Laughs.*)

Why do you say that?

NENYO: I think solar power is probably a better source. Solar and land and water is more natural and more abundant. Uranium power it seems like a big general waste; to go to produce electricity in a roundabout way. They know it's going to run out. It's just like anything else. Coal, uranium, it'll run out . . . oil . . . I think they're wasting their time. I think they could develop much better solar energy, wind power . . . .

How do the other miners feel about nuclear power? Do you talk about it?

NENYO: I asked several people about what they thought the energies were going for. What their job was, what they were doing. Since it doesn't really have to do with the atomic bomb, at least this mine. I think Homestake might, though. I think Homestake, [and] I know Kerr-McGee sells in the form of "yellow cake," now I don't know about the yellow cake, what happens to it from there, I don't know if they can turn it into what, U-238? Or whatever it is in an atomic bomb . . . I don't know. I was just told by someone . . . all the yellow cake from Kerr-McGee goes for nuclear fuel. . . . The rest of the miners, I don't know. They're there for the money. It's a job, if they don't do it, somebody else will. That's the general outlook I find in most workers here. It's good pay, and I think in the United States at this time, it's probably [one] of the better paying jobs around this area. It's too bad, [there will] probably be a great influx of people around here. But there have been plenty of jobs. I can go back to the mines anytime I want and have my job back.

Well now, in other parts of the country there are people who are trying to stop the nuclear power industry. Is that talked about here?

NENYO: This community is based on this, Grants, Milan. Even San Mateo [an old Spanish settlement near Grants] now, most of the people, these people here, have been living here for generations. I mean a long time. And they were dependent up until 1950 [on] agriculture so this is recent to them. But now, in this twenty years, they're completely dependent on the economics of the mines. Most of the men are, in some way or another, related to the mines or that industry of uranium. And if it fell through, no jobs. There would probably be a very great economic turmoil here if these jobs . . . they employ a lot of people. There's a lot of mines, three shifts, I can't exactly say how many people but I know that, say, Kerr-McGee's got eight mines, Homestake's got four; you know, there's, say, twenty pretty good productive mines here that employ a lot of people and the majority of Grants.

The only thing . . . San Mateo is concerned with, I guess, [concerning] uranium, is the mills. The mills. The mines don't seem to be a general concern. They're out, away from town. And no one hears the vent holes. No one hears, you know, the trips from the mines.

CHAPTER THIRTY-SIX

Pollution

THE PROBLEMS CREATED by the routine discharge of radioactive materials from powerplants seem to be on their way to solution. During 1972, the
Atomic Energy Commission, under pressure from the courts, its own scientists
and the general public, began to move to restrict emissions from powerplants.
This movement has been completed by the Nuclear Regulatory Commission.

Since the 1950s there has been heated public debate over the consequences of
small quantities of radiation in the environment. In earlier years, the source of the
controversy was the fallout from nuclear weapons tests, which created a small but
steady increase in exposure to radiation for the entire population. From that
debate, a number of scientific agreements and legislative standards emerged.
Radiation, it was widely agreed, could cause some damage at any level of exposure, no matter how small. This being the case, all standard-setting bodies
adopted the principle that there should be no unnecessary exposure of the population to radiation and that any exposure should be balanced by a corresponding
benefit. The International Commission on Radiation Protection recommended
that the general population receive no more than 170 millirems of radiation
exposure each year, roughly the amount they then thought was already being
received from natural sources (later research showed natural radiation averaged
only about one-half this amount). This limit applies to exposures *in addition* to
those from natural sources and from medical therapy. The underlying principle
of risk-benefit balancing and the 170-millirem limit were adopted by the federal
government and now apply to all radiation exposure from the civilian nuclear-
power industry.

There is a certain contradiction inherent in these principles and standards. If
radiation causes damage at any level of exposure, then *any* radiation exposure
should be prevented, since it carries a health cost, unless it is justified by some
greater benefit. The fact that a standard of 170 millirems had been set, although
no explicit balancing of costs and benefits had been done—or that at least the
balancing of costs and benefits was not generally acceptable—came to light when
Dr. John Gofman and Dr. Arthur Tamplin, then employed by the AEC, publicized the implications of existing radiation standards in a series of papers beginning in 1969. Adopting the theory that radiation does damage at any dose level, as
all groups engaged in setting radiation standards have assumed, Gofman and
Tamplin tried to calculate the effects on the US population if exposures of 170
millirems per year, permitted under federal regulation, were, in fact, to occur.
Using the experience unhappily available from the nuclear explosions at
Hiroshima and Nagasaki, records of individuals exposed to X rays for medical
purposes, and other data, Gofman and Tamplin estimated that "the present
population would experience sixteen thousand additional cases of cancer and
leukemia each year if exposed to the extent now allowed for thirty years."

That this is not a generally acceptable cost for nuclear power was demonstrated
by the storm of protest which followed the publication of the Gofman-Tamplin

estimates at nuclear-powerplant licensing proceedings. Because the two scientists were employed at the AEC's Lawrence Livermore Laboratory, their views carried a great deal of weight with the public and within the scientific community. The manner in which the estimates were presented and the uses to which they were put by citizens' action groups created a good deal of controversy. But because the Gofman-Tamplin calculations were based on widely shared assumptions, the controversy was short-lived. In public statements, Gofman and Tamplin called for a tenfold reduction in allowable radiation exposure; at least so far as the nuclear-power industry is affected, reductions will actually be far greater. The AEC committed itself to standards which keep radiation exposures at a level as low as is "practicable," and while this is subject to the sort of analysis promulgated by the NRC, the dollar value given to radiation exposure by the NRC may force radiation exposures to as low as one or two millirems, a hundred-fold reduction of present standards. A study of the upper Mississippi River Basin conducted by the AEC estimates that nuclear power in this region will result in an increase of about one-fifth millirem per year in average radiation exposure by the year 2000; similar regional studies in the remainder of the country are likely to produce similar results. These results would also be in agreement with a recent study conducted by the Environmental Protection Agency. While these studies may be optimistic, assuming as they do the proper functioning and maintenance of nuclear equipment yet to be tested in long-term service, they do predict that present technology can keep radiation exposure levels well below one percent of present standards.

As the political battle over radiation standards is almost over, the scientific debate, to the extent there was any, has also ended. As the chairman of the Federal Radiation Council pointed out in a letter to Senator Edmund Muskie in 1970: "Gofman and Tamplin, in reaching their conclusion that the Federal Radiation Council guidelines should be reduced . . . used an approach similar in principle to that used by expert advisory groups . . . in developing radiation protection standards and guidelines. This approach is based on the assumption of a direct linear and nonthreshold relationship between dose and biological effect."

These same assumptions were more recently used by a panel of distinguished scientists established by the National Academy of Sciences–National Research Council (NAS-NRC), which, in a report published in November, 1972, explicitly confirmed many of the estimates made earlier by Gofman and Tamplin. The report reviews, point by point, the assumptions and calculational techniques they used. It concludes that exposure of the present population of the US to 170 millirems for thirty years would result in "from roughly three thousand to fifteen thousand cancer deaths annually," which is reasonably close to the Gofman-Tamplin estimate of sixteen thousand per year. The report also somewhat confirms more recent predictions by Joshua Lederberg and Linus Pauling that such radiation exposures would increase the mutation rate, with a resulting increase in the general level of ill health. And finally, while avoiding any recommendation about changes in radiation standards, the report notes that "radiation exposure averaged over the US population from the developing nuclear-power industry can remain less than about one millirem per year . . ." assuming present technology performs as hoped. This estimate lends further weight to the proposition that radiation releases can be maintained at less than one percent of previous standards with "practicable" technology.

CHAPTER THIRTY-SEVEN

Millennia

BARRING AN ACCIDENT, only tiny and inconsequential amounts of radiation are released by powerplants, although such plants each produce enough radioactive material to eradicate whole nations each year. This material must be taken from the powerplant and kept in a safe place for many centuries. In theory this can be done, but in practice doing so has been a source of severe difficulties.

In this area, as in others, present problems are exacerbated by our having built a civilian industry upon a military foundation. In the 1950s, only the reactors designed for military use were available, and it seemed that only the military demand for plutonium would make them economically feasible. Accordingly, as the first private powerplants were built, private industry entered the plutonium business.

Plutonium is made in uranium reactor fuel as a by-product of the nuclear reactions. Some neutrons are absorbed by ordinary uranium, uranium 238, which is thereby transmuted into plutonium 239, the stuff of which bombs are made. This process goes on steadily in any reactor fuelled with uranium and is the basis of all plutonium production.

When power companies first examined the nuclear technology developed by the government, they concluded that by extracting plutonium from their reactor fuel and selling it to the government for the manufacture of weapons, the otherwise prohibitively high cost of nuclear powerplants could be justified. Plutonium sales and power sales together would support plants that could not be justified economically for either purpose alone. Reactors were therefore to be "dual purpose" in civilian hands. As the government's own single-purpose plutonium-production reactors grew obsolete, the entire massive military enterprise would be turned over to private ownership. The power industry would double as a defense industry.

Accordingly, it seemed clear from the outset that the nuclear-power industry would require, not only nuclear powerplants to make electricity, but chemical-processing plants to extract plutonium from the powerplants' fuel. The government had developed an effective chemical technique, after ten years of effort, for handling the radioactive fuel to extract plutonium and the unused uranium. In 1959, a group of five power companies joined with the construction conglomerate W.R. Grace and Co. and with American Machine and Foundry to build a private plant using government-developed technology to process fuel from civilian power plants. This consortium was called the Industrial Reprocessing Group; and Nelson Rockefeller, then governor of New York, was instrumental in organizing its first commercial venture. The state of New York, with Rockefeller's urging, created an Atomic Research and Development Authority to lure the fuel-processing business to New York with heavy subsidies. The federal government provided its own inducements to encourage the development of a private industry, in accord with the dictates of the 1954 Atomic Energy Act: since there were not yet enough private plants operating to justify a fuel-processing facility, the Atomic Energy Commission signed a long-term contract with the group, guaranteeing them the work of processing fuel used in the government's military programs.

With these inducements and subsidies, the nation's first privately-owned plutonium-production plant went into operation on the banks of the Cattaraugus Creek, in West Valley, New York, in 1966. The private consortium operating the plant had by then been rechristened Nuclear Fuel Services, Inc. (now owned by Getty Oil Co.).

The Nuclear Fuel Services plant at West Valley, not far from Buffalo in upstate New York, was a close copy of the facility built by the government at the Hanford Works in the state of Washington. As at the Hanford facility, which was designed when uranium was in very short supply, both plutonium and uranium from the spent fuel were recovered; the plutonium would be sold to the government for weapons production. This was to be the principal source of profits expected in the fuel-processing business; early estimates were that the government would pay fifty dollars or more for a single gram—a thirtieth of an ounce—of plutonium, in the 1950s still a scarce commodity. Uranium was to be returned for further processing in uranium-enrichment plants.

The residue of this extraction, the radioactive fission fragments created by the splitting of uranium, had little value. They were intensely radioactive, and the hot acid solution produced by the reprocessing plant was corrosive as well. The government stored such wastes in shallowly buried steel tanks at Hanford and Aiken, South Carolina; the West Valley plant was to do likewise.

The government waste-storage facilities did not perform well. The steel tanks, after ten or twenty years, began to crack and leak under the unremitting bombardment of radiation, heat and chemical attack. Hundreds of thousands of gallons of highly radioactive wastes seeped into the soil in Washington and South Carolina, close to the Columbia and Savannah rivers. The Atomic Energy Commission staunchly maintained these wastes were tightly bound to the soil and were moving only very slowly toward the rivers and ground-water supplies. The chronic leakage made it clear, however, that this was at best a temporary means of storing the wastes.

About eighty million gallons of radioactive waste were produced in the nation's military programs, and all of this huge mass of intensely hazardous material is stored in the tanks at Hanford and Aiken. These wastes must be stirred and cooled to prevent the kind of melt-through problem which plagues nuclear powerplants.

After several years, however, the intensity of radioactivity declines enough for the material to be left without further mechanical cooling; at this point, it is possible to consider solidifying the wastes in some inert form for permanent disposal. Because some of the military's wastes are now over thirty years old, experimental tests of waste solidification and storage have been undertaken.

After several years, it is possible to incorporate the radioactive wastes into glassy or ceramic materials. They then become hard, inert solids, highly resistant to attack by water or acids. Radioactive materials once trapped in this form should remain there, short of intentional efforts to extract them. The solids might then be buried very deeply—beneath a mile of granite, for instance, at the bottom of a well drilled for this purpose, or at the bottom of a salt mine from which geologists were confident water would be excluded for millions of years.

The Atomic Energy Commission went so far as to test the suitability of a salt mine in Kansas for such long-term storage, but public outcry and geologists' embarrassing revelations—the salt mine chosen was, in fact, riddled with drill holes and was not sealed off from sources of underground water—defeated the proposal.

It will be twenty or thirty years before such long-term disposal techniques are needed by the civilian power industry, and until the end of the century the overwhelmingly predominant problem will be the eighty million gallons of wastes produced by the military. While the technical problems of waste disposal do not seem insurmountable, the very long spans of time and the hazardous nature of the materials make these problems difficult to deal with in ordinary terms.

After several years, nuclear waste has declined in radioactivity enough to be solidified, but it is still producing heat and is still intensely poisonous. The dangerous combination—a poison *and* a source of energy to disperse it—is still present and creates problems in some ways resembling those of powerplants. If the waste is stored in a salt mine, so long as it is exposed to air, which can carry away the modest amounts of heat it produces, the material will remain solid and undisturbed. But once the waste is covered with salt, or the mine is closed, as it must ultimately be, there is no longer a medium to carry heat away from the waste except the salt itself, which is a very poor conductor indeed. The solid waste capsules may therefore melt themselves, and the salt around them for some distance. This can occur in granite as well. Some proposals have sought to take advantage of this phenomenon; freshly buried wastes would melt the rock deep beneath the earth and become dissolved in molten granite. After decades the rock would cool and harden, trapping the wastes within it.

An exotic proposal would use the heat of the wastes to bury them more deeply. Three scientists proposed, in the *Bulletin of the Atomic Scientists*, that fuel wastes should be placed on the Antarctic icecap. Over a period of years they would melt their way downward through the miles-thick icecap and come to rest on the continent beneath it, safe from any accidental disturbance. But others have wondered whether extensive melting beneath the icecap might not make it unstable.

Still more improbable proposals have been earnestly put forward. While James Schlesinger, later Secretary of Defense, was chairman of the Atomic Energy Commission, he said that the AEC was seriously examining an old proposal that these wastes be sent into outer space to fall into the sun. The difficulty with this proposal, aside from the expense, is the hazard of a rocket failure depositing large quantities of nuclear waste on, say, Kansas City.

Despite difficulties, the Atomic Energy Commission and its successor, the Nuclear Regulatory Commission, have maintained that deep burial of these wastes is a feasible means of permanent disposal. There probably would be no serious controversy on this point if it were not for the very long lifetimes of radioactive materials. Plutonium has a half-life of about twenty-four thousand years. Other radioactive wastes, more intense emitters of radiation, have half-lives measured in decades and centuries. (For material with a half-life of, say, a hundred years, after a century, one-half of the radioactive material will have settled into stable form; but it will take another hundred years for half the remaining material to subside, and a century for half the remainder again, and so on. For material with a half-life of a hundred years, therefore, it will be millennia before it is safe to release it to the environment.) Radioactive wastes will remain hazardous for geologic ages—forever, in human terms. It is therefore incumbent on those who produce it to show that it can be kept safely.

With infinite time, anything is possible; no means of storage can be said to be secure forever. But we make many hazardous materials with long lifetimes. When asbestos, mercury or beryllium is extracted from ore and collected together, we create a similarly long-lived hazard. Many lake and river bottoms in the United States, Japan and Europe have large deposits of mercury in the methylated form, left by paper mills and chlorine-caustic soda plants which have long since closed their doors and moved on. There are no records of many of these deposits. Yet methyl mercury is comparable in its hazard to radiation; it causes genetic defects and cancer, as well as immediate and horrible poisoning if taken in large doses. There are no plans to collect this material and store it in a safe place; but it will remain hazardous forever. Once radioactive materials have decayed past the point at which their internal generation of heat is a dangerous self-contained means of dispersal—in a few decades, or well within a single human lifetime—it is difficult to see in what way they are any more or less hazardous than other poisons produced by industry.

CHAPTER THIRTY-EIGHT

West Valley

THIRTY-EIGHT PEOPLE may have inhaled plutonium at the West Valley processing plant; thousands of workers there received radiation exposures ranging up to the maximum permitted by federal law; some may have received exposures in excess of what was permitted. The plant is now shut down for modifications and for expansion, but during the six years of its operation, from 1966 to 1972, it released more radioactivity to the environment and exposed more workers to radiation than all of the then-operating nuclear powerplants taken together.

There are at present no fuel-reprocessing plants in operation; it is possible that never again will there be any, even though the nuclear industry continues to expand. Fuel reprocessing to recover plutonium and uranium from wastes is a marginal economic operation. Its importance to the nuclear-power industry is not at all clear; now that the original military justification for its existence has been lost, there is some question as to whether it will be continued.

The process seems to be difficult to operate economically without exposing workers to unacceptable risks. Because of the intense radioactivity of nuclear fuel after it has been used to generate power, most of the operations in a fuel-reprocessing plant are done either automatically or by remote control. Even carrying the fuel from the powerplant to the processing facility is a hazardous business and contains a good deal of risk to the public at large.

While the West Valley plant was still in operation, it received fuel from civilian powerplants as well as from the AEC's military reactors. In a civilian plant, fuel which had been used until it was no longer capable of sustaining a chain reaction was lifted from the boiler. The fuel, now intensely radioactive because of the build-up of radioactive wastes, was lifted by a travelling crane and carried to a large pool of water for storage. After a period of weeks or months, when the initial intense radioactivity had somewhat declined, the fuel was loaded into a massive steel shipping cask and carried by railroad to the West Valley plant of Nuclear Fuel Services. These shipping casks were provided with their own internal cooling systems, since the fuel still contained enough radioactivity to melt itself and the surrounding containers unless cooling was continuously provided. There was a continual hazard that radiation might be released in any accident occurring during shipment.

At West Valley, the fuel was unloaded from its shipping cask and carried by a remotely-operated crane into the first of a series of sealed chambers behind thick concrete walls. Operators using remote-handling equipment chopped the long bundles of fuel rods into shorter sections and then dissolved them in nitric acid. The hot acid solution of plutonium, uranium and radioactive wastes was then pumped through a series of chemical reactions until the three classes of materials had been separated. Plutonium was drawn off in the form of plutonium nitrate, still in acid solution, for delivery to the weapons factories.

The federal government now has far more plutonium than it can ever use, however, and has stopped buying the material from private companies; the original rationale for fuel reprocessing has therefore been lost. The plutonium which is the chief product of the operation is simply put into storage; the Atomic Energy Act has been amended to allow private companies to own plutonium, and enough of this potential bomb material is stored at private facilities in the state of New York to arm a major nation. The plutonium will remain in storage until some means of using it in power plants or some means of disposing of it is devised.

Uranium extracted from the spent fuel, although contaminated with man-made isotopes, is near natural uranium in its U-235 content and so has economic value; it may be sold for further processing into reactor fuel.

The mechanical and physical handling needed to extract these valuable materials and to dispose of the residue of highly radioactive waste is carried on behind concrete walls by automatic machinery or remote-handling devices. However, the elaborate plumbing systems occasionally spring leaks; gaseous products must be vented to the outside air; and all of the machinery needs occasional maintenance.

At Hanford, the government's facility was designed to permit remote handling in repair and maintenance work. But at West Valley, to save money, men and women regularly had to go into highly radioactive areas to make repairs and to clean up spills.

The difficulty was that, once the military demand for plutonium was satisfied by our enormous Cold War build-up, the plutonium-separation plant was deprived of its principal source of expected income. The cost of retrieving plutonium and uranium from used fuel is now higher than the value of these materials as reactor fuel. In hearings on a proposed fuel-processing facility in Barnwell, South Carolina, the Nuclear Regulatory Commission estimated that in 1980 it would be paying $163 for the uranium and plutonium contained in one kilogram (2.2 pounds) of spent reactor fuel, but that it would cost $145 to process that fuel— leaving a small margin, about eleven percent, for profit. But the NRC figures included some subsidies; the market value of the uranium and plutonium would be no more than $107, according to Sierra Club testimony at the same hearings. Even accepting the NRC figures, the margin for profit is small and does not permit the kind of elaborate remote-handling facility the government used in its own plutonium-production plants; the inducement to cut corners is great.

At West Valley, unskilled laborers were used as a kind of living remote-control system. The plant had about a hundred regular employees who were occupied with skilled, remote-handling tasks. Whenever a dirty clean-up job had to be done directly by hand or a repair job was needed in intensely radioactive surroundings, the company would recruit hundreds of unskilled workers. These men (apparently few or no women were employed) were given a few minutes of instruction, dressed in protective clothing and then sent to do their job—washing a wall, perhaps, or loosening a bolt on a highly radioactive bit of machinery. Within a few minutes, or perhaps a few hours, each worker received the maximum radiation exposure permitted, even if he had followed all of the safety procedures necessary in dressing and undressing; after the allotted time, therefore, each group of unskilled workers was pulled off the job and replaced with a fresh group. The pay was three dollars an hour, and a worker was paid for a minimum of four hours, even though he might be able to work only a few minutes in the radioactive setting. It was the possibility of receiving a half-day's pay for a few minutes work which attracted employees; it was popular work, and the company had no trouble recruiting temporary laborers, although there have been some accusations that some of the recruiting was done in bars and in the local skid row.

An average of fourteen hundred workers a year were thus employed by Nuclear Fuel Services at the West Valley plant, although its permanent work force was only about a hundred.

During this period, there were also excessive discharges of radiation to the air and water surrounding the plant, which were the object of serious complaints and criticism from local residents, a group of scientists at Rochester, New York—the Rochester Committee for Scientific Information—and, ultimately, the Atomic Energy Commission. The plant was forced to make some improvements in its operations, but, by 1972, it became clear that adequate improvements to protect workers and the local environment would make the plant completely uneconomical to operate—it was already losing substantial sums of money. In 1972, therefore, the plant shut down. The company announced plans to expand capacity and improve radiation controls; the declared hope was that the added volume of work handled by a larger plant would reduce unit costs sufficiently to allow profitable operation (assuming continued government subsidy, however) with improved safety. Whether this will prove feasible remains to be seen; no date for the reopening of the plant has been announced, although the NRC is considering a request for a construction permit to allow the announced expansion.

If the nuclear-power industry continues on its past assumptions, it will require about half a dozen new fuel-processing facilities. There has been no rush to build them. General Electric some years ago undertook to build a new fuel-processing facility, near the concentration of nuclear powerplants around Chicago, to be called the Midwest Fuel Recovery plant. Construction began in 1969, when it had already become clear that the plutonium bonanza was over. This plant accordingly was designed to operate on a new process, the Aquaflour process, which was to be so highly automated that labor costs would be reduced to a manageable minimum. There was to be no automatic maintenance equipment, however; and when the company began testing this plant, it found that leakage rates in the plumbing and the general radiation level were so bad that manual maintenance was impossible, even on the human-discard system of West Valley. Sixty-four million dollars was spent on construction of the plant, but it will probably never operate; roughly an additional hundred million dollars would be required to redesign the plant, and it is more likely that it will be abandoned.

Since there are no processing plants in operation, used nuclear fuel is accumulating at powerplants, and there is a shortage of storage space. There has even been some speculation that the powerplants will be forced to shut down for lack of a place to send their used fuel. Spent nuclear fuel cannot simply be piled in a vacant lot; it is so hot that it will melt unless water-cooled; and it contains enough plutonium, mixed with highly radioactive wastes, to create many nuclear weapons. It therefore must be guarded as well as cooled, and there are very few facilities in which both can be done adequately for such massive materials.

It seems likely that some sort of interim storage facility will soon be built, simply to provide a repository for these used fuel rods until some way is found to untangle the reprocessing confusion. For the time being GE's aborted reprocessing plant is being used for such storage. The Nuclear Regulatory Commission plans to make a decision on the use of plutonium for powerplant fuel early in 1977. In the interim it is expected to issue an operating permit for the Barnwell, South Carolina, plant, now under construction and to be ready for operation before 1977.

This plant has been designed for a new objective. Plutonium is no longer needed for bombs—but in theory it is usable as reactor fuel. The Barnwell plant, far larger than either of its inoperable predecessors, is to produce plutonium oxide for fabrication into powerplant fuel. Its expanded size and modified design, according to the Nuclear Regulatory Commission, will allow greatly enhanced worker safety and reduced releases of radiation to the environment. But the NRC has also cast doubt on the fundamental plan to use plutonium as fuel. The future of the Barnwell facility is therefore still in doubt. The Sierra Club, which is opposing the application for an operating license, is likely to take the case to court if a license is issued.

Considering the difficulties inherent in the process and the very modest financial rewards to be expected, it is somewhat surprising that such intense efforts are being made to develop workable fuel-reprocessing systems. At first glance, it seems possible simply to store, and ultimately to dispose of, the reactor fuel without further processing of any kind. In Canada, where reactors based on natural uranium are used, precisely this is done. The Canadian reactor designs, developed specifically for civilian power production rather than as part of a military program, use ordinary uranium. There are no uranium-enrichment plants, and hence there is no reason to extract uranium from spent fuel for reenrichment. Nor is plutonium usable as fuel in these powerplants; and there never was a military market for plutonium in Canada. North of our border, therefore, while there is a civilian nuclear-power program, there are no fuel-reprocessing plants and consequently none of the hazards to workers and environment caused by such facilities. Fuel is being kept in storage for eventual deep burial—or sale to the US.

Furthermore, since plutonium is never extracted from the Canadian reactor's fuel, there is a very much reduced chance that it will be stolen for clandestine or criminal use. Ironically, however, the Canadian reactors, which resemble our plutonium-production reactors at Aiken, South Carolina, can more easily be adapted to military needs than US plants.

There are some disadvantages to this "throw-away" fuel process: the long-lived plutonium remains hot longer than most radioactive wastes and hence complicates the storage and ultimate disposal of the spent fuel; and, of course, there is the monetary value of the plutonium and depleted uranium, if only some means

of economically extracting it can be found. The sums involved are modest, however. A kilogram of fuel in a powerplant of 1980 should be capable of producing
electricity worth roughly $24,000 at the generating plant; the additional $163—
which the NRC estimates the recoverable plutonium and uranium will be
worth—is less than one percent of the power's value. It is hardly an important
factor in the economics of nuclear power.

Someday this situation may change. If nuclear powerplants continue to prolif-
erate at a rapid pace, the cost of providing increasingly scarce fuel for these plants
may rise to the point at which uranium and plutonium will be worth extracting.
But that time, if it ever arrives, is far in the future.

CHAPTER THIRTY-NINE

Stop

THE FRIENDS OF THE EARTH is an international organization of
thousands of members which is also, to a large extent, the personal instrument of David Brower, probably the most widely known conservationist in
America. Brower is an engaging man of about sixty, blue-eyed, white-haired,
with an easy and frequent smile. He appears to speak with very little calculation
and opens himself with surprising frankness to a reporter. A member of the
Sierra Club since 1933, and for many years its executive director, Brower led a
series of nationwide battles to halt construction of dams across Western rivers. A
veteran of national publicity campaigns, legal battles and legislative contests,
Brower has turned his considerable experience and resources, and those of
Friends of the Earth, to a struggle to halt the nuclear-power industry. A
layman—a college drop-out at nineteen—Brower is representative of the many
people active in conservation organizations who have come to oppose nuclear
power. Friends of the Earth has established numerous local chapters that
combine—as Brower does—a concern for preserving wilderness with a bitter
opposition to nuclear powerplants.

The headquarters of this effort is in San Francisco, a few miles from Brower's
birthplace in Berkeley. Friends of the Earth—or FOE as it pugnaciously calls
itself—has offices in a short commercial building surrounded by the new skyscrapers of San Francisco's manhattanization. Abutting the FOE offices is a
major substation of Pacific Gas and Electric, the power company that Brower has
been fighting for years; the noise of the transformers fills Brower's office. A
restaurant and bar occupies the ground floor of the small building.

A visitor takes an aging elevator from a tiny lobby to the crowded offices of
FOE—where there are no receptionists; a room, crowded with informally
dressed young men and women working at desks, is decorated with posters and
propaganda of dam and mining battles; its working space is heaped with books,
publications and files, and a visitor must make himself known as best he can.
Brower fortuitously emerges from a rear office, in shirtsleeves and open collar,
and conducts the visitor to a shabbily furnished office. Brower is entirely unself-

conscious in answering questions about his departure from the Sierra Club, and his assumption of the leadership of a new antinuclear movement.

You left the Sierra Club when you set up Friends of the Earth–right?

BROWER: Yes—or vice versa. (*Laughs.*)

The Sierra Club left you?

BROWER: Yes—the Sierra Club thought I should walk the plank, and I did. This was primarily a nuclear disagreement; that is, I always had a certain amount of inability to get things done, but that was constant for all my years there. It was a nuclear thing that sort of topped—topped it all and that was an argument over siting—simply siting—of the reactors now being built in Diablo Canyon on the California coast. There was a nip-and-tuck battle on the Sierra Club board. It was fairly evenly divided; I thought it was quite important that we not put reactors on unspoiled bits of coast, and I got a slate of directors to run with me to try to get enough on the board and get the right vote—and we lost.

So I left. And did start the Friends of the Earth to do some of the things that I hadn't done within the Sierra Club, but I wanted to and was getting more and more resistance. On the nuclear matter the Sierra Club position is now far beyond what mine was then. They don't want any new reactors anywhere until safety has been proved.

At that point, you weren't so much concerned about reactor safety?

BROWER: No—that was just siting; I thought that the thing to do was put reactors where there had already been a lot of major disruptions, and where there were transmission lines, and so on. Not being quite bright enough to understand that they didn't *want* to put reactors that close to developed areas for various reasons that are still not all out in the open.

Were you at that point opposed to nuclear power at all, really?

BROWER: No. I was just, I was getting shaken a bit but I had been a staunch advocate [of nuclear power] for a good many years. Some of my earlier testimony will show that, in the battle for Dinosaur National Monument, that I was advocating a nuclear alternative: let's not build any more dams, let's do it the right way. In the Grand Canyon battle I was beginning just to cite the nuclear alternative for the economic benefit it would show.

During the days when you were working for the Sierra Club you were fighting dams, you were fighting coast development?

BROWER: And loggers and miners and the usual bunch of rascals.

There was a division within Sierra Club?

BROWER: Oh, there had been differences of opinion, and those started way before my time; for example, in the Hetch Hetchy [Dam] battle, there was a very sharp division. And it just about split the club in two. There was [John] Muir who wanted no dam and then there were those who felt it was a very reasonable way to use the resources. There had been battles like that, there were battles about whether or not there should be wilderness preservation, and the Sierra Club had to evolve quite a bit. In the late '20s, for example, they were advocating roads across the Sierra through the wilderness and it was not until the thirties that there began to be a serious question about that.

How would you describe these divisions: as preservation of pristine wilderness versus conservation for use, or some formula like that?

BROWER: In some such measure. I think that the biggest [disputes] came on trying to get wilderness preserved, either in the national parks or in the Forest Service, where there was a very strong pressure to go about logging as usual.

When you left Sierra Club, you then set up Friends of the Earth. How would you characterize it in terms of this division? Is it a preservationist group?

BROWER: It certainly stresses preservation in the areas that need to be preserved. But we're broadening our concern about fighting the battle for preservation. You can get wilderness forest preserved if you can improve forest technology. You can lessen the burden on wilderness in upstate New York if you make the cities less uninhabitable. The concern then broadened, about what was happening to resources. And certainly now our great concern is what is happening to energy, and the misuse of energy is really at the root of almost all environmental ill we can think of. So it's spread out into other concerns than just the preservation of a given beautiful piece of land.

Friends of the Earth is now an international organization and a very major factor. How do you go about setting up such an organization?

BROWER: All we did, really, was to set it up in the United States. And then, since I was—in the Sierra Club—concerned about conservation activities abroad, for slightly different reasons I was anxious to see how we looked from there and to try to see our problems in the eyes of others, so that we could perform a little bit more responsibly: we, in the United States, as the world's principal polluters and resource-users. But, to set up Friends of the Earth, what we did was just find, a few of us, would just go around from country to country and see who sort of had the kindred spirit operating. The first two, or really three—France, Britain, Sweden—were set up largely on the cooperation of Edmund Matthews who's an attorney, a Harvard Law School man then. . . . He and I went around and set up meetings at some clubs, such as the Travellers Club in London, and Sweden, and invited in maybe twenty or thirty of the principal environmentalists, and said that we thought we had something that would help. We were not seeking to compete in any way with what they already had, but we thought that our combination of legislative activity, as appropriate within any country, of being ready to sue, of publishing, of being activist, would—would help. And invariably there was a little resistance in the beginning, but they got over it.

How did you get set up in the United States?

BROWER: Various people who had been my supporters in the Sierra Club thought they could also support me somewhere else. These people, a good many became our original board of directors for Friends of the Earth, and a good many of them still serve.

The chairman of Atlantic Richfield was one of that group, wasn't he?

BROWER: Robert Anderson—Robert O. Anderson—was one of the people who came in with some initial support for the John Muir Institute, never for Friends of the Earth. Altogether he put in about seventy thousand dollars to help its program and to help some Aspen conferences, at least the first one. And then he lost interest when Friends of the Earth sued to stop the pipe [the trans-Alaskan pipeline, in which Atlantic-Richfield was a major partner]. Along with the Wilderness Society and others.

We had some financing from McCall Publishing Company. They wanted to pick up the series of big books that I had been doing [for the Sierra Club] and get me to do some for them, and we had a contract for twelve, in which they were advancing something like $16,500, which we divvied up for the authors and kept the other half. And then they paid us $30,000 because we are such nice people, to help us get started. And that helped us get into the direct mail, and we had a pretty active direct-mail program.

At the outset, Friends of the Earth was not focused on the nuclear issue?

BROWER: We were not going to be very far away from it, because I was still deeply concerned and had already, in the Sierra Club, begun to lose my enthusiasm for reactors in any form. But, near the end in the Sierra Club, I remember being quite critical of Fred Eisler who was down near the Diablo Canyon area. He lived in Santa Barbara. And kept telling me about the various safety threats, and had I seen what the evacuation plan was. I was a little disdainful at his protests and said, "You shouldn't worry about that, they've got people who are really experts and they can take care of this sort of stuff." (*Laughs.*) And how right he was and how wrong I was. But we really didn't take very long to get excited about nuclear reactors as a whole, and I do know that it was right after the founding of Friends of the Earth, which was July of '69, the next month I was at this convention in Chicago with [AEC Commissioner James] Ramey. He didn't want to give a clean breast of what was happening. And then, the following January, we had our first international meeting. We had people who were getting interested in Friends of the Earth in Paris. And it was there that we began to get really quite excited about the nuclear threat because of what the French and others were fixing to do. And Esther Peter Davis, the wife of Gary Davis of the international passport clan, was really a tower of strength on that. And she came into that meeting shifting rapidly from German to French to English as necessary without losing a syllable. And getting everybody kind of worked up about the threat. The genetic threat, among other things, and [she] was one of the first to promise to write a book for us. I don't think it's out yet. (*Laughs.*)

But she got a very good antinuclear movement started in France and in England, and indeed in the Unites States. We brought back the resolution from that meeting that we didn't want any new reactors until the safety had been proved. Pretty much the position that is involved now in the California initiative.

Now is that the point at which you began to oppose nuclear powerplants per se?

BROWER: Yes, from there on it was pretty easy. One of the things that I turned up early on was the series by Gene Bryerton, who then worked for the Eugene [Oregon] *Register-Guard,* who did the book *Nuclear Dilemma.* And it was a series of, I think, maybe eight or ten long pieces in the *Register-Guard,* and I got all of those and read them all and thought he had done a beautiful job of drawing right down the middle, trying to say, "the advocates say this and the opponents say that," and by the time you got through with it, being as objective as he possibly could be, you knew you didn't want any [nuclear plants] around. At least I did. I got permission to put it out as a Ballantine paperback and that was one of the early ones that we did put out. And from there on my respect for reactors went steadily downhill.

In the Bryerton book, what was it that impressed you?

BROWER: I think I was pretty impressed by the Price-Anderson [federal accident insurance] discussion. I didn't know anything about the imminent problems in the emergency core-cooling system, didn't get into that until I talked to Henry Kendall later on. And really got ready to worry about that part. But it just did seem to me that we weren't getting a fair shake on insurance, and to this day we're not. If it's so good, why can't you insure it? And then wastes began to become a very troublesome thing. Now, I had been reassured in the Sierra Club by Lawrence Moss, Larry Moss, who is good on every other subject except nuclear, he's a nuclear engineer from MIT. I was assured way back in the mid'60s that that's going to be all right, because they would take the waste and concentrate it and encapsulate it in glass or in ceramic and put it way down in the salt mine and you wouldn't hear about it any more. Then the years went by and they still weren't doing it.

The industry says, in fact, that they will do that, and that they can do that when it seems like a suitable thing to do.

BROWER: Well, they were saying that ten years ago and they still haven't done it. And I got into a discussion with a physicist whose name escapes me now, this guy said he was worried a great deal about what migration [of radioactive materials] might happen in the salt. There are all kinds of things you didn't know about when you started playing around with atomic structure. And I began to have just a healthy skepticism.

Do you think it's just not going to be technically feasible to store the waste in the way they describe?

BROWER: I think it isn't. There is no way they really can handle anything for that period of time . . . from five hundred thousand years, up.

Suppose they do embed this stuff in glass or ceramic. And it's buried, oh, a mile down in a granite or a salt formation?

BROWER: You still have to ship it from the reactor to the site. And we don't have that good a record of shipping anything at any time. And I still like the line that the Mafia [has] the capability of intercepting any shipment at any time it wishes in the United States.

So. . . . If the stuff were vitrified and deeply buried you would not be concerned about it. It's the process of getting it there that worries you.

BROWER: Well, the process worries me and the story about what's going to happen down there is worrying me, too. That is, the longer I have been associated with this controversy, the less respect I have for the advocates [of nuclear power]. They just don't seem to be willing at any time to adopt any other course except the instant denial of problems. And if they would even *admit* that they had a serious problem I would begin to find them believable. But they don't seem capable of that. It's just all: "We got it all fixed, it's a technical fix. We can put it down and it will be all right."

There are a number of points of discussion right now in the nuclear debate, and I wanted to kind of go over them and see how important the separate issues are to you. One that's been a continuing thing is reactor safety. The problem of accidents. Do you feel that that's a decisive issue?

BROWER: I feel that it's a very important one because the chance of accident is far greater, I think, than Rasmussen's [reactor-safety] report suggested, and I think that was indeed a bought report.

It was a what report?

BROWER: A bought report. . . . The initial concern that I had, the most important one, was on waste management. What do you do for future generations, what have you done, what have you left them? For what advantage, for what temporary advantage for us, what have you left? And it just seemed to me as immoral as you could get. My stress was on waste. Then I got into my discussions with Henry Kendall, who I had known for a long time, finally found his concern about the emergency core-cooling system [ECCS] that was the accident *he* thought was the thing that needed to be discussed because it would catch people's attention here and now. If one of these things goes, it's—you don't have to wait five hundred thousand years for a problem, or thirty years for your cancer, you're in trouble right now. And there was an immediate attention grabber in the ECCS problem. And I read his first statement back in March in '72, I remember it was before Stockholm [United Nations environmental conference]. Read that on an airplane on the way home. He gave me a copy. I stopped by at MIT, and he told me a little about his very grave concern about that; and by the time I read that, I was really quite alarmed—wrote all kinds of notes on the cover and along the margins and everything else. We finally published it, a revised version of it.

Brower then described the problem of theft of plutonium and the possibility of clandestine nuclear weapons, which he had learned of through a profile of weapons designer Theodore Taylor, written by the *New Yorker*'s John McPhee.

BROWER: I began to see that we should really—if we just wanted to be completely US-oriented and not worry about what happened anywhere else—we should avoid jeopardizing our own security by having these reactors put around where, as we later found out, an MIT chemistry student who isn't all that bright could devise . . . something. And then you hear the things [about potential sabotage] you don't put out in print. The crowbar and two pounds of explosives [that] can give you a meltdown. For Christ's sake, how many reactors do you want around this country, where a few people get out of sorts, to be as euphemistic as you can, [and] say, "Well, let's cripple the country; let's get twenty pounds of explosives and ten crowbars and put this place out of action"? Now, somebody in the Defense establishment should be thinking a little bit about that. But, that's a completely new worry. . . . It is an extreme vulnerability, and you finally get around to thinking, and for what? And then, after *those* concerns—it moved from waste, to accident, to diversion and sabotage, to an understanding of the toxicity of plutionium and now, now it almost all becomes academic if you look at the economics of it. They've got a loser, and I hope they get into an honest line of work.

The business people I have talked to seem to feel, quite earnestly, that nuclear power is cheaper right now than anything else that's available.

BROWER: They just haven't done their sums, God damn it. They're sitting there dreaming their dreams. In our little paper, *Non-Nuclear Futures*, the introduction to that book, Amory Lovins alludes to the "envelope exercise." I don't know whether you've seen that yet or not.

No, I haven't.

BROWER: He said that if they would simply do their sums on the back of an old envelope they would know how crazy it is. So I put in as an illustration what that, what the back of the envelope looks like if you do those sums, and it comes out really absurd.

[The "back of the envelope exercise" is an illustration in the Friends of the Earth newsletter *Not Man Apart*, headed "Question: What would a 'safe' 'clean' nuclear future really cost?" The exercise begins by assuming the world's energy demand will grow at 5.6 percent per year, far above historical growth levels; even if one nuclear powerplant were built every day for the next twenty-five years, at an estimated cost of nine trillion dollars in 1975 money, nuclear power would supply only one-eighth of the world's energy in the year 2000 at these rates of growth. The calculation hardly seems to answer the question which is asked or to support the broad assertions made by Brower in conversation; whether or not nuclear power is to replace all other energy sources is a very different question from that of nuclear power's place in the electric-power industry in the United States.]

Now, are you talking here about the actual cost to a power company of a reactor?

BROWER: Right now I'm just talking about money. If you're going to put in nuclear power at the rate the AIF [Atomic Industrial Forum] and others say we need it, then you're going to have to put in a hell of a lot of it. And if you do, you come up with a deficit.

Well, there are some fairly bright guys in the power business. How come they haven't figured out that this is losing them money?

BROWER: I think they should be fired for not even having given it a thought. If they put these reactors in at that rate, looking at the reliability factor, the capacity factor, looking at the cost, as they are now going. Using present dollars, not throwing in inflation. Building a reactor a day, or commissioning a reactor a day, you come out with a major deficit in energy and you're burning fossil fuel about five times as fast as you were when you started. Now this is in a growth period, and they're saying that they gotta have that dough. And the Chase Manhattan [Bank] is crying because it's [the rate of energy growth] only four percent. So you go up to five and a half or whatever it is, then you start getting all these doublings. I was talking to a Rockefeller Brothers [Fund] environment man the other day; we had this thing out. He was going over the exercise and he was going to send it on over to Chase Manhattan and scorch them. They're coming up with these crazy ads talking about how we have got to have this growth. I don't know whether you've seen this series, full pages in the [*New York*] *Times* and in the [*Wall Street*] *Journal.* And they have no concept in those ads about the availability or the necessity of having resources. You just form capital—it just condenses out of space. Resources: forget it, there is just a great big blind spot. Your man in AIF probably hasn't even looked at the price of yellow cake [uranium ore] lately or what's happening to that. Then they say, Well, it's going to be such an insignificant part.

Well, they do say that. That electricity costs are very insensitive to uranium fuel costs.

BROWER: Well, I think they have to look a little harder.

If there were a breeder program, would that remove the resource constraint on the program [breeder reactors would produce plutonium to relieve the shortage of uranium fuel]?

BROWER: No, because you're then making these tons and tons of plutonium. And that makes a greater resource constraint than everything else. There goes damn near everything in the interests of security. Everything that you were trying to get power for goes out the window while you try to keep plutonium out of people's lungs and bones. We just don't need that. With about a million people who die in the Northern Hemisphere now, because of plutonium from atmospheric [weapons] testing.

If we could just kind of summarize this up to now, tell me if I'm wrong, you're saying that all of these points are decisive issues. That safety and waste storage and plutonium diversion and radioactive contamination of the environment are all decisive issues as well as the shortage of uranium. And that any one of these would be sufficient to prompt you to oppose nuclear power.

BROWER: The shortage of uranium, at this moment, interests me least. The shorter the better. (*Laughter.*)

Well, let me ask you again the kind of question I was asking about economics. How come you're so smart, Mr. Brower? There are all these guys—armies, almost literally, of scientists and engineers working on these problems—and along comes David Brower and says, you are all wrong on the science, on the engineering, and on the economics. You're just wrong.

BROWER: Simply that I know a few people that hadn't been trapped by the industry into shutting up. Look at the people who finally began to open up when they found a brave enough man in Henry Kendall to start telling what they knew to. One of the worst kinds of pollution has been the pollution of the free flow of information. There just is this compulsion to look the other way when there's truth facing them. . . .

On technical issues now, like emergency core-cooling safety. There are some people who say that the things are going to be very reliable, and there are some people who say that they are not going to work hardly at all. How do you know which of them is right?

BROWER: Well, I would, I would certainly look for some judgment that had some independence. This is General Motors passing judgment on its Chevrolet. Since they [the reactor manufacturers] did not want to risk a reactor by seeing what would happen if the emergency core-cooling system had to work, they tried to compute all the things that were going to happen and they worked out their codes and made the computation. And sent those—let's see how, what was that route—General Electric sent them to the AEC to check and the AEC sent them out to Idaho Nuclear and Idaho sublet the review to General Electric. Now somebody should have gone to jail for that: "Do you like your work?" "Sure I like my work, it's the best work I ever saw." Any questions? (*Laughs.*) It would almost be funny—it would be funnier, I suppose, if they were just doing it for kicks and weren't trying to build real reactors.

On the economics, I don't consider myself an economist and I will consider it the dismal science as much as anyone else will. But when you start getting some of the approaches that come out of the Council on Economic Priorities, or what's happening here in this book that's coming up, how much energy deficit we should get into in a rapid growth of reactor building, you just wonder where these other people were. And when we look at Chase Manhattan and their energy department, they've got a bunch of very bright people, but what have they done? They've said we're going to need all of this enormous amount of capital to accommodate the growing need for energy around the world. And they post that number, and everybody starts to go "See, well, Chase says so, it must be true, where are we going to get it?" And I think that in itself is going to be a trigger of a massive inflationary movement all around. We can't get it in the market, you can't pay the interest rates so you gotta get it at the [gasoline] pump. That shot prices way up. I was talking to one of the Chase people on that question and was trying somewhere along the line to get a meeting with David Rockefeller [chairman of the Chase Manhattan Bank] to say, Hey, why don't you stop these silly ads.

What happened to the meeting with Mr. Rockefeller?

BROWER: I haven't had it yet. I just haven't really concentrated on trying to meet with him somewhere along the line. Not that I can outexpert any of his experts, but I can probably outgeneralize some of them. I think that the generalists have a function at long last.

What's going to happen if the nuclear industry continues on its present path? What kind of accident do you think is going to happen?

BROWER: I don't know. I don't know which one would happen first. One caused by sabotage and terrorism is probably the one that would happen first.

What would be the consequences of such an accident?

BROWER: Sabotage the very easy way; you don't even have to get into the inside security of the reactor. We've got all of these constipated reactors now, with their spent fuel lying around, while they wait to see what they're going to do with it. They don't know, nothing is ready, really, to handle it in [the] volume that they're talking about. And all somebody needs to do is void the coolant [drain the cooling water in which the spent fuel is stored]. Not in the reactor, just outdoors.

What would happen?

BROWER: We'd have a meltdown.

Okay, then what?

BROWER: You'd get a massive release of radioactivity in the environment.

If there really were a bad accident, what would the consequences be?

BROWER: Well, the really bad one would be Indian Point [New York] with the wind going the right way. The numbers vary all over the lot. Rasmussen was accepting some evacuation capability and I don't think he spent any time in New York City. He never tried to get a cab in the rain or tried to get out of town from 4:30 to 7:00.

What sort of consequences do you think of when you think of an unacceptable accident? Do you think in terms of numbers?

BROWER: I think in terms of numbers; I would be disturbed to see somebody, just one person, unnecessarily killed. Because this kind of stupidity, we shouldn't be into it. But numbers make it worse, and the thing that makes it really worse is, of course, the genetic damage. And this we don't know enough about. We get all kinds of denials that I'm—I'm sure there is genetic damage, quite likely.

The Rasmussen report purported to find that, in fact, not too many people would be injured in most such accidents.

BROWER: Their figures have been shot pretty far apart by the American Physical Society, by I think almost anybody except the Rasmussen group. The UCS [Union of Concerned Scientists]-Sierra Club study found a lot of faults in it. I haven't read it all, but I just look, again, at the source and that's what worries me there.

What do you think such an accident would be like? How many people would be hurt?

BROWER: They could get up into the millions if it were in New York. Beyond that, you'd put a large area out of action for a long time. You can't go back to it and you can't grow things. I suppose that I start backwards on that, but what conceivable, sensible, sane reason is there for a risk of anything *approaching* this magnitude, if there are alternatives? And there certainly are alternatives and the one we start teaching and hollering about most is the simple alternative of using a lot less. I think that's frightfully important that we do that. And that we not share for a moment Senator Tunney's idea that if we cut our energy twenty-five percent there would be riots. I don't know where he got that one.

You say "conservation": will that do it; is that all we need to do?

BROWER: That is all we need to do, plus doing a little bit better with the alternative technologies we have. If you then take what solar could do, solar and wind, by the end of the century you're in pretty good shape. It's not only important to think about alternatives, but to move toward them swiftly, because if we start committing the vast amounts of capital required for nuclear and rapid [extraction] of oil, doing those two things, we'll have precluded our ability to finance the alternatives. There are two rapidly diverging paths and we have to jump to one or the other. A foot in the boat and a foot on shore.

If I can ask you a very general question: All of the problems, or nearly all of the problems, that we have been talking about arise in military nuclear programs and have, in fact, arisen over the years in military nuclear programs. Military nuclear power requires reactors and waste storage and transportation and shipments abroad. And this has been going on, really, for thirty years. Does the civilian program really add that much more hazard?

BROWER: I think that George Wald's point that the US is building three hydrogen warheads a day, that it has become a lethal society, is a very valid point. Of course the military use of the atom is one that is at the root of the trouble and must somehow be banished from the world. And so must war. "How?" is, of course, the question. What can be done in changing attitudes and getting a little bit more equity around the world and a little less envy—envy being the principal fuel of war, I suppose. This is going to be man's challenge, humanity's challenge from here on. If we don't find a way out of that, then we will indeed not survive. Wald will be right on target—he's saying that he doesn't see how we're going to get much past the year two thousand. I'm not quite that despondent, because I think possibly there are some people who are getting a little bit impatient of the futility of humanity's course in the last century—what the industrial age did to us, full of lovely toys and conveniences but at frightful costs to the earth, and to our likelihood of staying on it. I think that there is a fairly rapid change coming now, if you watch the number of people who are seeing . . . the harm that growth does, who are trying to reappraise what "progress" might be. Then you see hope. I still see hope. I think that the magnitude of what we have created with our cleverness is possibly the one thing that can calm us into a rational approach to the future . . . . But all I can do there is sort of just whistle in the dusk.

Do you find any reassurance in the fact that nuclear power has been managed for thirty years, in the military program, without the kinds of disasters that we've been talking about?

BROWER: I call it luck, and I still like the [John] Gofman line there: "When you built a house last week, you don't brag yet because it hasn't burnt down" . . . . The military has been pretty lucky, they've lost a submarine or two, we hear about these things, and sometimes possibly we don't. We spilled a little plutonium on land in Spain, and we've—a little slip here or there, the two guys in the experimental reactor at Arco [Idaho] probably didn't like being splattered around in it [three men were killed in the explosion of a military reactor in 1961]. We've been extraordinarily lucky.

Those are the questions I had prepared. Was there something I should have asked you about that I didn't?

BROWER: No, no, I don't think so. What, what do I think is going to happen . . . I think it's going to wind down, I'll go along with Nader: I think that in five years we'll be phasing it out as fast as we can. I'm perfectly willing to see, if the AIF and others who made the enormous investment in that will come off it, I'd be perfectly willing to see society as a whole help pick up the cost. They did what they did because they believed in it, even as I believed in it. And I just wish they would see the error of their ways. It took me fifteen years; where were they when I was learning all this?

The interviewer thanks Brower for his time, and leaves, vainly trying to imagine the meeting at which David Brower explains to David Rockefeller that the staff of the Chase Manhattan Bank has not done its sums.

CHAPTER FORTY

Problems

CARL WALSKE, seated in a plushly carpeted office across the continent from David Brower, could not be more distant from him in other respects. Walske is calculated, cautious in his choice of words, and much on guard with an interviewer, frequently checking points of fact with public-relations man Eugene Gantzhorn, who accompanies him. The slang of professional science still clings to his speech: where a layman would say "approximately" or "very roughly," Walske says "zero-order approximation." Without pausing in his speech he translates a probability of one in seventeen thousand per reactor per year to a probability of about one in three hundred for the fifty-odd reactors now in operation. His only major hesitations seem to be in taking care to avoid using his knowledge of classified military activities. Perhaps because of his experience of the most secret of all secrets, the design of nuclear weapons, Walske takes exaggerated pains to explain that what he says is based on public statements of the Defense Department or the Atomic Energy Commission, not on personal secret knowledge.

As president of the Atomic Industrial Forum, the nuclear industry's trade association, Walske might be expected to take a hard line concerning nuclear power; but, in fact, he acknowledges its problems, as he views them.

One issue that has been at the center of discussion about nuclear power is the question of the storage or disposal of high-level wastes. How much of this waste was produced in the military weapons-production program?

WALSKE: I don't happen to have any quantitative measure of it. Well, we know this: I think that at Hanford [Washington] a total of nine reasonably large reactors were running at one time or another, and in Savannah River [South Carolina] five. Now, whether those are larger or smaller in the amount of plutonium produced than power reactors has never been said as far as I know. So, you could probably get a rough idea . . . .

I'm asking because we evidently have some experience over a period of thirty years with the waste from the military program. Have there been problems with those wastes?

WALSKE: Well, I think your question implies you know something about the subject. They put those in steel tanks, as I understand it—and I probably have the same information that you have, what I read in the papers—but, they put it in tanks. There were leaks in some of the tanks, and there was inadequate monitoring, apparently, in some of the failures of those tanks. They had a system set up to do the monitoring, as I understand it, but the human part of the system wasn't adequately pursued. Therefore, there was some leakage in the soils up in Richland, Washington. I've also heard the statement—these are government statements, I don't have my own independent information—that there's no threat from any of the leaks that occurred, that they were relatively small, there's no threat of these migrating into the ground water, something like that. That they've taken corrective action, and their monitoring program is now jacked up and they'll know sooner, rather than later, of any recurrence in the future. I guess I should point out that these [waste storage tanks] are not as good as what would be used—what will be used—in the civil program as interim storage. They're not a permanent solution anyway, as storage, but as an interim solution for a few tens of years or so I guess they're okay.

There's been some decades of experience in handling military waste and that's experience you'll be able to use in the civilian program.

WALSKE: Yeah, improve on, I hope, since there were some leaks at Richland that were not adequately monitored. On the other hand, that's not the end of the story. There are also a number of reactors operating right now, in the US: fifty-five. How many abroad, Gene?

GANTZHORN: Oh, about seventy.

WALSKE: Now, spent fuel out of those reactors will have to be disposed of in some way. Either it'll be chemically reprocessed or there will be some kind of fuel disposal, where you figure out a way of throwing away the fuel elements without separating [plutonium and uranium from] them chemically. In which case, you have to worry about environmental matters anyway. So that it's not a question of whether or not we will handle radioactive waste—when you combine the military waste that exists plus the civilian waste that's being created—but *how* we will handle it, that's the question. And how *much* we will handle—because you would obviously not have the option of saying, Let's stop right now. That's an important point; I know [former AEC chairman] Dixie Lee Ray likes to make that point. It really says that you don't have the option of going back to square one. We really have to go forward [because of wastes already produced by the military].

Now, then, there are a number of options to going forward. The knowledge we have of alternative ways of disposing of radioactivity [is] a combination of laboratory experiments and military experience, as you cite—the research laboratory-level type thing, and not demonstration projects, if you like. Those are slated . . . we would like to have had them now, so that we would be utterly convincing to people who think about these things. Not having them represents a public-relations problem the industry has.

Is there a problem of safeguarding the plutonium produced by reprocessing plants?

WALSKE: Well, there are two separate [situations]—one is the military and one is the civil. And in the military you have weapons and you also have naval reactor fuel. I was once the Assistant to the Secretary of Defense for Atomic Energy, and very much involved in that, the military side of things, [but] I'm not—it's really not proper for me to give you my impression of what I know, based on how things were when I left two years ago. I haven't a clearance today. . . . I mean it would be better to talk to my successor in the Pentagon, Don Carter, who is now the Assistant to the Secretary of Defense. Just to stay off of that. I would say this, though, that the amount of material in terms of weapons and reactor fuel for naval reactors, and so on, in the military side is much, much greater than separated plutonium or highly enriched uranium on the civil side. On the civil side we have a certain amount of fuel . . . which contains highly enriched uranium, for high-temperature gas-cooled reactors. We have a little bit of plutonium oxide which is being recycled on an experimental basis back into reactors. I'm not sure there's any of that right now, at the moment, but there has been. We have separated some commercially in the past, but we are not now, as I indicated a moment ago, separating any plutonium commercially, so that it will be a year or two before we start again. At best. So that the problem today is rather small. [In fact, large quantities of plutonium have been separated from civilian reactor fuel at West Valley, New York, and are in storage facilities maintained by private power companies.]

Now, against that small problem today the AEC has been making improvements. They made improvements in the fall of '73 in the regulations; they put them into effect, got really going in the spring of '74. They issued another set of regulations, I think, earlier this year [1975], on antisabotage protection around reactors. As I recall, it was earlier this year. I'm sure they're still studying the problem and we'll get more improvements as people see things that can be fixed up and made better. I've given speeches myself on safeguards—in October '74 and just recently, early in June [1975]. In each speech I list what improvements could be made in the safeguard system to make it better. . . . Safeguards is an important problem. It should be dealt with right, and there's no point in trying to sweep it under the rug. It's something we're all going to live with for a long, long time and we better do it right. Military and civil both.

Walske proposes to establish new federal security agencies to transport and guard plutonium and to retrieve it when stolen. On June 18, 1975, he addressed the annual meeting of the Institute of Nuclear Materials Management, in part as follows:

> The principal need for change is in our communications and reaction system. We should deploy a federally operated, high-frequency [radio] network connecting fixed installations and shipments which involve special nuclear materials. Demonstrations have shown that such a network can be highly reliable in maintaining radio contact with all parts of the US.
>
> Reports from this network should feed into a federally operated command center which would be organized to coordinate the response to any attempted diversion or sabotage involving special nuclear materials. Perhaps NRC should be the operator of this system. ERDA [Energy Research and Development Administration] could equally well be the operator.
>
> In any case, the responsible federal agency will have to effect agreements with local and state police, the national guard and federal armed forces for the rapid provision of reinforcements to meet emergencies. Where there are capable local law-enforcement agencies, these are probably best suited for responding to incidents at fixed installations. State and federal forces will usually be most effective for incidents involving material in transit. The agreements will involve state governors for the use of state police and the national guard. They must necessarily involve the president for the use of federal armed forces. Perhaps it would be possible to arrange to pre-position presidential authority with a duty officer in the White House situation room.
>
> In the event that a diversion attempt were successful [and plutonium were stolen], retrieval would become our objective. It appears to me that one or several special recovery forces should be organized and trained by the federal government for this purpose. Such forces would need access to such intelligence information as may exist. Probably they should be a part of the FBI, under the direction of the attorney general.
>
> This proposed federal communication network, command center, reaction or recovery force and central direction is now missing from our anticriminal safeguards. In its place are the presently required communications and arrangements with local law-enforcement agencies. While these may be adequate for the present circumstances, where our exposure is minimal, they are not sufficient for the future. The federal government should start now to implant a national system for communications and coordination of response forces.

Shipments of special nuclear material represent the most vulnerable point for attack. I believe, therefore, that the federal government should assume custody of such materials in transit and should be directly responsible for the associated security arrangements. The guards accompanying such shipments should be under governmental control and authorized to act under government orders, as necessary. The guards may be government employees or contract guards under government orders. Similarly, the mover may be private under a government contract, or the shipping may be handled entirely by government employees using government equipment. The main point is that the federal givernment should be responsible for and should directly control shipments of special nuclear material.

The protection of shipments involves physical measures, as well as guards. I am already on record as favoring the use of armed vehicles with immobilizing features for shipments of concentrated special nuclear material. The armor should be such that it deters penetration sufficiently long for reinforcements to arrive. When such armored vehicles are not used extra escort forces should be added, specifically, more than are presently required.

At fixed installations, guard and security forces should be under the supervision of the plant management, with one exception. The exception is those guards charged with the use of armed force. These, I believe, should be governmentally organized and authorized, at least as regards their duties which require the use of armed force.

Currently, the Nuclear Regulatory Commission is conducting a study mandated by the Congress of a National Security Agency for overseeing the protection of special nuclear materials. What I have just sketched represents my ideas for the responsibilities that the federal government should assume.

The problem of safeguarding plutonium—would you characterize it the way you do the waste-storage problem? That is, it's a problem we've got, regardless of whether there is a civilian nuclear industry?

WALSKE: Well, you sure got it on the military side that way. On the civil side, I believe that also [is the situation] because civilian nuclear technology is not a one-country item any more. As we indicated before, there is a tremendous surge toward nuclear power [abroad]. At a time, in the last year or so, when we were cancelling and deferring plants in the United States, there has been a great expansion in nuclear powerplant orders all over the world. And Gene has a nice little set of figures on that he can probably give you.

GANTZHORN: It's up a third.

WALSKE: And that's not a point fully appreciated by everybody in the States. Where we have lost momentum on our new orders for nuclear plants, for the time being, the momentum has never been stronger abroad. And I guess it's because of the fuel crunch; the energy crunch is more severe in countries that have fewer resources than we. You get these countries with no petroleum and no coal, like Japan, they're really squeezed. And I could say the same for France, and less so of Germany because they have coal, still less so the UK because they have lots of coal and a little bit of oil. The Germans still order [nuclear plants]; in Germany they are ordering about half nuclear and half coal. In France they are ordering all nuclear. In the UK, inasmuch as they are ordering anything, they are ordering nuclear, still rather slowly. Spain is all out for nuclear. It's a smaller country in terms of energy requirements.

There has been some concern about the orders for nuclear powerplants in the less developed countries. It's being called nuclear proliferation now. Is that a problem for private industry to worry about?

WALSKE: Yes, you know we want a good posture across the board. When people are concerned about any aspect of nuclear power, we should be interested. The main posture, though, of private industry is—vis-à-vis government—is for government to lead and for industry to cooperate. The senior guys that I know in the industry who have been involved in sales abroad feel rather strongly that they have acted responsibly. For example, I know that the people at General Electric who provided the Tarrapur [electric power] reactor to India felt that they worked very hard with government people, in order to make sure that the Indians accepted the right kind of safeguard system. This was as of that time, you know, when it was sold to them. Some changes have evolved in people's thinking about what constitutes adequate safeguards over the years. But at that time I think GE went all out to do it. I don't have a lot of details about other countries, but they seem to have that attitude, that they acted responsibly and I don't have any reason to think that they didn't. . . . There are limitations on how much we can do, as Americans, to dictate to the world what they should do. Secondly, there are limitations on the attractiveness of using power reactors as a way to get a nuclear-weapons program. [In his June address, Walske had pointed out that most countries could obtain plutonium for weapons more cheaply by building or purchasing a simpler experimental reactor similar to those marketed by Canada—this was, in fact, the means by which India obtained its first nuclear explosives.]

I also advocate having fuel-reprocessing facilities or uranium-enrichment facilities be multinational—sometimes I say "regional". And that's simply a way of insuring that they are open enough to more than the country where they are located, so that we get signals if the people appear to be diverting material, or technology, or whatever, from those facilities to a weapons program. That's only a way of sounding a warning, it's not a way of preventing something, but that's of some value.

Do you think there will be more nations with nuclear weapons?

WALSKE: Oh, I think so. Yeah.

Do you think that would be so with or without the civilian reactor program?

WALSKE: Yes, I think so. But not necessarily at the same rate. I think that the spread of civilian nuclear technology will have some influence on it, it's hard to say how much. But I don't think it will double the rate of spread. In other words, if we get five more countries [arming themselves] without nuclear-power reactor programs around in the next 25 years, I don't think we'll get ten with it. It might increase . . . something more than one but less than two. (*Laughs.*) I don't know. It's tough to say.

I think you get a little insight into this when you read my argument which essentially says that the best way to get a military program is not to spend a few hundred million dollars on power reactors, but the best way is to spend a few tens of millions on a natural-uranium, heavy water–moderated type reactor which could be called a test reactor, and which is bought cheaper and could be done outside the safeguard system—perfectly legitimately so.

What about the hazard of accidents in powerplants?

WALSKE: Contrary to our public-relations problem on safeguards and on waste disposal, the public as a whole is considerably less concerned with reactor safety today. There was a day when it was relatively *the* most important problem. But I think that the things that have contributed to easing that in the public mind are, first of all, the continuing good safety record of the plants as far as injuring anybody is concerned—they certainly had some problems like the Browns Ferry fire [in control wiring] and things like that, which aren't beneficial. (*Laughs.*) But on the other hand, after all, the Browns Ferry plant suffered what was really a very annoying accident that didn't hurt anybody. [Two TVA reactors were shut down for more than a year after a fire in control wiring immobilized normal and emergency cooling systems.] And that's probably neither a net plus nor a net minus. The fact that it happened and it was major is a minus; the fact that nobody got hurt is a plus; it might be about a washout in the public mind. I don't think it's [getting] that much follow-up attention. The Rasmussen report, which was really a pretty sincere effort to do a decent job of analyzing all the possibilities for accidents in the reactor types they were looking at, it went I think a fair ways. It's not out [in] final [form], yet, but it's gone a fair way. Reassuring people. We had nothing, nothing like it before that which was as scholarly or as well done. However much the critics might criticize the Rasmussen report, it's so much better than anything that has ever been done before that it's got to be called a step forward and acknowledged.

As I read it, the report says that meltdowns are not extremely improbable events, but that it concludes anyone being injured as a result of a meltdown is extremely improbable.

WALSKE: That's right, the meltdown plus a leakage of the radioactivity from the containment *after* the meltdown is exceedingly improbable. Yeah.

The meltdowns are not *exceedingly improbable, though. Do you think a meltdown, by itself, would be damaging to the industry as a whole?*

WALSKE: Well, it depends a little bit on when. I wouldn't want to have one now (*laughs*), you know, from the point of view—

GANTZHORN: He didn't say they were probable, either.

No, but we're not talking any more about one in a billion . . . .

WALSKE: I can't remember his numbers.

One in seventeen thousand per reactor per year.

WALSKE: If your number is right, it's about one in three hundred per year for the number of reactors that are now operating. So you have one chance in three hundred. That's pretty improbable, but it's not so statistically improbable that you would bet your life on it, you know, unnecessarily. You're not betting your life on it. You're betting money on it [a meltdown]. I think it would have a negative impact. And just how negative I don't know. You see, the situation is—nuclear power has problems, we've been talking about some of them. Every one is being worked on; you know, the components are solvable, soluble. But there are enough problems there so you would not have any nuclear power if you didn't need it. You know, it really comes down to that. And the fact of the matter is that we really need it.

Now, there is another thing. The other alternatives. Even though we're going to exploit them fully, like domestic gas and oil, coal and so on. They all have their problems, and that's one of the reasons why nuclear power, from some points of view, looks pretty favorable. That's why utilities pick it. In the first place, it's cheaper. People have said, Well, that's phoney economics. Those of us who are in the business and are really talking to each other, you know, in closed rooms with no reporters or anything present—we really believe nuclear power is cheaper, lots cheaper, like a factor of two cheaper. We also acknowledge that the capital-investment part is higher than the capital-investment cost of the alternative forms. So that's a disadvantage in terms of getting it started. Once you get it going, the operating costs are so low at present that it's great.

But then, you know, look at the problems with the other stuff. Well, I don't have to tell you. It's off-shore drilling for oil, it's pipelines for oil, it's diminishing supplies of gas. There, you're just running out. Coal, it's a whole series of environmental problems. Also capital-investment problems. Labor problems, environmental releases, land consumption. So what I say again is that if it weren't for the fact that you really need nuclear power, and if it weren't for the fact—that the other sources can't provide enough anyway; they have their own things that are problems and are holding [them] back from being developed further—probably people would say "I won't go to nuclear power, at least not now, not for a long time." But it isn't like that.

V. CONTROL

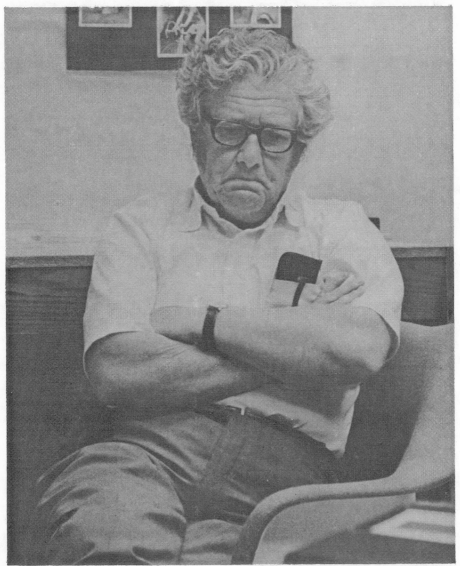

Peter Stern, vice-president for planning, Northeast Utilities
Corporation (Chapter 42)

David Pesonen, San Francisco attorney who helped draft the California
Nuclear Safeguards Initiative (Chapter 48)

Gus Speth, staff attorney, National Resources Defense Council (Chapter 50)

Representative Mike McCormack (D/Wash.), member Joint
Congressional Committee on Atomic Energy and chairman of energy
subcommittee of the House Sciences and Astronautics Committee (Chapter 49)

Madeleine Adamson, coordinator, The Movement for
Economic Justice (Chapter 51)

Tim Sampson, head of the Citizens' Action League (Chapter 51)

CHAPTER FORTY-ONE

A Refusal

THE CONSOLIDATED EDISON COMPANY of New York is the in-
heritor of Thomas Edison's flagship station at Pearl Street. The company
made an early and strong commitment to nuclear power and at one time proposed
to build a nuclear powerplant in the East River between Manhattan and Queens,
in the midst of some of the most densely populated real estate in the country. In
1974 the company fell upon hard times, however, and for the first time was
forced to omit a dividend to its stockholders, sending tremors throughout the
utility industry. It became clear that the company would not be able to finance
the nuclear powerplant it had under construction, nor did it have any immediate
need for more power-generating capacity; the state of New York, itself hard
pressed to raise money, was persuaded to purchase the nuclear plant from Con
Ed for $350 million (the state also bought an uncompleted oil-burning plant for
$228 million in cash); because of the near collapse of New York City's finances,
high interest rates will be paid by the taxpayers of New York for the bail-out of
Con Ed when bonds to finance the purchase are finally issued.

Con Ed's insistence on the merits of nuclear power still has not diminished.
The assistant to the chairman of the company is a former FBI agent named John
T. Conway, who joined the company after serving as staff director for the
security-conscious Joint Congressional Committee on Atomic Energy. Conway
now deals with the press for his boss, Charles F. Luce, the chairman of Con Ed,
but Conway will not grant interviews to people he thinks may be critical of
nuclear power or Con Ed. He explains that he has spent "hours" talking with
reporters about nuclear power and does not find his views reflected accurately in
what they write. All of this takes time from his work, he complains, so that he
must work at night and on weekends just to make the time to talk to people, and
he feels it is just a waste of time.

CONWAY: Critics of nuclear power have been irresponsible—and there are
people in the nuclear industry who have taken irresponsible positions. Critics of
nuclear power have made irresponsible, uninformed statements. Some things
that could have been of benefit to the American people have been lost.

*Can you be more specific? Were there powerplants not built because of irresponsible
statements?*

CONWAY: I would have to look into that more carefully. Some money that
would have been invested has not been invested.

Some financial problems have resulted?

CONWAY: Yes. It is irresponsible to close down coal-burning units and convert them to oil and put us in the hands of OPEC [Organization of Petroleum
Exporting Countries]. There has been a definite inflationary impact.

Are you concerned about the lack of a consistent national policy?

CONWAY: No, consistency is not important for its own sake. But when somebody passes a law saying that you can't use a certain fuel, that has an impact. *Certainty* is important. We had long-term contracts for oil at two dollars a barrel. Then a [New York City] ordinance makes these contracts null and void. We have to use low-sulfur oil, and this puts us in the hands of the OPEC nations. Now, there may have been environmental reasons for doing this, and you may think it was right.

Now you are being urged to burn coal?

CONWAY: It has turned completely around now.

All of this is said in the course of refusing a request for an interview. Conway enlarged on his reasons for not discussing the objections to nuclear power which have been raised.

CONWAY: The critics speak from ignorance. It seems that a decision has already been made, and they look for arguments to support it, first one thing and then another. There was this thing about very low levels of radiation, then it switched to safety, and then it switched to sabotage.

Is there no substance to any of these objections?

CONWAY: Certainly, there is substance to some of the criticisms. No industry is perfect. No question that waste disposal is a problem. But it can be worked out. No question that it's a difficult problem, but it can be worked out.

Is there anything else of substance in the criticisms?

CONWAY: I'll be glad to send you written information and answer any questions about the material. The written material will show what nuclear power means to *us*. It's been argued and discussed so long—I don't think I can add anything. I just want to save your time and mine. There's no need for you to come all the way down here.

Thanks very much for your time; I look forward to receiving the written material. May I quote the conversation we have had?

CONWAY *(laughs)*: Can you quote it verbatim? Sure. As long as you quote me accurately, no problem.

In due course a press release arrived, describing the high prices Con Ed would have had to pay for imported oil if it had not relied partly on nuclear power. There was no information about the half-completed nuclear plant Con Ed had sold to New York State. The Federal Trade Commission and the state Public Service Commission are investigating Con Ed's claims for nuclear power after the nonprofit Council on Economic Priorities disclosed the figures released by the company did not include all capital costs for the existing nuclear plants, and that a coal-burning plant would have produced cheaper electricity.

CHAPTER FORTY-TWO

Decisions Are Made

T HE ELECTRIC-POWER INDUSTRY has a long tradition of promoting executives from within a company's ranks. Chief executives therefore tend to be elderly; there are very few scientists employed by power companies, and Ph.D.'s are extremely rare among the managers of the industry and its lower-level employees. The typical vice-president or president has a masters degree in electrical engineering. Reactor operators rarely have even a bachelor's degree and are qualified by a few weeks of training offered by the reactor manufacturers (utilities rarely are able to train their own reactor operators). Others of the operating staff of a power company have even less formal training. Like other radiation jobs, operating a nuclear powerplant is routine production work. Power companies generally do not conduct their own research—nor do they, of course, attempt to design or even to supervise the design or construction of nuclear plants. Such work is left to the government, and to the private vendors of atomic power, General Electric, Westinghouse, Babcock & Wilcox and Combustion Engineering. Utilities spend less than one-half of one percent of their revenues on their own research, and this work, such as it is, often concerns what the industry calls "beautility"—how to make transmission lines less ugly—or efforts aimed frankly at public relations (the Union Electric Company of Missouri contemptuously funded a modest study of the mortality among birds resulting from collisions with transmission lines; company officials have a tendency to snicker when discussing the "bird-wire collision study").

It is something of a puzzle how companies so short on technically trained personnel are able to make the complex decisions involved in choosing nuclear power. The executives of power companies do not often care to discuss internal procedures of their companies; most share the attitude expressed by John Conway of Con Edison: talking with reporters, let alone directly with the public, is a waste of time and interferes with more important business. Dealings with the public are usually left to public-relations officers who have little actual knowledge of the company's operations, or to outside organizations like the Edison Electric Institute.

If power-company executives do not like talking with reporters or the public (and very few businessmen do), they strenuously object to efforts of those outside the company to intervene in company decisions. Gregory Vassell, vice-president of American Electric Power Service Company, the management unit of the nation's largest private power system, is one of the few industry officials who has been willing to enter discussions with critics of nuclear power, and he is painstakingly courteous.

VASSELL: Our main problem is overregulation. I mean we have now too many regulators, too many people who want to help us *(laughs)* by adding another layer of regulation. We feel that we, and the public, would do much better if, maybe, we would somehow find a way—still, of course, the industry needs to be regulated, because it is by its very nature a semimonopoly industry and provides a vital service; this has to be recognized, but—find some way where we can be regulated to make sure that we meet the public-interest test, but at the same time give us a chance to *do* it.

We have too many cooks, now, in this situation. What I would like to plead for—and you may want to pursue it to some extent—is that we just have to find some reconciliation, and some understanding, of the goals of the environmental movement and goals of economic prosperity, economic development, and full employment, and the over-all human environment that includes a lot of things other than nature and so on.

I think that we too often forget that we all, really, are humans, we all are rooted in the same characteristics and traits and reactions and what not. And the tendency often is to assume that somebody in business is more likely to be selfish than somebody in publishing or in the environment, or whatever. But the facts are that we all are humans, we all have a certain degree of selfish interests. We might as well recognize that. But we all also find satisfaction in public service. And I know that we very often just cannot get the point across, but I know many people in the industry, and certainly in our company, and I know that most of us have the best intentions. We are trying to do what is right for the people, for the public. And we are trying to do the right thing as far as the environment is concerned, also, in terms of balancing things. And we believe that, maybe, slowing down on some of these things is *exactly* the right thing to do. And we believe it, because we are trying to do the right thing for the country, not because of some selfish reasons. Now I don't know how to bridge this gap. I mean, it was there for some time.

May I ask you how you would respond if, say, a member of Congress said to you: Well, we're all good, well-intentioned people in government, so we're not going to have elections anymore. We're just going to run the country and you guys trust us.

VASSELL: What you are saying is that a certain degree of contention and advocacy is necessary, and I agree. I agree, a certain degree is necessary. That's what this society is all about.

Don't you think there should be more outside inputs into management of something as important as the electric-power business?

VASSELL: Well, I think it's good to voice the views, and I think we should listen to the views and consider them and react to them. I do think that we need to keep sight of who, in the final analysis, carries the burden of responsibility for certain things. In other words, in our society, we do have to specialize in some way or assign various jobs to different people, groups or categories of institutions, or whatever. We elect, as you mentioned, we elect the president to run the executive, we elect congressmen to legislate, we elect a mayor to run a city. We expect various businesses to run their particular businesses. We expect publishers to keep us informed and stimulated and so on. And while we need this exchange of views, at the same time we need to keep in mind, in the final analysis, the institution that was made responsible by society to perform certain functions; it is, in the final analysis, responsible for performing those functions, and you have to leave them some leeway and you have to rely eventually on their judgment in doing it. If you don't, then you can't expect them to be responsible for what results in the end. And that's why, while I fully support exchange of ideas and fully support the need for inquiries, I don't support the idea of everybody doing everybody else's job. In other words, if I spend all my time, for example, specializing in the area of electric-power supply—analyzing it, looking at it, projecting it, determining what is needed, trying to optimize the whole thing, so that it's available at the lowest cost. It would be . . . damaging if we would say, Well, let's select five people—you know, one environmentalist, one legislator, one this and one that, and all of them together are going now to do the planning for me. And then, five or six years hence, when the time comes to really see whether the things were done right, the five gentlemen walk in different directions, and I'm here to hold the bag as to whether it's right or wrong.

Isn't that how, in fact, nuclear power was first developed? Elected people in the Joint Congressional Committee on Atomic Energy and experts from other parts of government and the military got together and, in fact, made this kind of decision that tax money would be spent developing nuclear electric powerplants. And then, I guess, they all walked away and left the power companies to deal with this. Do you think that was bad?

VASSELL: Well, I was not particularly thinking of nuclear power. Of course, we do have the functions for legislative committees, we elect officials, and once we elect them, presumably we like them not to be hounded all the time, but rather to do the job for us, even though we have to look over their shoulder, help them along, and so on. And certainly, convey to them our views as they go along, and the same applies to all the other areas. It's a difficult thing to really define. I don't know, really, I don't know the answer.

Another official willing to talk about how decisions are made in power companies—and about his dislike of outside interference—is W. Donham Crawford, now president of the Edison Electric Institute and a former vice-president of the Consolidated Edison Company of New York. In the early 1960s, Con Ed was among the first private companies to announce their decisions in favor of nuclear power, and none was more aggressive in its commitment than Con Ed. The company's first nuclear plant, a small experimental plant at Indian Point, thirty-five miles north of New York City on the Hudson River, went into operation in 1962. Before the year was out, on December 10, 1962, Con Ed announced plans for a much larger nuclear plant—to be built in the heart of New York City, in the Ravenswood District of Queens, just across the East River from Manhattan's East 72 Street, and about a mile from United Nations headquarters.

This proposal was eventually withdrawn, but for a few months there was an extraordinary uproar. The New York City Council began to consider whether it could ban nuclear powerplants from within city limits; citizen groups were organized, and a great many people were exposed, for the first time, to expressions of doubt about the safety of nuclear power. There was probably little or no chance that the federal government would have permitted construction of a nuclear powerplant in the heart of New York City, but Con Ed could hardly have made its own commitment in stronger terms.

The abandonment of the Ravenswood scheme (an oil-burning plant was built at that site) did not end Con Ed's efforts to place a reactor within New York, however. In 1968, the company made further announcements: it said it planned to build a whole complex of nuclear powerplants on David's Island, just beyond New York City limits in Long Island Sound (within the suburban community of New Rochelle). This scheme, too, was unrealistic. The reactors were never built, and by April, 1976, Con Ed was trying to persuade the town of New Rochelle to accept David's Island, a gift that had already been rejected by national and state park services and private conservation groups because of the cost of building a bridge or providing ferry sevice to the island. Toward the close of 1968, the company again affirmed its interest in building a nuclear plant in the heart of New York City: on November 7, the *New York Times* carried a front-page news story which quoted the company's chairman, Charles F. Luce; while nothing was definite, Luce hoped that in "planning for the use of Welfare Island, the City will leave a portion of the island or plan to use a portion of that island in such a way that we can put a nuclear reactor beneath it." Welfare Island lies in the East River between Manhattan and Queens and extends southward from the Ravenswood site toward the United Nations; its southern tip is about a mile from Times Square.

Plans for a nuclear reactor on, or under, Welfare Island were ultimately abandoned in the face of strenuous public opposition. The company has so far built and operated only one nuclear plant, in addition to its first experimental unit. The first full-size nuclear plant in Con Ed's system, Indian Point II, went into operation well north of the city, in 1973. Efforts to build another full-size plant at the same site, plagued by delays and construction problems, ultimately were abandoned in Con Ed's fiscal crisis, and the plant, still incomplete, is in the process of being taken over by the state of New York. For a commitment of almost twenty years, and many hundreds of millions of dollars of its stockholders' money, Con Ed can show only one full-size nuclear powerplant and a reservoir of public ill will and hostility to nuclear power that could not have been more successfully created if the company had set out with that objective.

In 1968, when Con Ed announced its extensive plans for nuclear expansion within New York City, the company's vice-president for administration and the head of its nuclear-development program was William Donham Crawford, a graduate of the Columbia Military Academy and the US Naval Academy, with a master's degree from the California Institute of Technology (1948). Crawford, tall and gray-haired, at fifty-two is very young to be a senior executive in the electric-power industry. In his office at the Edison Electric Institute in midtown Manhattan, Crawford answered questions about the early decisions on nuclear power at Con Edison.

At Con Ed, did you guys sit around and say, well, there are problems with wastes, or radioactive emissions, or whatever, and balance them against the benefits? Was there a process like that?

CRAWFORD: Well, yes, there is a process like that. First of all, the kind of thing you are talking about is *not* an inherent part of a specific comparison. Because the company, much before they would get to such a point as to contrast a coal-burning plant with a nuclear plant, would have satisfied itself that these problems were not great problems that could not be overcome. Of course, the company would understand the questions about radiation levels, about safety aspects, about public acceptance, and so forth. But they would come to a conclusion that Yes, these are problems, but there are lots of virtues of nuclear power, as well. One is that you don't have stack gases coming out of the plant. Another certainly is economy. So, you conclude that nuclear power is a viable option, on kind of a general basis. Then, when you get to the point where you have to add a powerplant, you look at what your alternatives are, and, of course, what you do is see how much your load is growing, how fast it is growing, when you need a new plant, and then what kind of a plant should it be and where should it be located. And then you have to look at the question of close-in plants, and the lack of transmission expense, versus far-out plants and transmission in. The generating cost, the fuel cost, as best you can predict them, the environmental costs that you might be exposed to; and then you put all these down and take a lot of time and a lot of pain to do it and you finally come to a conclusion as to which one is the better course of action.

There is a generic decision at some point that nuclear power is one of the options, and then decisions proceed from there.

CRAWFORD: I would say that, but then I would also go on to say that when you make a concrete decision on a plant, that you would review these matters again, and say, Yes, these conclusions are still intact.

I know these things are hard to pin down, but very roughly, when would you say that Con Ed made that kind of generic decision with respect to nuclear power?

CRAWFORD: I would say that it was made in—now this was before I was with Con Ed—but I would think that it was made in the late 1950s, because Indian Point was started, construction of Indian Point I was started in approximately 1957 or 1958.

So, Indian Point I was not viewed as a test of nuclear power's suitability, it was a result of a decision that this was an option?

CRAWFORD: Yes. Because, the test of whether nuclear power was viable would have been a predecessor to Indian Point I, namely, the Shippingport [Pennsylvania] plant [the first nuclear powerplant, which began operating in 1957], the submarine reactors, and the whole background of nuclear power. So, by the time that all had been acquired, the company would have concluded that nuclear power was a viable option and they decided to build Indian Point I.

Let me ask you specifically about safety, and the catastrophic kind of accident that has often been at the center of discussion: was that considered an event that had to be evaluated, or was it simply not considered a possibility at all?

CRAWFORD: As to whether there could be a catastrophic event? Bear in mind that I was not at Con Edison at the time [in 1957].

Well, but even in the. . . .

CRAWFORD: In the general term?

Yes.

CRAWFORD: I would say that it would not be considered to be impossible. Nothing is impossible. But when you look at the kinds of studies that are made, the kind of charges that are made [by critics], I think you have to try to focus on those as to just what they say. And here is an instance where I think some of the critics of nuclear power really do a disservice. . . .

The disposal of radioactive wastes seems to be acknowledged as an unsolved problem at this point, although there is some disagreement about the severity of the problem. In any case, there is no federal program right now that defines the ultimate disposal of the waste. Should the utilities be concerned about that?

CRAWFORD: We are concerned about anything about nuclear power that is a fundamental consideration, and so, yes, we are concerned about that, but simply as a part of an over-all problem. But the way I look upon the storage problem is that it is much ado about nothing. Wastes have been stored for thirty years—on a temporary basis, true, but wastes can be stored, they can be stored safely. The fact that an *ultimate* decision has not been reached as to what to do with the wastes does not say that it is an unsolvable question. The question really is, of the options that either have been developed or can be developed, which is the best one? We are talking about a very small volume of waste; it's not a problem to deal with it on a temporary basis until you decide what the ultimate answer is. But the fact that the government has not concluded on what is the best of the ultimate answers, it seems to me, should not be an obstacle to continuing nuclear-power development.

Because of the very long lifetime of the wastes, some people consider this a kind of separate and unique problem. Do you think that this is a moral problem, a legacy to future generations?

CRAWFORD: I do not think it is a moral problem. I don't see the morality of the problem. We are talking about a way to meet people's needs for energy. Part of meeting that problem in this way, which is a desirable way to meet it, has to do with disposing of the ashes of this particular form of fuel. And, yes, the lifetimes are long, but they can be dealt with. Presumably we'll continue to have a viable society for as long as anybody can foresee. I don't see that it is asking a lot, a whole lot, to believe that succeeding generations can watch the salt domes, or whatever they are, as well as we can.

What is the option? The option is, if you are going to provide energy and you don't do it this way because you think it is immoral, what do you do? You provide it some other way. The other ways also have their problems. I can go into some of those if you care to, but there are problems with any way that you produce energy.

The question of a meltdown in a reactor is a real question. Mr. Walske [of the Atomic Industrial Forum] accepted the estimate that for the reactors now operating, there was one chance in three hundred per year that there would be a meltdown [equivalent to one chance in ten over the thirty-year lifetime of the fifty-six nuclear powerplants then operating]. Now, I think it would be clear to anyone that is not a very high probability, but it is not the "chance in a billion" either, it's not negligible. How do you feel about that, the chance of a meltdown without injuries; is that a real problem for nuclear power?

CRAWFORD: A meltdown without injuries, is it a real problem? Well, I don't know that you could say that is not a real problem. You have to persuade yourself that it's something you can deal with. . . . I think you have to look at the safety aspects in the beginning. And that is, How do you design a reactor? There is no other industry that's ever been developed where you deal with safety in the same way that you do with nuclear power. You assume that the worst possible thing will happen, and then you design to compensate for it or to prevent it. So, the question really is, whether a meltdown could ever occur, and if it did, what you could do to contain it or to control it. And it seems to me that the great number of redundant systems that are built into the reactor to prevent the meltdown in the first place, or to contain any consequences of them—everything has some risk, but it looks to me like you are doing so much to guard against this, that it's the kind of risk that society is justified in taking.

I was not asking you whether this was a decisive issue, I am asking whether this low, but not negligible, probability of a large financial loss is something that is taken into account in making decisions about nuclear powerplants?

CRAWFORD: I would say that the potential losses of a plant are taken into account when we make a decision to build a plant. It's certainly a factor that must be considered.

At Con Ed was this quantified in any way? Did someone try to get the present value of the possible risk or anything like that?

CRAWFORD: No, I wouldn't say so.

So, it is handled intuitively?

CRAWFORD: Well, first of all, you have the Price-Anderson protection [of federal liability insurance] and, ah . . . .

But that doesn't insure the company against its own property losses.

CRAWFORD: It does not insure the company against its own property losses, it insures liability. But, the conclusions that were, I am sure were drawn at Con Ed on this point was that if there were any realistic expectations that something like this would eventuate, they would never have built a reactor in the first place.

Is there commercial insurance that covers the possible loss of a plant? Do companies carry that kind of insurance?

CRAWFORD: There is property insurance provided by the pools [of insurance companies to provide coverage for large risks] up to a certain level.

Do you think that would cover the complete loss of a nuclear powerplant?

CRAWFORD: No. I don't remember what the exact amount of the limit of liability, of the coverage, is, but if you were to conjecture that you would lose the entire plant, I'm sure the insurance is not that much. But if you did that I am sure the expectation would be that you would not lose the entire plant.

What do you think would be the effect on the industry as a whole, if there were a meltdown, even if no one were injured?

CRAWFORD: Well, that is a very good question, I think. I've heard people say that if you ever had a meltdown or a very large accident, that that would shut all the nuclear plants around the world down. I don't believe that, because I think that it's very clear—and not just to the United States, but to the other industrialized nations of the world—that you can't meet your power requirements without it. And, while there would be great concern and great distress and great outcry about a major nuclear accident, I think that would not shut the rest of the nuclear plants down, because you have to have them. Now, it might mean cutting back on the power levels or operating them in a much more conservative way or something like that, but I do not think it would mean the end of nuclear power, if you had one accident.

Do you think the accidents involving large numbers of injuries, which I think everyone concedes are of low probability—are such accidents a factor in the power company's cost-benefit analysis? Do you try to get some sort of discounted value of that?

CRAWFORD: My own feeling is that they are so remote that you should not try to quantify them.

At Con Ed, did anyone ever do the kind of cost-benefit analysis that the National Environmental Policy Act now requires? Adding up health effects, environmental effects, economic effects, and balance them for the different kinds of powerplants?

CRAWFORD: Well, I think I would answer your question "yes," but it is a matter of the detail in which you specify it. If you are talking about an alternative being to burn coal, and having certain adverse medical effects from that, and you try to quantify that, I'd say we didn't do that. But to the extent that you could do it on a reasonable and fairly confident basis, those factors are all considered.

So you would not include, say, the health effects of sulfur dioxide or the long-term effects of radioactive pollutants, but you would include economic. . . .

CRAWFORD: The things that you could feel reasonably confident about quantifying, these things would normally be taken into account.

If serious accidents are too improbable to consider, then there is no reason, from a power company's point of view, why nuclear plants should not be built in the heart of New York City. In discussions with power-company executives, one encounters little or no evidence that nuclear powerplants are thought to pose problems different in kind from other plants, or that they require any different decision-making process or management. Peter Stern, vice-president for planning of Northeast Utilities, a large holding company that operates power systems in Connecticut and Massachusetts, discussed the planning of nuclear plants at great length and described the factors which induced his own company to make a heavy commitment to nuclear power—a commitment that has so far been more successful than Con Ed's. Stern is another of the young executives brought into positions of authority in recent years; he is particularly unusual in being an executive brought into the power business from entirely outside the industry— Stern is a professional planner.

The brick building which houses Northeast Utilities is on a campuslike hilltop, overlooking the rolling countryside of Connecticut's suburbs, but approached from the New York City direction, south, it is not marked by any identifying sign, and a visitor must drive some distance past warnings against trespassers before reaching the building. Peter Stern's windowless office is approached through a lobby guarded by a woman receptionist; a guest is escorted to an elevator and met at the proper floor by Stern's own secretary, a woman who guards another reception area.

Stern himself is a cordial, rumpled, plump man of about fifty, who talks rapidly at great length and with little inhibition. He represents his company at frequent public hearings, as the utility's extensive nuclear plans have received equally extensive public opposition.

What environmental information goes into deciding what kind of powerplant the company will build—assuming that there is a choice of plants?

STERN: We certainly wouldn't be authorized today to build an oil-fired plant for base-load [steady operation]. The PUCs [public-utilities commissions] wouldn't let us. And it would take, at this point in time, an awful lot of guts to build a coal-fired plant.

Why?

STERN: Supply, and particularly the transportation system. This is a situation which occurs nowhere else in the country. We have, in fact, lost our railroad transportation system in New England. Oh, yes, there's a shoreline railroad running from New York to Boston. But you should look at that track some day. The reason we can't have any Metroliners from Boston to New York, as opposed to New York to Washington, is because the track's so bad. And that's the main line; anything beyond the main line—and I'm exaggerating, but only to make a point—it's an absolute mess. You cannot cross the Hudson River into New England anywhere between New York City and Albany by rail.

Is barge transport not a possibility for you?

STERN: A possibility for shoreline locations only, not for inland locations. And I would postulate here, as I said earlier, I don't see any Connecticut shoreline sites available for nuclear at this time. There's too many competing uses. You know, we can't build any more state parks in Connecticut, so why should we have any nuclear powerplant sites? Connecticut shoreline—I've forgotten the statistics, and I wouldn't want you to use an inaccurate number—but less than four percent of the Connecticut shoreline is open to public access. Just to give you an idea of the competing uses.

So, in the light of that, where are you going to get a thousand acres today of shoreline? That's where the barge problem comes in. Now, yes, barges can go up the Connecticut River, and do. We have Connecticut River plants which are [fueled by] oil delivered by barge. The main problem with putting plants on the Connecticut River—and I can assure you, due to Mr. [Senator Abraham]

Ribicoff's [Dem., Conn.] Gateway National Park proposal, which did not turn out, we now have a state conservation compact on the lower Connecticut River which is trying to accomplish what Ribicoff did not with National Gateway—there is a host of problems on the Connecticut River, too.

I would say that to the extent that nuclear powerplant sites, or *any* major powerplant sites, are possible in Connecticut, other than the one in a central city—we have a site, an old site—they are not accessible to rail or to barge.

Just to summarize, you feel oil is not available because of the federal policies based primarily on foreign policy. . . .

STERN: Why, hell, we barely avoided being ordered to convert oil to coal on July 1 [1975] by FEA [the Federal Energy Agency, which attempted to persuade utilities to reduce oil consumption]. It so happens, thanks to our large nuclear commitment, that we managed to talk Mr. Zarb into exempting us from conversions.

And coal is not an option because you can't bring it here?

STERN: Can't bring it here, and because we have been out of the coal market so long. We would have an extremely hard time obtaining any contractual, long-term coal commitment without really committing to opening up mining [inaudible], which means an investment in coal mines.

The logical geographical distance for coal, for us, is still Appalachia—for us to bring [low-sulfur] coal from Wyoming makes absolutely no sense—it means high-sulfur coal, and that would mean scrubbers, and we would rather have the Midwestern [coal-consuming] power companies go through experimentation with scrubbers, before we have to commit to a scrubber.

So we'd rather not, frankly; we'd rather not.

"You'd rather not" is a little different from not having a choice.

STERN: Well, we'd rather not, but I think we can make a rather convincing argument, at this point. Not just to the nuclear regulator, who obviously is just concerned with the nuclear plant. If we had a regulator who regulates federally—as may very well happen under federal law—any powerplant siting, if we started to make a comparative analysis of economic, environmental, logistical, national-interest factors—you'd find that coal would look pretty sick.

Then did your company make a generic decision to build nuclear plants?

STERN: No, not quite. I would say there was a generic decision in the early 1960s to be early, or ahead of the pack, essentially ahead of the country, to go into nuclear then, without any preconception that that decision would mean *only* nuclear decisions thereafter. In other words, the fact that we built Connecticut Yankee [a nuclear plant] in 1960 and signed with GE for Millstone I was not, in itself, [a decision that from] then on it was going to be all nuclear. On the contrary, at that time, obviously, nobody knew what the operating experience [would be] and the economic evaluation made by TVA in 1967 was just like that *(holding fingers together)*, a stand-off. You know. There were many, many things that changed.

So I would say it was an early commitment to *try* nuclear because of the historic disadvantage of New England in terms of distance from fossil-fuel [coal and oil] supplies, and hence the traditional higher cost.

When there were still options as to what kind of plant to build at a particular site, would you have been involved in the discussions at that point?

STERN: Sure.

What sorts of environmental considerations would go into that decision?

STERN: There were two decision episodes that were fairly recent. One affects Millstone III [in Connecticut] and the other affects the Montague plant [in Massachusetts]. The Millstone III discussions were carried on in the area of 1969 and 1970.

All the far-sightedness, the interdisciplinary nature of environmental research, did not lead to the acquisition of the Millstone site, yet except for some menhaden kills we haven't had any major environmental problems there. [The site was acquired from an estate in the middle 1950s and was kept in reserve simply because it was available at a good price and had access to the ocean.] The Millstone III decision [as to what kind of plant to install] was made, I think, in 1969 or '70. And the honest comparison, both internally for management purposes, and for environmental-report purposes, the comparison was a third [nuclear] unit at Millstone versus a fossil-fired plant in Stamford, Connecticut. Stamford, Connecticut. You passed through Stamford today. It's the first, largest city as you enter Connecticut. It's really near the load center [of the utility's system]. Not only that, it's a little west of the load center. You could, potentially, relieve some of New York's problems. The site was occupied for fifty-five years by an old coal-fired plant which we tore down in '68, or around there—I can't remember whether it's '68 or '69. In the middle of Stamford, right in the heart of it. It clearly would not meet AEC criteria [for limits on population near a nuclear plant]: Stamford has a hundred thousand plus. But for an oil-fired plant, [it had] a deep-water harbor, potentially good site for an off-shore oil-unloading platform and pipeline; proximity to the transmission grid, power-system grid; probably underground connection to the grid in town, you know. And water. And compare that with Millstone. Where the environmental analysis. . . . We had baseline data from Millstone I and II, and we hired outside help to do a biological reconnaissance assessment of the waters off Stamford. Terrestrial ecology was not a problem in the middle of a city. Radiation monitoring—not for an oil-fired plant. So it was really a question of trying to look at an optimum location for intake and discharge [of cooling water for the oil-burning plant] in the harbor. The problem was not so much upsetting the natural biological community, but the fact that Stamford's sewage treatment plant was discharging into Stamford harbor very close to our powerplant site. Unless we were careful with the location of our intake pipe we would find ourselves piping the sewage through our plant *(laughs)* and into an area that had not previously been affected by the sewage discharge. So it was that kind of physical studies, hydrographic studies, primarily, on a reconnaissance basis, however. And as a result of that, we at least preliminarily located a discharge and intake structure—at least in such general location within the harbor that we could make economic comparisons with Millstone, where there was no need for a remote intake or discharge.

That summarizes the comparison. The economic comparison, at that time, came out very favorable to Millstone even though we were still getting four-dollar [a barrel] oil. Based, of course, on what was *then* the assessment of capital cost of Millstone, which I forget what it was. It was certainly nothing like what it is today.

So that Millstone makes, today, even more sense than it made in that comparison year, but for completely different reasons. Oil prices have gone out of sight, capital costs of *both* [fossil fuel and nuclear] have gone up; today, to make a comparison, if we had the same situation today, Millstone III versus the other, we would be comparing with a coal-fired unit—possibly at the same site [in Stamford], but then we would have to include scrubber costs, because we couldn't get low-sulfur coal—I wouldn't know exactly what the comparison costs would look like.

The decision to build Millstone—nuclear—was economic, and it was environmental, because this was just a year before the amendment to the Clean Air Act [in 1970, imposing stricter controls on sulfur emissions], it was clearly coming; we knew then that low-sulfur oil—we were burning, believe it or not, we were burning 2.2 percent [sulfur] oil; I came from TVA where they were burning coal with 5 percent sulfur. [These are now considered extremely high levels of sulfur, ten times what is now permitted.] So there was an environmental reason that affected the economic comparison.

You didn't do the kind of cost-benefit analysis that is required now under the National Environmental Policy Act?

STERN: We did it more intuitively and less explicitly. Because it wasn't required. But I would say that the internal economic justifications, in which environmental costs were—insofar as possible—quantified, were done.

To what extent, even intuitively, were you considering health effects and long-range problems with pollutants?

STERN: Neither those from the discharge of sulfur and particulates, nor those from discharging radiation. But with respect to radiation, it has always been a company assumption that as long as we could design and operate a plant that performed under what were then permissible releases—which are now "as low as practicable" [under federal regulations]—we would, in fact, have no health effects.

The executives and directors of Northeastern Utilities, like those of other power companies, made no independent evaluation of the safety or long-term impact of nuclear power; nor did they have any particular scientific or nuclear-engineering expertise. By and large they relied on the federal government for assurances that the plants were safe, that wastes would be disposed of and that radiation releases would not cause sickness or death. No company is known to have taken seriously the possibility that its investment in a nuclear powerplant might be lost, or that such a plant might do injury to the general public. The comparisons between nuclear and fossil-fueled plants were and are made solely on economic terms, although environmental effects are counted to the extent they have an impact on economics; in short, nuclear power is a business matter decided like other business matters. Its advantages in this regard seem clear to most industry executives, and there is no reason to suppose that they have not done their sums correctly. The only large power system to choose not to adopt nuclear power in any significant degree is the American Electric Power Company which owns extensive coal reserves.

Neither those who choose nuclear power nor those who choose coal take into account in their decisions the manifold issues of public safety and national interest which play so large a part in the public debate over nuclear power. Power-company officials have taken the safety and desirability of nuclear power for granted for so many years that it is evidently difficult for them to give serious consideration to the complaints of the critics of nuclear power.

CHAPTER FORTY-THREE

The Battle against Socialism Begins

SAMUEL INSULL'S companies were extraordinary creations; as corporations, they were mantled with privileges the courts had fashioned to shelter private property. As monopolies, they were not subject to the competition which is supposed to regulate private economic activity. Protected from public and private interference alike, the power companies assumed a place in the economic system of the nation—alongside those other private utility monopolies, the railroads—as entities of enormous power.

Long before the utilities achieved this position, Insull anticipated the public opposition which would result. Ten and twenty years before, the agitation against railroads had become serious enough to raise the threat of public takeover, and even the Supreme Court, in deciding the Granger cases, seemed to give ground to the arguments of socialism: businesses, the Court said, conducted for a public purpose should be subject to public control.

The power business was peculiarly vulnerable to such arguments, for many power companies were already owned by governments—the municipalities. Samuel Insull himself had been the salesman who persuaded, through various means of inducement, corrupt municipal governments to purchase the powerplants then being sold by the Edison companies. By the 1890s, there were hundreds of such municipal power companies; there were also many publicly owned street railways which generated their own power, and water companies which generated electricity for their pumps. In many communities where the power companies were privately owned, there was agitation for municipal takeover.

When the railroads earlier had been faced with the threat of public ownership, they had accepted the compromise of public regulation of privately owned monopolies. In the great court battles of the 1880s and 1890s, and in the better-publicized legislative battles, a compromise had been reached. State regulatory commissions were established to oversee local freight and passenger rates; the Interstate Commerce Commission, the federal government's first venture into economic regulation, was established in 1887 to oversee interstate operations of the railroads.

Insull, anticipating public opposition to his monopoly scheme, began advocating a similar compromise for electric power almost from his first days as president of a power company. In 1898, when he first publicly called for monopoly control by private power companies, he coupled this with a plea for acceptance of state regulation.

Insull seems to have been almost alone in his industry in seeing the implications of monopoly at so early a date. He seems to have conceived the entire pattern of the industry in a single inspiration: the use of low industrial rates to create a monopoly, and the use of state regulation to forestall municipal control or ownership; the economic efficiency of large central plants was to be used to sell the scheme to the public.

Insull's audacious arguments were those of the capitalist, the hated builder of trusts, but they were also the arguments of socialism: competition breeds duplication and waste; the efficient means of production are centrally owned. The difference was, of course, that Insull proposed that he should own the monopoly; the socialists preferred government ownership. From 1898 onward, Insull's speeches contained an open plea for the advantages of monopoly, but he coupled this plea with an awareness of the dangerous ground he has treading. Monopoly was not only highly unpopular, it was oddly near to the schemes of the socialists; it was widely viewed, in fact, as merely a step toward socialism. Once competition had been ended, the justification for private ownership was gone. A monopoly might just as well be owned by the government, over which the public had some control. Insull, therefore, became the first public advocate of regulation of the power companies, at a time when they had not even been recognized as utilities in need of regulation. State regulation was a necessary concomitant of monopoly; the only other possibility, Insull believed, was public ownership and socialism. In 1898 as well as today, regulation was seen as a means of preserving the private ownership of power companies.

When Samuel Insull first laid his scheme before his fellow power-company executives, he stated the question in an address under the heading "Public Control and Private Operation":

> A subject of growing importance to a number of our members is the question of the public ownership and operation of the undertakings now operated by electric-lighting companies. The agitation in connection with this subject has called forth a great deal of discussion, partly by interest in it simply with a view to extending the influence of political parties, and partly by serious disinterested thinkers who believe that the best interests of the greatest number are to be obtained by the creation of a municipal socialism, which, if carried to its logical conclusion, must ultimately result in municipalities performing, with others, such public-service work as we are engaged in, and also in producing the food we eat and the clothes we wear . . . .
>
> The fallacy of the so-called reformers' theory results from looking only at what the public calls the injurious effects of corporate management without taking into account its indisputable benefits. . . . It appears to me that a correct division of power and responsibility requires political government merely to control private industrial management. Where political government and in-

dustrial management are merged into one interest, the power of control is seriously impaired, since a political administration cannot be reformed without overturning the party in power.

I cannot bring myself to the belief that the citizens of this country are in fact opposed to large aggregations of capital in corporate form, as such aggregations are absolutely necessary to the operation of all great undertakings by private enterprise.

Samuel Insull pressed for the doctrine of state regulation for more than twenty years. In the early 1900s, political-reform movements in some states, notably those of the Midwest, echoed his arguments; the reformers were moved, as Insull himself was in part, by the corruption and bad management of the municipal governments, whose periodic sale of utility franchises had become a fabulous source of graft. In 1907, Wisconsin enacted the first power-company regulation statute, which established a state commission to set rates for privately owned power companies. It was patterned after the earlier railroad commissions and was limited by the same strict safeguards on behalf of private property. In many states, the old railway commissions were simply enlarged to take responsibility for all public utilities. (Texas was a kind of missing link, or living fossil, in the development of public utilities, with a railway commission, but no public-utilities commission until 1975.)

Very few things about the electric-power business are more striking than the degree to which Samuel Insull's message of 1898 has been absorbed, repeated and taken as the motto of the industry, emblazoned on every surface available to the industry's considerable publicity organizations. It is a message which is taken to heart by both private and public power officials, by company stockholders and by the federal officials who regulate the companies. Here, for instance, is Joseph Swidler, then the chairman of the Federal Power Commission, speaking on February 4, 1965, to the Chicago Law Club, very near to the spot on which Insull had so many times delivered similar addresses:

"The American system of public-utility regulation plays a vital role in the national economy by making possible the dominance of private enterprise in the operation of the industries which are of a monopoly character. Of all the basic industries, only the mails are exclusively government-owned and operated. The railroads, the airlines, the motor-transport industry, the maritime industry, the telephone and telegraph industries, the radio and television broadcasting systems, and the electric-power and gas industries, are all either wholly or dominantly in private hands, despite the fact that they are the foundation of our industrial economy and by their nature not subject to the same forces of competition as other commercial enterprises. In the United States private enterprise operates a larger share of these vital industries than in almost any other country because of our balanced system of regulation by public authority."

CHAPTER FORTY-FOUR

Ownership and Control

RALPH NADER is the acknowledged leader of the efforts to shut down the nuclear-power industry; he is also the man the industry's leaders accuse most frequently of wishing to injure the private-enterprise system. A reporter asked Nader about state regulation and ownership. His answer is not much different from Insull's.

NADER: Well, they've done an absurdly, abysmally poor job, obviously. But I haven't yet concluded that they can't do a much, much better job. . . . If the federal government does it, and you get a reactionary president, that's it for four years or eight years. Whereas, at the state level you spread your risk. However, there's going to have to be a federal presence, whether it's overlapping or unilateral.

Should the states have more legislative authority in the nuclear field?

NADER: I think they should have a veto power, yes. I think the Vermont statute [reserving the final decision on nuclear powerplants to the state legislature] is a good step in that direction. Because one of the most basic functions of state government is the public health and safety. To strip it by a [federal] preemption doctrine is an absurdity. However, legal arguments can be raised. The point is that the reason why the states have . . . the safety function, is because they're closer to the scene of the massacre, or they're closer to the scene of the disease.

When the existing state commissions are confronted with a request from a company to build a nuclear powerplant, they are basically being asked to decide on whether or not they should interfere in the management of the company. Which is quite different from setting the rates, which has been their traditional role. Do you think that the state governments are properly, should properly get into these kinds of management decisions?

NADER: Yes.

And should they decide on economic grounds whether or not a company should build a nuclear powerplant?

NADER: Most definitely. They're being asked that now, in reverse. They're being asked to permit automatic pass-throughs [to consumers of costs of] construction work in progress, automatic pass-throughs of shutdown costs, so they're being asked, see, they're being asked to make those decisions in favor of the utilities, and it's quite clear that it's got to work both ways.

Would it not be very much easier if the state government were to make management decisions, if they owned the company they were trying to manage?

NADER: Not necessarily. The whole quality of state regulation depends on the quality of involvement by the consumers. I hasten to point out that the American Public Power Association is as gung-ho for nuclear power as is the Edison Institute.

I mean that, just in terms of the difficulty of imposing their decisions on the company, if a regulatory commission in Missouri wants to stop a company from building a plant, they have to fight that decision through the courts, by and large, and through hearings and appeals for a period of years. Whereas, if they owned the damned thing, they could simply decide which way to go. Isn't that an important difference?

NADER: Yes, it is, except if they decide to go the wrong way, then who is going to fight the government? If the utility goes the wrong way, the government is at least in a position to fight the utility. But if the government goes the wrong way, who is going to fight the government, under our present system?

CHAPTER FORTY-FIVE

Propaganda

SAMUEL INSULL may have been right about the importance of state regula- tion in maintaining the private ownership of power companies. But he was wrong to say that it would be sufficient. Many power companies were already publicly owned by municipal governments; others would come to be owned by states and by the federal government. Today there are about three thousand publicly owned power systems and only four hundred privately owned com- panies. About two-thirds of the publicly owned systems are municipals—small distribution systems serving small towns; most of the remainder are rural electric cooperatives; and there are forty federally owned systems. Nevertheless, about seventy-five percent of all power-generating capacity in the nation is privately owned, and eighty percent of all customers are served by private companies. Many of the publicly owned systems are merely distributors of power purchased from private companies.

In terms of generating capacity, a quarter of the power industry is publicly owned—a percentage that has not changed greatly in recent decades. The Los Angeles power system is owned by the city government; there is no private power at all in the state of Nebraska; since the 1930s all of the state's power production has been state-owned. The largest power-generating system in the country, the Tennessee Valley Authority, is owned by the federal government. In short, the municipal ownership which was widespread at the outset of the power business remains a significant factor, and the pressure for increased public ownership has continued during the twentieth century.

The power industry has reacted in a variety of ways to this pressure. Few, perhaps, saw the importance of the question as early as did Samuel Insull or were as liberal in their response. In general, the industry reacted with simple and total opposition to government interference in any form. Public ownership was viewed and characterized as the beginning of socialism in America, and it was fought with corresponding vigor. In discussing municipal ownership in 1891, the president of the National Electric Light Association, C.R. Huntley, told a conference of his association, "I believe the most conclusive answer we can make to the sophistic arguments of an ill-disguised socialism, presenting itself in this municipal-ownership scheme, is to give the very best possible service at the lowest rates compatible with a fair profit."

But by 1894, language at the trade association's annual meeting had become somewhat stronger: "This unholy, this infamous, communistic warfare on these enterprises—not alone upon us, but upon kindred enterprises—is something that is distinctly antagonistic to the principles of our government, to the principles of our institutions, and to the principles of fair-minded business equity; and I tell you whenever such principles are suffered to gain a foothold in any part of this continent, whether it affects a small plant in some village or town or a big company in a large city, it strikes at the interests of all alike."

Despite the strength of language being used, there was no suggestion, at this stage, that the power companies should join in any concerted opposition to public ownership; speaker after speaker at the trade association annual meetings, and one trade publication after another from this period, insisted that the superior efficiency of private companies would quickly defeat the corruption and inefficiency of municipal governments; that the lower rates and more reliable service claimed by private companies would triumph.

But for whatever reasons, municipal ownership did not disappear; rather, it enlarged and multiplied. General Electric and Westinghouse were still selling generators to municipal governments to create a market for their products, and the municipalities were competing directly in many cases with private companies. Henry L. Doherty, who would later organize one of the great consortia which came to dominate the power business, seems to have been the first to suggest that a concerted public education campaign should be the industry's response to this menace. In 1902, he told the National Electric Light Association: "Ninety percent of all the municipal plants that have been installed have proven rank failures. Many of them have been abandoned, and many others of them would be disposed of if the taxpayers really knew the results obtained in these plants. It is idle to suppose that a municipality can ever operate an electric plant as economically as can private capital . . . [W]e must meet the municipal ownership fallacy with hard facts and not with wind. Our opponents generally deal only in wind, but pose as friends of the people. If we deal only in wind we are at a disadvantage, for we must necessarily appear as foes to the people. To meet both kinds of competition described above [municipal take-overs and municipal competition] we must educate the public."

A policy espousing state regulation, as Insull had urged, was adopted by the trade association in 1909, and, by the First World War, many states had followed the lead of Wisconsin and New York in establishing regulatory commissions, agencies which still have almost exclusive jurisdiction over the electric-power companies. But this was clearly not sufficient to halt the pressure for more public ownership and control. Efforts to sway public opinion were evidently needed, as Doherty maintained. Samuel Insull became a convert to this new cause.

From 1914 to the conclusion of the First World War, Samuel Insull, a former citizen of Great Britain, devoted most of his efforts to mobilizing public sentiment in favor of American entry into and prosecution of the war; at the war's conclusion, he turned the propaganda machinery which he had built over to the electric-power industry's trade associations, which by now were completely persuaded, as was Insull himself, of the correctness of Doherty's proposal. The public must be educated. The result was the greatest public-relations campaign the United States had ever seen. So enormous was the effort that when its extent was revealed, the public reacted with anger. By the time the Depression struck, public revulsion against the utilities' propaganda program was so great that the power and gas companies became the scapegoats of Congress. The massive public-relations campaign on behalf of free enterprise was itself a strong stimulus for greater public control.

In 1928, the Federal Trade Commission began investigating the utility conglomerations which Insull, Doherty and others were erecting. Its investigations continued over a period of eight years, and its reports filled eighty-four volumes, twenty of which were devoted to the propaganda campaign in behalf of private power. The summary of this portion of the investigation, despite its evident effort to affix blame, is a useful and concise description. Much of the FTC's evidence was later used in an unsuccessful effort to convict Samuel Insull, his younger brother Martin, who managed some of Insull's companies, and other utility executives of fraud:

"The record in this investigation establishes conclusively that the electric and gas utilities, since about 1919, have carried on an aggressive country-wide propaganda campaign. In it they have made use not only of their own agencies, but have enlisted outside organizations in active, and often secret, aid. In it they have literally employed all forms of publicity except 'sky writing,' and frequently engaged in efforts to block full expression of opposing views. The record shows that this propaganda had for its objective the disparagement of all forms of public ownership and operation of utilities and the preachment of the economy, sufficiency and general excellence of the privately owned utilities. The record establishes that, measured by quantity, extent, and cost, this was probably the greatest peacetime propaganda campaign ever conducted by private interests in this country. The record establishes that the activities were carefully considered and planned by responsible heads of the industries. Numerous declarations and excerpts from minutes and committee reports show clearly that the character and objective of these activities were fully recognized by the sponsors and planners, and the director of National Electric Light Association, the leading propaganda organization, boasted that the 'public pays' the expense. Often methods of indirect approach were employed and injunctions of secrecy given. All of these facts are established by the records of the utilities themselves. They are drawn, not from adverse or conflicting testimony, but from their own documents and declarations . . . .

"The record establishes that these propaganda activities were carried on chiefly through the National Electric Light Association, the national association of the electric-light industry, comprising in its membership over eighty percent of the industry, and by the American Gas Association, the national association of the gas industry, which comprised over ninety percent of that industry. The National Electric Light Association divided the country into twelve geographic divisions and had one such division in Canada. It also maintained State associations in many states, in some instances several States being grouped into one association. Both geographic divisions and State associations had set-ups under the constitution and bylaws similar to the national association. Their particular function was to carry out locally the work nationally planned.

"In addition to such geographic divisions and State associations, there were also set up the so-called 'State committees' or 'bureaus on public-utility information,' which were, in fact, set up for and solely devoted to propaganda. At one time there were twenty-eight of these, covering thirty-six of the most populous States. . . .

"Acting upon the theory that the two greatest public-opinion-forming agencies of the present and future generations are the press and schools, the utilities, in planning their publicity and propaganda activities, gave most consideration to contacting and exploiting these two agencies. Officials and State directors were selected for their experience, acquaintance, and ability to make contact with newspaper men and school men.

"As to the press, the publicity obtained ran the gamut from harmless utility news items to reprinting of 'canned' editorials supplied by the utility writers.

"In addition to the publicity obtained through the press generally, the utilities carried on propaganda through a a number of subsidized agencies and also sent out large quantities direct. Full advantage was taken of the good will induced by advertising expenditures. In a number of instances newspapers, or a controlling interest therein, were acquired.

"The influence with school men was obtained in numerous ways. Some were invited to speak at utility meetings, others were engaged for vacation jobs, often at no small remuneration. Others were invited to sit in on committees with utility men to plan courses for utility studies. Others were paid to make studies or to write articles. Direct money payments, some quite large, were made to many educational institutions, including several of the leading universities. Surveys were made of textbooks in use and pressure brought to bear on textbook publishers to eliminate matter deemed unfair or prejudicial by the utilities. The statement was frankly made that the pressure was first to be brought on the largest publisher for the effect it would have on the smaller ones.

"The organization chart of the National Electric Light Association shows that it also had various committees for contact and cooperation with other industries and with many associations. In this way agencies such as the United States Chamber of Commerce, Kiwanis, Rotary, Lions Club, women's clubs, and even at times the church and clergy, were utilized to aid the utility program.

"Repeated attacks were made upon every outstanding public project, whether in existence or contemplated. The Ontario hydroelectric system, which was the largest and most outstanding, drew much fire. In fact, a series of books was printed to disparage it. The accuracy and fairness of each publication was challenged by the Ontario Hydro authorities but, regardless of that fact, each succeeding book quoted the prior challenged works as though they were accepted authorities. Much and varied propaganda was also directed against the proposed Muscle Shoals and Boulder Dam government projects.

"A favorite method of attack was, as indicated in a memorandum by the assistant director of the Illinois committee, not to meet the public-ownership argument but to 'pin the red label' on their proponents. Thus advocates of the right of the people to own and operate their own public utilities were labeled as 'Bolsheviks,' or 'reds,' or 'parlor pinks.'

"During the period of these activities, it has been common knowledge that in only a few instances was there any approach to effective State regulation. Yet the utilities proclaimed the sufficiency and complete effectiveness of such regulation as a foil to any further federal or local regulation or to any form of public ownership and operation. In many states the utilities engaged in direct political activities against any project of a public nature and in favor of men and measures agreeable to the privately owned utility program.

"So, while engaged in all of these activities to disparage public or municipal ownership and operation, the utilities pursued their ultimate objective of creating a halo around all their practices, including financing.

"Thus was attempted to be created a firm belief by the general public in the soundness and value of all security issues of privately owned utilities. How fully this was attained appears from the harvest that came from sales of their security issues in the years 1928 and 1929, as shown in the various financial reports and studies now in the record of this investigation. It is true that these were boom years in all stocks. But, relatively speaking, no other industry put out such a quantity and a variety of issues as did many privately owned electric and gas utilities and their holding companies. Additional super–holding companies were also created which issued securities with little or no regard for the underlying soundness of or necessity for such issues—in a field devoted to public use and supposedly regulated as local monopolies. . . .

"Much of the selling . . . was accomplished by actual door-to-door peddling by employees of the subsidiary operating companies. In other instances customer ownership and other high-pressure, local sales campaigns were planned and carried out. They called to mind real-estate development and sales methods to be employed in public utilities with their dedication to public use and their supposed limitations to their necessities, and regulation as to charges and returns. Besides these, there were also the 'baby bonds,' which in some instances amounted to trading a bond for a ten-dollar bill . . . .

"In all this, the years of propaganda activity undoubtedly proved a powerful aid in having made the general public [to be] utility conscious. The facts as to such issues and sales will appear in the report under the financial phases of the investigation.

"To measure what the investing public, as distinguished from the speculators, lost on these issues is impossible. First of all, no one has assembled, and it is likely that no one can assemble, the varying prices and amounts paid. In any attempt to measure the loss it would be impossible to segregate losses due to the depression in general, and likewise it would be impossible to say how much these campaigns for utility stock sales had to do with the unsound rise of the whole investment structure which finally collapsed. However, it would probably not be overstating to say that the losses to investors in utilities securities attributable to utilities campaigns for selling and the increases in price effected thereby were in the billions. This would seem to be true because the activities of the utilities in these campaigns must have had a powerful effect on the entire upward price trend, and because in those instances alone where the entire holding group or corporate structure crashed to bankruptcy, the losses ran into hundreds of millions."

CHAPTER FORTY-SIX

Informing the Public

THE ELECTRIC-POWER INDUSTRY finds it difficult to take seriously the charges made by critics of nuclear power, and so it is not surprising that officials are tempted to find ulterior motives in the opposition they face. This temptation is much strengthened by the industry's history of red-baiting and its fears of public ownership. A fair segment of the industry's management probably would agree privately with the sentiments of the author of this letter, a vice-president of Portland General Electric, replying to the question of a customer concerning the debate over nuclear powerplants:

> "I really wish I knew what the reasons are for groups . . . to come out with overstretched, half-truth statements [regarding nuclear powerplants] as they do and posing as factual. Sometimes— and this is my own personal opinion—[I think] that they belong to a different political philosophy than Americans do and are doing things like preventing needed electrical energy [in order] to destroy our country. If I were on the other side, I'd follow their footprints of destruction exactly. First, I'd get all our kids to use drugs and dress like tramps . . . then I'd start a campaign to convince the populace that nuclear power is a killer. With youth and dwindling energy resources, we'd be ripe for destruction."

There are also private accusations that critics of nuclear power have been subsidized by the coal industry. The National Coal Association takes the public position that nuclear power will be severely limited by shortages of uranium but has refrained from any substantive public criticism of the industry. The United Mine Workers, which represents most coal miners, for a time was more explicit and virulent in its opposition to nuclear power, an opposition which severely damaged the union: its District 50, a conglomerate of industrial unions which had, in turn, organized some atomic workers, split off from the parent union, severing the Mine Workers from its only prosperous and growing segment. (The seceding unions joined with what became the Oil, Chemical and Atomic Workers Union, which now represents the uranium miners of New Mexico.) In the late 1960s the corruption of the Mine Workers union was widely rumored (its leader, Anthony Boyle, later was convicted of the murder of Joseph Jablonski and his family), and it was not difficult to imagine the union conducting secret campaigns of opposition. There is very little evidence, however, that any criticism of nuclear power was inspired by the Mine Workers or any other interest group, and in recent years, under the spotless regime of Arnold Miller, there is no longer any reason to think that the union would be pursuing its former self-destructive course.

Of all that has been said and written in criticism of nuclear power, only a single, privately-printed pamphlet by a physicist was secretly subsidized by the coal industry, via the National Coal Policy Conference, a trade organization which includes the Mine Workers union, the coal operators, the railroads and the utilities. In that case, the secrecy was probably employed to circumvent the power-company members of the organization as much as to deceive the public. The pamphlet bears no indication of its sponsor, but is otherwise an unobjectionable piece of literature written by a competent scientist.

Against these feeble efforts, the power industry, the nuclear-equipment manufacturers and the federal government have campaigned vigorously for twenty years to secure public acceptance of nuclear power. It is hard to avoid the impression that, as in the case of Con Ed's efforts to secure approval of a nuclear powerplant in downtown New York City, the efforts to persuade have themselves summoned up much of the opposition. For better or worse, however, the electric-power industry continues the elaborate public-relations efforts of the kind described almost fifty years ago by the Federal Trade Commission's investigations. The Edison Electric Institute, operating through a number of subsidiaries, including Reddy Kilowatt, Inc., still distributes copious numbers of comic books, to which member companies can add their names in appropriate blanks before donating them to local schools, where they can be given to students. Current numbers, with drawings in the style of the comic books of the 1940s, are "Our Spaceship Earth—Needs More Fuel!," "The Atom, Electricity, and You!" and, "The Story of Electricity," all of them rather heavy-handed pieces of propaganda. Another EEI subsidiary distributes a book proclaiming the virtues of nuclear energy, written by Ralph E. Lapp, a widely-respected physicist and independent advocate of nuclear energy, but which is copyrighted by Fact Systems, "a public education division of Reddy Kilowatt, Inc."

Donham Crawford, president of EEI, argues that its publicity campaign on behalf of nuclear power is in the public interest:

What problems do you see in the nuclear-power program, if any?

CRAWFORD: Well, I see a number of problems in the nuclear-power program. Let me start off by saying that our very strong feeling is that there is no way that the power requirements of the nation can be met without nuclear power. We feel that electric energy based on coal and nuclear must be the way that we can provide the power and evergy requirements of the future, thus decreasing our dependence on oil and natural gas. So, you can't get from here to there without using nuclear power. Now, the question is, What are the impediments, what are the hurdles you have to get over to be able to bring nuclear power in? I think there are several problems. One problem, certainly, appears to be public acceptance. This is not true everywhere, but there certainly are areas of the country where there is significant resistance to nuclear power. Maybe not in numbers of people, but certainly in the extent of the opposition, the volubility of the opposition that you see in the press, and that sort of thing. So, that is a problem, to get the public generally to come to a point of greater acceptance of nuclear power.

How do you do that? Well, I, for one, feel that the government has been remiss in its own efforts in this regard, in terms of educating the public about the merits of nuclear power. For many years, it seems to me that the government has been very reluctant to speak out in defense of nuclear power, to educate the public about it. We have a much lesser effort, it seems to me, in this country than in other countries where nuclear power is coming along.

You mean a smaller public-education effort? Or a smaller nuclear program?

CRAWFORD: No, a smaller public-education effort. I don't mean a smaller nuclear program. It just seems to me that the AEC for many, many years was very timid about speaking out and educating the public about nuclear power, and I think a lot more needs to be done in that area. At the present time, because of the change in the AEC structure, and the changes at the Joint [Congressional] Committee [on Atomic Energy], there's really nobody in the government defending nuclear power. The question is, Is nuclear power in the national interest, or isn't it? Of course, we feel very strongly that it is. And if it is, and if the government concludes that it is, then it seems to me that the government has a responsibility to try to further the acceptance of nuclear power. On the other hand, I don't see very much of that at all. Therefore, it has been necessary for industry to try to fill this gap and to try to provide the leadership and the educational and information efforts itself. We try to do some of that, the Atomic Industrial Forum tries to do some of that, and a lot more of that needs to be done, it seems to me.

So, that's one general problem area, the question of public education and public acceptance of nuclear power. There are other problem areas that are brought up from time to time, and I guess they all really revolve around this question of public acceptance. One would be the safety aspects of nuclear power. We think there is a very good story to tell there. Not a very good job is being done of telling it, but this is a matter that is of concern to people, and I think it's understandable that it is. We feel quite confident about the safety of nuclear power; otherwise, we would not be building the plants. But this is an issue that has to be resolved in the public mind, that there is *not* danger to the public from building and operating nuclear plants. The question of radiation emissions, here again we think there is a very compelling story to be told in favor of nuclear. It is not really being told, but it's a question that has to be resolved in the public mind.

One of the real problems that we have is the length of time it requires to get a nuclear plant licensed and constructed, and here we think there are many things that could be done to expedite the administrative procedures. That's a real problem. I think that the government has its heart in the right place in trying to reduce the amount of time required from ten or eleven years to a lesser amount. But despite their good intentions, very low results are being shown.

You'd say licensing delay is a "real" problem in contrast to the public-acceptance problem?

CRAWFORD: No, I'd say it's a more, oh, *concrete* problem. It is easier to get your hands around than the broad amorphous problem of public acceptance.

I think you have answered this question already, but let me ask you specifically: do you think that the critics of nuclear power are well informed?

CRAWFORD: No, I really don't. I think that the leading critics of nuclear power have been led down the primrose path by opponents of nuclear power. I think Mr. Nader, who leads this group, is not well informed about it; I think he is on the wrong issue. I think that what he is doing is a disservice to the country, because nuclear power must be a way to provide our power requirements. He and his disciples are making it more difficult to arrive at that position, and while I don't want to be broadly critical of him, because in some areas I think he has done some good things, I think he is wrong about this one, and that he ought to change his position on it.

You say some of the leading critics have been led down the primrose path, is that just a way of saying they are mistaken or is someone doing the leading?

CRAWFORD: I won't be specific on that, but there are well-known people who have their pet theories about explosions and effects of radiation, that for years have been going around the country making a living out of making speeches along this line, and I think that they have had their sway with Mr. Nader and he has accepted their arguments, and he has ignored arguments on the other side.

Do you feel that the other scientists who have spoken out against nuclear power are similarly mistaken?

CRAWFORD: No, I am not going to say that there is no room for argument here, because obviously there is. But it seems to me that there are some of these leading scientific critics of nuclear power who are making a crusade of this thing, who I think really are not—I won't say that they are not informed, but certainly I would disagree with their conclusions.

The Edison Electric Institute distributes a public-relations handbook to member companies which strongly urges each company to undertake a local public-relations campaign and which offers various publications and services to assist the member companies. All of this activity, of course, is paid for by the customers of the power companies, since dues to Edison Electric Institute are allowed as an operating expense in most states (advertising costs are increasingly being disallowed); the rate-payers are therefore subsidizing the efforts to persuade them of the correctness of the power companies' activities—an aspect of these campaigns which created the public furor of the 1920s and 1930s.

The public-relations manual distributed by EEI claims that two out of three people in a survey "didn't even understand the terms 'private ownership' and 'public ownership'" and therefore recommends that power companies work to extirpate these terms from the English language. The manual suggests that for "private ownership" the companies substitute one of the "terms that people understand" given in a lengthy table of terms relating to public ownership. A part of the table is the following:

WHEN REFERRING TO AN ELECTRIC COMPANY:
People *Do* Understand These Terms:
Invester-owned electric utility company
Shareholder electric company
Investor-owned, taxpaying electric company

People *Do Not* Understand These Terms:
  Private utility
  Private company
WHEN DESCRIBING POWER SUPPLY BY GOVERNMENTAL
BODIES AND AGENCIES IN GENERAL, AND
PARTICULARLY FEDERAL GOVERNMENT PROJECTS:
People *Do* Understand These Terms:
  Government-owned electric power
  Government power project
  Government power
  Government operation
  Government power program
People *Do Not* Understand These Terms:
  Public power
  Publicly owned electric power
  Publicly owned plant
  Publicly owned electric industry

The campaign to obliterate the semantic differences between public and private ownership has been very successful. Even Senator Lee Metcalf and Vic Reinemer, in their bitter attack on the industry, *Overcharge*, which documents the more extreme public-relations efforts of EEI in recent years, conscientiously use the "correct" term—"investor-owned utilities"—instead of saying, "privately owned power companies."

The emphasis of the public-relations manual on the terminology of public and private ownership reveals the industry's continuing preoccupation with this issue. And the managers of the power business generally still seem to view themselves as locked in mortal struggle with some sort of conspiracy. One power-company vice-president agreed to talk, off the record, about management attitudes.

The power business has a reputation—its management has a reputation for being somewhat backward, somewhat bigoted.

EXECUTIVE: Backward, bigoted—yes, I would say so.

For instance, are there any women in executive positions in your company?

EXECUTIVE: No. No women executives, and no women engineers. There are no applicants. There are no black executives or black engineers, either, because there are no applicants. We would hire them if there were any candidates. About seven percent of all the employees are black, and we would hire more if we could.

Management also has a reputation of, frankly, being somewhat anti-Semitic. Even Samuel Insull's biographer admits that he was anti-Semitic.

EXECUTIVE: Sure, just look at the low percentage of management who are Jews. Why? I don't think it has anything to do with Insull, he was no different, probably, than other businessmen of the time, in the twenties. But there is something self-fulfilling about this. Look at the intervenors [in nuclear licensing hearings]. There are three independent groups intervening [in a particular plant's licensing] and each of them is headed by a (*giving an apparently Jewish name and shrugging*). They aren't related to each other. You find that all the time.

CHAPTER FORTY-SEVEN

Ideals

THE FIGHT FOR PRIVATE ownership of the power business during the 1930s and 1940s is too recent to have been forgotten, but it is easy, even for those who lived through that period, to forget how and why the fight seemed important. Electricity was not just another product; it was a new age. For Left and Right, rich and poor, electricity was the answer to social problems. Lenin was supposed to have said, "Communism is socialism plus electricity"; the Soviet Union probably took more pride in and devoted more effort to electrification than any other peaceful effort. The United States boasted of and planned the benefits which power would bring. Here is how it looked in 1936:

"In all the range of problems with which President Roosevelt is called upon to deal there is probably none with which he is more concerned, or to which he gives closer study, than those relating to electric power.

"The President's thinking goes first to government—democratic government—and after that to economics. As it relates to power and what it can do for the nation, his thinking runs far beyond electrical conveniences and release from drudgery. It is concerned primarily with democracy, and then secondarily with economics as the means of sustaining and building up democracy.

"Any utility man can tell you that this is totally haywire. Generating and selling electricity is a business, and the supplier either follows the rules and makes a lot of money, or takes chances and very likely goes bankrupt. However, remember that many things are upside down nowadays, as proved by the fact that in the nineteen twenties they were in just the reverse of their present position. And remember, too, that electricity is a mysterious force which already has worked a number of miracles that even hard-headed money-makers feel it safe to talk about. Then consider the thought, attributed to the President by his close associates, that electricity can do much to build up and revitalize democracy . . . .

"The fundamental problem of democracy, as the President sees it, is to raise the level of the physical well-being of men, and with it their intellectual level. This done, democracy would be able to cope with the new conditions and the new problems that the onrush of modern science, coupled with an intense but disorderly industrial development, and the concentration and irresponsibility of financial power, have laid upon our doorsteps.

"Like a great many other Americans, the President is not satisfied with the nation as he finds it. He sees overproduction and underproduction generating the same evils of want and of joblessness in a world wherein science constantly makes more riches available. He sees millions of people, in metropolitan slums, in small cities and even on farms, denied education and opportunity, living in homes that are primitive and that breed disease. They are deprived of the greater part of the necessities of life. He sees tens of thousands of tenant farmers, existing without either security or substantial hope, and other millions of people trying to live on incomes so meager that the pall of family disaster hangs over them day by day. It is decidedly significant that Roosevelt is discussing and attacking these conditions instead of becoming a cheer leader for another boom pretty certain to be followed by another bust.

"In the prodigies of effort he put forth to lead the country out of the bogs of depression he therefore sought, and seeks still, more than what he has termed 'a purposeless whirring of machinery.' It is important that every man have a job, that every factory have orders to fill and that business as a whole earn profits. 'But,' as he said in his annual message to Congress in January, 1937, 'government in a democratic nation does not exist solely, or even primarily, for that purpose.' The factory wheels 'must carry us in the direction of a greater satisfaction in life for the average man. The deeper purpose of democratic government is to assist as many of its citizens as possible—especially those who need it most—to improve their conditions of life, to retain all personal liberty which does not adversely affect their neighbors, and to pursue the happiness which comes with security and an opportunity for recreation and culture.'

"This growth of well-being and security, insofar as it is attained, will make men more tolerant toward their fellows, and so will make for more of kindliness and of social peace. It will help to increase the number of 'men of good will' among whom can be achieved the 'era of good feeling' toward which the President says we are moving. Well-being and security will also make for more intelligent and more capable citizens. Unless our political and economic processes can produce such men, and produce them not alone for leaders but for intelligent followers and participants, democracy will be unable to cope with the host of infinitely complex problems that modern science, which has already outstripped economics and politics, has raised up for us. There is a need, the President said in his inaugural address of 1937, for moral controls 'to make science a useful servant instead of a ruthless master of mankind.'

"Accordingly, and considering now the ultimate rather than the immediate aspiration, it is safe to say that Roosevelt's ideal would be a nation, or perhaps a world, of people sufficiently well equipped physically and mentally and morally to endow them with a sanity of outlook and enjoying enough leisure to enable them to study and to think.

"To suit the President's standards and ideals best, such a people would have to live close to nature—to live on the land. They might constitute a social group somewhat like the ancient Greeks, but with myriad economic advantages which for that people were inconceivable.

"Even a tentative and halting approach to such an ideal requires some animating, vital economic force, serving as a social catalyst, or even a social prime mover. *And that is where electric power comes in* [Italics in original]. Roosevelt sees it as the answer, perhaps the God-given answer, to this problem of making men healthier and happier and especially of making them more fit for the democracy which must try somehow to keep up with science. Consequently power seems to the President as essential to the new civilization as, one might say, the light and heat of the sun. It may slowly—remember that this is an ultimate aspiration— transform men and make them intelligent rulers of their own political destiny.

"Hopelessly visionary? Yes, but provided again that one is convinced that generating and distributing electricity is a business, and must always be that, and always be nothing more than a business. If one believes that, unless an investment is recovered with a profit—except when bad judgment, social change, gambling, lust for empire, hard times or some other misadventure intervenes—the undertaking is a failure.

"But if the business concept is subordinated, it is not hard to see at least a little of the bricks and mortar of the President's conception. That electric power can transform industry, making it vastly more productive with a phenomenal lessening of human labor, no man can dispute. It has transformed a great part of industry already, electric motors replacing the tangles of belts and cogwheels in more than three-fourths of American factories. More than the beginnings of social transformation in cities are an accomplished fact. Devices which less than a century ago would have been thought miraculous have become so commonplace that they are never realized dramatically until the power goes off, as it did in New York City within recent memory, and quick communication ceases, horizontal and vertical transportation comes to a standstill, lights flicker out and almost all the other normal features of big-city life are abruptly reversed. The low-income groups are beginning to benefit. PWA [the Public Works Administration] is providing electric cooking and refrigeration in slum clearance and low-rent housing projects in many cities. Private manufacturers make the equipment. Private utilities supply the energy in wholesale quantities at about one cent a kilowatt-hour. EHFA [the Emergency Housing Finance Administration] finances similar equipment for other homes, mostly served by private utilities.

"Outside the metropolitan centers, cheap power is working its varied miracles. Mason City, Washington, for example, designed for eight thousand people identified with the building of Grand Coulee for the Federal Government, is a town without chimneys. All its heating is done with electricity—bought by the construction firm from a private utility. The rate charged the consumers, by the way, is three mills [0.3 cents] per kilowatt-hour. Washington State College is using the town as a field laboratory for studying the practicability of electric house heating. Down the coast in Seattle, Ross's City Light is working on an electric heating machine which it hopes to place in the reach of the average home owner. This device [a heat pump] is a reversed refrigerator, collecting heat from the outdoor air, however cold, and distributing it through the house.

"These things are just a few of the obvious manifestations of electricity's place in the modern scheme of things. They take no account of the birth of great new electrochemical industries, foreseen by authorities as a consequence of abundant cheap power, of the coaxial cable carrying two hundred and forty simultaneous two-way telephone conversations, of air conditioning in all seasons, or of the robots with electrical 'brains.' To all these must be added the infinitudes of the photoelectric cell or 'electric eye,' which sorts eggs, turns on lights in the henhouse or city street when daylight dims, counts cans or automobile radiators on a swiftly moving conveyor belt, grades coffee beans, and starts the auxiliary ventilating fans of the Holland tunnel when the smoke gets too thick. Then there are television, possibly the transmission of power without wires, and a host of other present and prospective developments and staggering potentialities that have emerged recently from the laboratories.

"The future of electricity is unclear, but that is in no small measure due to the very fact that its promise is so dazzling. About its revolutionary potentialities there is not even any noteworthy disagreement. Edison declared shortly before his death that he was even more positive than he was forty years earlier that 'the electrical development of America has only well begun.' The Socialist leader, Norman Thomas, sees in electricity 'the slave on which mankind must depend to conquer poverty.' And Herbert Hoover, with a difference merely in his angle of approach, exclaims that 'mankind has never before grasped such a tool.' A Federal Power Commission of the Hoover era styled the possibilities in electric power 'beyond human conception,' very much as Stuart Chase has declared that 'its future is limitless' . . . .

"Write off ninety percent of this as visionary if you like. Even then it would remain a fact that [there has been a] swift, surging increase [in production] of billions upon billions of kilowatt-hours of electricity in the last two years . . . .

"That is one reason why Roosevelt, undismayed by the wails of waste and staggering losses as federal workers wrestle with the mighty Columbia, tumbling out of the Canadian Rockies and snaking across the state of Washington ready to be put to work producing the greatest supply of power in the country, bids the engineers hurry up and confronts the prophets of doom with a grin. Another reason for his firm confidence, of course, is that vision of his of a strengthened and more vitalized democracy.

"In agriculture as in industry and urban living, new forces and new standards of life are in evidence or in sight. They will manifest themselves increasingly as rural lines are built, appliances are perfected, and costs are slashed radically. And for Roosevelt, a countryman, these changes in rural life and the far greater changes that appear feasible, are profoundly and tremendously important. Roosevelt has a conviction, dearly held, perhaps rooted in the instincts of a man who comes from the land but anyhow held with all the tenacity of a true believer, that much of humanity will be emptied out of the great cities into the country, and that there it will be immeasurably better off.

"The President believes that, as Dr. O.E. Baker of the Department of Agriculture says, 'the land is the foundation of the family, and the family is the foundation of the State.' He would like to see, adapted to American social patterns, 'a continuity of family proprietorship in farming,' as China has known it for forty centuries and Germany and some of the other European countries for more than two thousand years. But the President would have many of the industrial and the white-collar workers live on the land, too.

"For this to come about it is necessary to decentralize industry, to undo the work that Watt did with his condensing steam engine which chiefly brought about the first industrial revolution. It has been said that Watt and his engine did more to change civilization than Caesar, Napoleon and Charlemagne. It has also been said that Watt was socially blind, like hundreds of innovators before him and after him. Certainly many of the social by-products of his innovation could not have been intended, and Roosevelt is one of those who most wholeheartedly deplores them.

"Decentralization, if and when it takes place, will break up the great conglomerations of people in sprawling, dirty, noisy slums and brutalizing sweatshops, lift the swarming hordes out of tenements and subways and street-cars, and put them back on the land. Electric power, a vast unseen ocean of electric power that will run factory machines, light the countryside and bring relief from drudgery to the homes on the land, is for Roosevelt the seemingly certain instrument of this decentralization. It will transport people to places where they can work naturally, live decently, breathe deeply and see the open sky. No child will, as did one in

New York's tenement districts a few years ago, draw a picture of the sky as a flat covering over the earth like the lid of a kettle.

"Something of the President's vision and aspiration in this direction may be found in his address to the World Power Conference in 1936, in which he recalled how 'workers had to go to the steam engine, whose energy could not be divided into parts and sent out to them.' He said then: 'Now we have electric energy which can be and often is produced in places away from where fabrication of usable goods is carried on. But by habit we continue to carry this flexible energy in great blocks into the same great factories, and continue to carry on our production there. Sheer inertia has caused us to neglect formulating a public policy that would promote opportunity for people to take advantage of the flexibility of electric energy; that would send it out wherever and whenever wanted at the lowest possible cost. We are continuing the forms of overcentralization of industry caused by the characteristics of the steam engine, long after we have had technically available a form of energy which should promote decentralization of industry . . . . It is not irrational to believe that in our command over electric energy a corresponding industrial and social revolution is potential; that it may already be under way without our perceiving it' . . . .

"To effect, or even to achieve in creditable part, such economic and social transformations as are here envisioned, electric power must be abundant, cheap and widely distributed. To assert that it is already, in the face of its niggardly rationing to most householders at from four to twelve cents a kilowatt-hour, is, in Herbert Agar's words, to 'dishonor the dream by saying that it has been realized, that it lies all about us today.' There must be a materialization, or perhaps only an approximation, of the wizard Steinmetz's vision of electricity supplied so cheaply that it would not pay to meter it, and so plentiful that men would work only two hundred four-hour days in the year.

"Roosevelt is determined to make power almost infinitely cheap, as well as all but infinitely abundant. Those who have talked with him most about electricity say that he would like, if he could, to give it away, as water is given away to all who come with their jugs at the public fountains of Spanish squares. To attempt to inflate and pyramid profits in the production, transportation and delivery of such an element, and to attempt then to legalize and even sanctify the profiteering, so that the nation is helpless to curb it, is, for the true believer in electricity as the agent of thoroughgoing social transformation, almost on a par with the old custom in some European countries, of taxing sunlight and air by levying an impost upon windows."

This dream of the 1930s—the extraordinary social changes which would be brought about by electricity "too cheap to meter"—persisted almost to the present day. Nuclear power, at the end of World War II, was widely believed to be the technical breakthrough which would make all this possible. With a kind of religious faith, many people, particularly those of the political Left, believed that electricity "too cheap to meter" would move mountains and raise up the most humble; that it would usher in a pastoral utopia without smoke or slums. But nuclear power, as the private power companies well understood, held no such promise. Nor, perhaps, would abundant, free electricity from any source have worked the kinds of changes which were hoped for. But the utopian quality of the visions of public-power advocates easily led to a feeling that enormous benefits were somehow being withheld by the power barons of private industry; when nuclear energy appeared, there was a concerted effort to wrest free these enormous benefits for public use. The private power companies, which quite plainly did not share this vision of utopia, saw only a plot to deprive them of their property.

CHAPTER FORTY-EIGHT

Initiative

CALIFORNIA is not like other places. It seems to have no history; Californians speak of five years ago as the distant past. In fact, of course, the state has a history as long as that of most parts of the United States, a history of native, Spanish, Russian, Mexican and American communities, but, perhaps because most present residents have recently arrived, and perhaps because earthquakes and fires have obliterated many physical monuments of the past, the past is not much in people's awareness.

A visitor can experience a kind of disorientation; nothing is fixed, not even the landscape. Religions are invented as they are needed, political parties have no fixed identity and fission or coalesce; dozens of newspapers are on sale everywhere, but none seem to give very much news of the outside world. Ordinary newspapers are jumbled together with the most extraordinary pornography. There is a fashionable prejudice that the proliferation of cults and parties is limited to the southern part of the state, but, in fact, cosmopolitan San Francisco is afflicted with quite as extensive a profusion of sects and causes. The one issue which seems to unite Californians, however, is their dislike of the local power companies. On a recent visit to San Francisco, the *Bay Guardian*—a newspaper whose lead story that day was "Lunaception: Can Pregnancy Be Prevented By Sleeping With A Light On Three Nights A Month?"—carried an attack on the large corporate stockholders of Pacific Gas & Electric [PG&E], the local power company; the *Berkeley Barb*, whose front page featured a medieval astrological chart and the headline "Psychic Politics," carried sixteen pages of ads for massage parlors ("complete satisfaction at your request") followed by an article charging that PG&E was keeping secret its attempts to seed clouds for rain to fill its reservoirs ("How PG&E Plays With The Weather"). The two conventional newspapers that day carried stories about efforts in the state legislature to force the utility to provide special low rates for poor customers (an effort which ultimately succeeded).

On June 8, 1976, the depth of Californians' antagonism toward their power companies would be tested, for on that date the state would vote on the nuclear-safeguards initiative placed on the ballot by petitions carrying more than three hundred thousand signatures. The initiative was a complex piece of legislation submitted directly to the voters. California's legislature had repeatedly refused to pass bills limiting nuclear development in the state; opponents of nuclear power therefore took advantage of the initiative process written into the constitution of California and many Western states during the years of Populist rebellion against the corrupt rule of the Eastern railroads. The voters of California would be given the chance to vote directly on the question of nuclear expansion.

The outcome of the vote was very much clouded by confusion over the initiative itself, however. Opinion polls showed that many of the people who would vote for the initiative had the mistaken idea that its passage would enhance nuclear power development; many opponents of the measure believed, incorrectly, that they were expressing opposition to nuclear power itself. The confusion was exacerbated by the ambiguous popular name, the "nuclear safeguards initiative," given to the measure by its supporters.

The lack of information about the nuclear initiative was not surprising. It was a long and complex legal document, and even its supporters had difficulty in describing its purpose and effects.

One aspect of the nuclear-safeguards initiative was that its operational effect, if it passed and was allowed to stand, would almost certainly be to shut down the nuclear-power industry in California within one year of passage. For this very reason, the initiative, if passed, would likely be held void by the courts as an unconstitutional infringement of federal authority, and therefore would have no practical effect at all. Despite the likelihood that federal courts would strike down the initiative if passed—as they have already struck down other attempts of state governments to regulate the safety of nuclear power, an area of commerce which the Congress has reserved to itself—a majority vote for the nuclear initiative would deal a strong blow against nuclear power by putting a majority of the citizenry on record as opposed to its continuance.

But the voters had rejected in 1972 an environmental initiative which explicitly would have halted the construction of nuclear powerplants, and opinion polls showed that the voters would vote in favor of nuclear power again, if asked to do so. The nuclear-safeguards initiative, therefore, was framed simply as a measure purportedly to make the industry more safe. While the measure is extremely complex, it has salient features. Nuclear powerplants would have to be phased out of operation within one year of passage unless federal limits on liability were removed or power companies voluntarily assumed unlimited liability for accidents; and the state legislature would have to certify, by a two-thirds vote, the safety of nuclear powerplants within five years of passage.

To understand the importance of the first provision, one has to examine the peculiar nature of hazards created by nuclear powerplants. An accident in such a plant could, in theory, do damage with no practical upper limit. Tens of thousands could be killed and many billions of dollars of property destroyed. While no one associated with the nuclear-power industry has ever conceded that such accidents are a serious possibility, they cannot be ruled out entirely; there is some, still unknown, chance of catastrophic accident. This was the position in the 1950s, when the federal government first began to press private power companies to take on nuclear powerplants. The companies replied that they could not build plants unless insurance were available. And the private insurance companies testified before Congress that, no matter how small the probabilities, risks as large as those shown to be possible in the nuclear industry were simply not insurable. No private company can undertake a risk which exceeds its ability to pay.

The solution to this dilemma was passage of the Price-Anderson Amendment to the Atomic Energy Act in 1957, providing for $500 million in insurance to be supplied by the federal government at nominal cost for every nuclear powerplant. The manufacturers and operators of nuclear powerplants—the utilities and their suppliers—would not be liable for any injuries to the public in excess of the federal insurance plus whatever insurance private companies could provide. There was an effective limitation on liability, therefore, of $560 million, since private insurance companies throughout the world, joined in two large pools, would provide only $60 million in insurance for any one reactor. (The total is still $560 but about $100 million is now provided by private insurance for any one plant.) No private company or person would be liable for damages to the public beyond that sum; and private insurance companies, all of whom participate in the nuclear-insurance pools, further covered themselves by excluding from all ordinary insurance policies, such as homeowners and life-insurance policies, damage caused by nuclear accidents. In the case of a catastrophic nuclear accident, it would be up to Congress to provide additional disaster relief.

The Price-Anderson Act was renewed for another ten years in 1975, two years before its expiration; it passed both Houses by substantial majorities and was signed into law by the president, despite intensive lobbying by opponents of nuclear power.

Power companies and nuclear manufacturers lobbied strongly for renewal of Price-Anderson, but they claim that the limitation on liability was no longer necessary to them. Donham Crawford told an interviewer that most power companies would be able to proceed without federal insurance and limited liability, as would the manufacturers of reactors, because in their judgment catastrophic accidents were simply too improbable to be considered. According to Crawford, however, small companies that supply parts—valves, meters, instruments and so on—would probably refuse to sell components of nuclear plants, since without limitations on liability their existence could be endangered by an accident which exceeded the private insurance available. The withdrawal of all small suppliers would bring the industry to a halt.

One need not take too seriously, perhaps, Crawford's assertion that private power companies are willing to proceed even without federal insurance and limitations on liability. Like Con Ed's willingness to build nuclear powerplants in Manhattan, this position is consistent with the industry's insistence on the safety of the plants, but is not a practical possibility. No private company can assume risks that exceed its assets without being challenged in courts and in legislatures; the outcome would be uncertain, at best. Company executives might even have to consider their own personal liability for injuries in case of an accident.

For all these reasons, it seems likely that abandonment of the Price-Anderson protection would mean a halt in the construction and operation of nuclear powerplants. And although Congress has strongly reaffirmed its support for this law, the California initiative would slowly phase out all existing plants unless the law were repealed or private companies voluntarily assumed unlimited liability.

Even if this were not the case, however, the initiative's requirement that the state legislature certify the effectiveness of a list of safety measures by a two-thirds vote is another direct challenge to the federal government. Neither is likely to be permitted to stand, a century and a half after John Marshall first declared that the federal government would stand between private corporations and such acts of the states. Nuclear safety was an aspect of commerce whose regulation Congress had explicitly reserved to itself, and a long, unchallenged line of Supreme Court decisions beginning with Marshall's had held the states powerless to act in areas preempted by federal action.

David Pesonen is a young attorney who helped to draft the text of the nuclear-

safeguards initiative. He is involved in litigation against PG&E (a libel suit resulted in a verdict against PG&E for nearly eight million dollars, but the verdict
was reversed and a new trial ordered in January, 1976). Pesonen is well known in
California for his leadership of an earlier and successful fight to stop the construction of a nuclear powerplant at Bodega Bay, north of San Francisco. He insists
that the initiative is not a ban and that nuclear powerplants, if suitably insured
and approved by the state legislature, could continue to operate in the state.
Pesonen, however, does not conceal his own opposition to nuclear power, which
he says stems from the earlier battle against a plant at Bodega Bay. In 1962,
Pesonen was a staff member of the Sierra Club. When PG&E announced plans to
build a powerplant on the beautiful shoreline spot north of San Francisco called
Bodega Bay, Pesonen, a biologist at the time, tried to persuade the Sierra Club to
oppose the plant. Lacking sufficient support, Pesonen failed and resigned from
the club, working at part-time jobs for the next several years as a laboratory
technician to support himself and his wife while he carried on a full-time fight to
preserve Bodega Bay. Midway in this struggle, Pesonen now claims, he under-
went a kind of religious experience on his way to an evening meeting in Sonoma
County: from fighting to preserve a patch of shoreline he was converted to
opposition to nuclear power.

"It was a beautiful evening, a touch of fog. I had a feeling of the enormousness
of what we were fighting; that it was antilife. I had an insight into the mentality
of it, I began to see it as the ultimate brutality, short of nuclear weapons."

After that "emotional turning point," Pesonen read everything he could get his
hands on pertaining to nuclear energy, and at a time when there was almost
nothing written for the layman on the subject other than the propaganda booklets
of the industry, Pesonen succeeded in gaining a firm grasp of the technical issues
involved in nuclear power. Through the organization he created, wide publicity
was given to the discovery of earthquake faults under the proposed site, which
was close to the San Andreas fault system that caused the great San Francisco
Earthquake of 1906. PG&E was ultimately forced to retract its application to
build a nuclear plant at Bodega Bay, and Pesonen allowed his organization to
dissolve and began attending law school. Now as an attorney, he has again taken
up the fight against nuclear power. Pesonen organized the umbrella group—
People for Proof—that sponsors the nuclear-safeguards initiative and claimed to
have raised all of its funds; he was responsible for bringing into this effort another
organization called Project Survival, which, with some affiliates provided both
the manpower and the funds for the campaign to get the initiative onto the ballot;
Project Survival continued to supply the backbone of support for the initiative
during the election campaign. More limited assistance was provided by the Sierra
Club, the Friends of the Earth, and other conservation groups.

The initiative was being opposed by a coalition of trade unions, primarily
construction trades, and nuclear-power companies, primarily PG&E and
Bechtel, the architect-engineering firm that built many plants throughout the
country. This coalition, working through an organization called Citizens for Jobs
and Energy, secured overwhelming support from organized political and union
groups. Project Survival, on the other hand, drew its support almost entirely
from white-collar and professional people and housewives of the same class. In
February, 1976, three nuclear engineers in middle-management jobs at General
Electric in California resigned, revealing that they were members of Project
Survival and would work for passage of the safeguards initiative.

William Burch, president of Project Survival, talked about its nature and goals:

And how long has Project Survival been around?

BURCH: Since April [1975].

And you have been president since its formation?

BURCH: I had joined it as a member, and then took over because I was working full time for it. There are seven men that are presently working for Project Survival, all on a voluntary basis, without pay.

They're all working full time?

BURCH: Full time.

How are you able to manage that?

BURCH: Well, some of us were able to take leave of absence. I can just run down the list. There's a schoolteacher who's taken a one-year leave of absence. There's myself, I resigned from an advertising-agency job after twenty-three years and had enough money to, I think, keep my head above water, so I decided to work full time on a voluntary basis. And I first went to work in April of last year, for a group called the Creative Initiative Foundation, an educational-research foundation, as their president, and they were involved in the study of nuclear power. In fact, you really have to go back to that to really say how we got involved in it.

Creative Initiative is an educational foundation, and some of us went to hear a man by the name of E.F. Schumacher, who had written a book called *Small is Beautiful*. He made a statement at a lecture at Saga Corporation about nuclear power, that it was the most hideous thing that mankind had ever come across. And we talked a little bit about that, and we said, "I wonder what he means by that?" He didn't really go into it, and so we assigned a couple of our people, a lawyer and another one, to look into the problems of nuclear power and report back to the board of directors of the foundation. What they reported back was very scary, that it really was, indeed, a serious problem and that we had been totally unaware of it really, going our own way. Because the foundation dealt mostly with human relationships and was more or less directed in that sphere; so we did more exploratory work. We talked to scientists that we knew that worked in the nuclear-power industry. Then we had some presentations from GE, they heard that we were interested. And I think the presentation that they gave us probably did more to convince us that there was a problem than anything. We were treated to such arguments as "Well, people have always been concerned; they worried about electricity because of the wires running up the walls." And we began to see that that was the kind of argument that was being given to people when they really seriously questioned, and some of our scientists in Creative Initiative just couldn't believe what they were hearing. And we continued to probe, more and more. Then we decided that the public was like us. They were just totally unaware of the situation and we really did a lot of research, we talked to [Henry] Kendall, we talked to [Norman] Rasmussen [author of the AEC's reactor-safety study], we talked to anybody we could lay our hands on. And the more we dug, the more concerned we got. So we decided, again as Creative Initiative, to put on a series of twelve presentations in the [SF] Bay Area.

And the women in Creative Initiative were deeply concerned. They said, We've got to get this story out. And they said, We will go to the women of the state and put on presentations and issue a call for information. And so the women who make up Creative Initiative, under the name "Woman-to-Woman: Building The Earth For The Childrens' Sake" began to do presentations. They went first to Fresno, they'd go in, maybe two or three hundred of them, and walk, make a march in the city, and invite people to a presentation on the concerns of nuclear power, show a film, and then get signatures on a Call for Information, asking the governor and the energy commission to conduct a full hearing and report back. Now let's see, I should back up. I'll back up to January [1975], when, in the midst of our concern, we learned about the nuclear-safeguards initiative.

They were trying to qualify it for the ballot, and that's when we heard about David [Pesonen] and that whole thing, People for Proof. They came and we called a meeting of the Creative Initiative people, under the name of education, to hear what they had to say. David talked, and we voiced what we had learned in our explorations. And then we said, Okay now, to the people of the Creative Initiative Foundation. We said: This meeting is over. You go outside, you have lunch in silence. Think about the implications of what we've seen. And if you *want* to, come back into this auditorium after lunch as a volunteer, to go to work for People for Proof, but the meeting of Creative Initiative is over. There was a very definite reason for doing that, because we're a nonprofit tax-exempt group.

Was it customary for you to have lunches in silence?

BURCH: Yes, we'd done that before, because we are basically a religiously based group. Not in any institutional way, much more eclectic. But the people have been to seminars, they are attuned that way, to a deep concern for life, and it was in that concern that they were approaching the nuclear issue.

Can I ask you a couple more questions about that? Because I'm afraid I've never heard of this organization.

BURCH: It has never publicized itself at all. It's been around for a long, long time—as a formal group, by that name, since 1968. But it's very involved, in, I would say, human growth. And the way to change people from being apathetic and uninvolved, to becoming functional and really committed people.

How did it get started?

BURCH: It grew out of—a long time ago—a thing called Sequoia Seminar Foundation, which was devoted to the teachings of Jesus. And it grew from that and, let's see, over the years emerged. It conducts seminars, conducts discussion groups. I would call it, for a long, long time, a horizontal activity that people came to, experienced and went their own ways. But along about 1962, a group of women who had been to one of the seminars, the wives of some of us, got together and said, There's got to be more. It was a time that the world was considering bomb shelters, and a lot of that, and we said, We've got to take responsibility and begin to move out with what we know. And so that was the essence of what you would call a movement today. But it has never been an attempt to reach out to huge numbers of people, it has been more of an in-depth experience.

What was the size of the group roughly at the beginning of 1975?

BURCH: Creative Initiative? Oh, I'd say fifteen, eighteen hundred people, probably in the Bay Area, primarily, with groups in Los Angeles, Sacramento, Portland, Seattle, Denver, back East a few here and there. The ones back East mostly [are] people transferred by jobs. The other ones springing up, with one or two couples that would then begin the discussion-group process in their own areas.

You were mostly management-level people, I guess.

BURCH: I would say so. Some construction, but very little. Mostly professional people: doctors, lawyers, engineers, businessmen.

Then in January, 1975, you met with David Pesonen?

BURCH: In January, we had the meeting with David, and the people came back in and said, We want to go to work, we want to help them get that initiative on the ballot as volunteers for People for Proof. And a lot of those people went out and worked to get signatures and we got, oh, I think roughly 200,000 to 250,000 signatures in four weeks, five weeks. We gave [People for Proof] some money, again as individuals—Creative Initiative gave no money—and we went to work. When it had qualified, and in the process of qualifying, many of us could see that there was going to be an uphill fight, that we were going to have to work hard and that that got us into the political arena which we could not get into as Creative Initiative. And so some other entity had to happen. And we began to talk of the possibility of something called Project Survival, that we would join and that we would encourage others to join. So that's how it came about.

In the meantime, those women were out. They went to Fresno hoping to get 15,000 signatures on that Call for Information, and ended up getting 30,000 in two days. They went to thirteen cities throughout California: Fresno, then Bakersfield, Stockton, Modesto and Los Angeles, San Francisco, Sacramento, Monterey, you know, around. All leading up to, culminating, on May 21 in Sacramento when they presented 365,000 signatures on this Call for Information, asking the governor and the [state] energy commission to conduct a full hearing and report back to the people by January 1 of 1976. And at that time, there were 4000 to 5000 people in Sacramento that day, and I would say maybe, oh, 800 of them were the women of Woman-to-Woman: Building the Earth, and the rest were members of Project Survival who had joined as they had gone around the state. And the women, who had collected those calls for information—they collected some of them and Project Survival members collected some of them— the women, in effect, handed them over to Project Survival, and Project Survival presented them to the government. And I would say that those women would probably say that at that point they turned it over to Project Survival. We've done our job, we've gone out for a Call for Information, now you, Project Survival, you take it over from here.

So, has the governor responded to that?

BURCH: The governor [Jerry Brown] was not there that day. You know, the governor is a very hard man to pin down. He doesn't show up where he doesn't want to show up and, of course, nuclear power is a very controversial issue, particularly with his father [a former governor of California] on the other side: Pat Brown is leading the fight against the initiative. So he was not there, but the chairman of the energy commission was there to accept the signatures on the Call for Information and, in fact, there will be hearings conducted here in the state during October and November. So, in effect, that fulfills our request.

Project Survival has emerged as an entity to electioneer in favor of the initiative?

BURCH: We are registered with the state as a proponent of the nuclear-safeguards initiative, so we are a bona fide political organization. And we are going around the state, continuing to do an educational job on the concerns of nuclear power, [in] support of the safeguards initiative. People for Proof is also listed as a proponent, and I believe there's two organizations registered on the other side, the Council for Environmental and Economic Balance, and a group called Citizens for Jobs and Energy.

Do you feel that these are industry-inspired?

BURCH: Oh, absolutely. The initiative as it was put up was called a "nuclear-safeguards act." The [state] attorney general, in writing his summary of it, called it the "nuclear-powerplants initiative," and the people who were against it called it the "nuclear-shutdown act." But it's obvious that they [the industry] view the initiative as a total fight for survival. That they feel that if it passes, they're done in. They're telling everybody, the banks, all the people that they come in contact with, that it's your fight and if it wins, it's all going to go under. Economics is going to go to hell, and the jobs, and everything else.

You are, in fact, working for an initiative that would, at least for some period of years, halt the construction of nuclear powerplants?

BURCH: Not really. That is what they're saying, in effect. Because what they're saying [is], We cannot meet the requirements of the initiative. The Initiative does not say, if passed, it halts powerplants. It says, if passed, it sets up requirements that must be met, and the first of those is the insurance requirements. With the idiocy of Price-Anderson, which a lot of people admit—$560 million of liability against a possible accident of $17 billion, and that doesn't even include people, that's just property damages—is idiocy. So, the first thing it says is that you've got to give complete coverage within one year, or you can't build any more plants and you must operate the ones you've got at sixty percent of their rated power for the next four years. And while that's going on, it gives you five years to prove reactor safety and methods of handling waste storage; [prove] to, first of all, an advisory board and then the legislature itself, which must, by a two-thirds vote say that it's okay at the end of that five-year time. It gives them a three-year escape. At the end of three years, they have to vote that there's a reasonable reason to expect that they're going to arrive at a decision—so if the industry is dragging its feet, or something else is going on, they don't have to wait the whole five years.

The name of your group seems to convey some urgency about this. Do you feel that civilian nuclear power is a matter of life and death?

BURCH: Yes, I do. I definitely do. I feel, myself, that it is only the first step to a whole, total change of life. I read an article by Barry Commoner that thinks that energy might be the issue over which the world does change and face up to its realities. We've got glimmers of it in the ecology, we've got glimmers of it financially in the monetary crises, but all of those seem to be somebody else's bag. But then you come down to energy, it touches everybody. And I do believe that unless there's a major change in the attitude of the world, we won't make it. A new sense of cooperation, a new sense that we are all in this together, and that we can all make it together or nobody makes it. And that the energy could be the *breakthrough* into that new insight in all the people. And if *we* don't do it—I think that's why the group that we are, [who] are supposed to be the affluent ones, or the ones who have it made, or the privileged—if we don't seize the privilege we have, the time, the money, the opportunity, if we don't lead through to some kind of a new way and say, "No, that's not the end," who can we expect to do it? Because everybody's trying to be like America and have what we had and it's not that. That's not the end. We haven't got the answer for the world.

What do you think the alternative to nuclear power is?

BURCH: First of all, conservation, without a doubt. It's got to be. We're doing it in our own lives. We've cut our energy uses *way* back: twenty, thirty percent easily, some of us more than that.

Are you speaking of the members of your own group?

BURCH: Yes, and [we] are able to do that. And I find that people, if they really understand, are more than willing to. They have been told that power was cheap, initially, that—leave all the lights on and have a beautiful American home with lights pouring out of every window. And also they don't trust—they think the energy crisis was manipulated. But if they really understand, if they say, Well, forget who manipulated it, the problem still is that if you keep demanding electricity, they're going to keep building it for you, the only thing to do is to show that you can get along with a lot less. They [the public] are more than willing to [reduce energy use]. But we go back to things that were sold to us, like instant-on television. You don't have to wait seven seconds for the tube to warm up when you walk into the room. Well, nobody told you that the set, in order to be instant on, was in fact *always* on and that it was drawing energy twenty-four hours a day to give you that little convenience. Because power was cheap. Or, the same thing with a frostfree refrigerator. The fact that it uses eighty-five percent more electricity than the other one, that you don't have to bother with the ice cubes anymore, that's a tremendous price to pay for convenience when it means building more powerplants. And I know those are little, infinitesimal things, but they all add up and multiplied out by millions, they make significant changes.

What is it that gives such urgency to this struggle?

BURCH: I think the fact that we may be running out of time. That the world really is in a crisis. Nuclear-weapons proliferation is everywhere. It's going to take some kind of a shock wave through the world to make a change, and I think if we don't do it, we may well have had it as a species on the planet. We seem to be not making any headway at the reduction of nuclear proliferation of weapons. Perhaps, if there is a shock wave of people taking another look at power, it can happen. I even ran through a scenario that says, You can take America from a nationalistic point of view—really wanting to be the top dog and the hell with everybody else: if the economic projections that we look at, which are that nuclear power will run us bankrupt in terms of capital . . . then if America were to make a decision to reject nuclear power and really put heavy emphasis into other forms of energy, such as conservation, such as solar, such as geothermal, really go all out for them, and let the rest of the world pursue nuclear, it could find itself in ten or fifteen years on top of the heap. So even if you pursue a purely nationalistic, self-interested scenario, that makes the most sense. No, I don't promote that, I would hope that by its example, America could lead the rest of the countries to say that this is the way to do it and we offer it to you free. . . .

In what you're saying there seems to be a note that civilian nuclear power is symbolic in a way. That it's a chance to kind of turn things around. Would that be fair?

BURCH: That's true. The point on which we're hanging the initiative is that the people should have a right to decide: knowing all of the prices, knowing that it's not just a case of a little pellet that's magic, but what is really implied in pursuing this—people [will] make the right decisions.

Suppose everything else in the world were okay. And there weren't nuclear weapons. There weren't armaments and we weren't in danger of warfare, society were organized in small manageable units. Would you still oppose nuclear power?

BURCH: You mean, with the concern about waste storage and all the rest of that, if all of those things were solved? I don't know. I know that the idea of unlimited uses of energy with which to turn around and diminish all of the other resources of the planet doesn't make sense. We've got to realize, we live on a finite planet and that everything must be recyclable eventually. And that we're running out of things, and that we need to use things and return them and not deplete our resources. You can't legislate that. That has to come about through an attitude of a human being who suddenly realizes where he fits in the total scheme of things, and I think that's where we have come from, through this religious process. I don't mean to use the word "religious" in an institutionalized sense, but in a sense of the oneness of everything, and the interdependence of everything; and knowing that and having discovered that for ourselves, then we relate that to the problems. We say, That's where we have to begin to work. And I think as we work in Project Survival we try to communicate that to other people, that there is a beauty and a joy, there is finding the meaning for your life, in realizing that you are interrelated with everything. And that it's only as you work to get things in the right relationship do you find the meaning for your own life.

I'm talking to some people who make their living in the nuclear-power industry, who get very angry at talk like this, because it means they will lose their jobs. Have you folks thought about that?

BURCH: Yes, we have and the Ford Foundation report [on energy policy] said that there would be changes in jobs. I think if you're going to say— obviously, if the guy's been in nuclear and you're not going to do nuclear, he's going to lose his job, he'll change to something else. Chances are he was an engineer at something else before that. We need engineers, solar will need engineers, geothermal will need engineers.

Of course, many of the people who are angry are not engineers, they are miners and construction workers and metal workers and pipefitters.

BURCH: I really believe that there's jobs for everybody. And I think the figures are there to prove it. If they want to really get a scary feeling, all you've got to do is say, All right, let's move forward with nuclear to the point where it will be sucking [in] forty to fifty percent of all the capital in the country, and your housing starts will be down, and a lot of other things will be down—then if you're worried about jobs, what's going to happen? I believe, and we're developing quite a bit of data, that the economics [of] moving ahead are even more frightening than the possible economics of not [moving ahead with nuclear power].

Just to personalize it, if I may, although this is happening all over the country, which is why I'm asking about it—your group is basically upper–middle class, well-to-do. Your opposition is, in many instances, blue-collar trade-union people who are fighting for nuclear power. And while the committees that oppose you here in the state, you say, are industry-inspired, yet they certainly have the support of much of organized labor.

BURCH: I think they've been able to reach out to organized labor with an approach that says, You're going to lose your job Friday night. I think that they would appeal to somebody that takes a very short-range view and, indeed, there are many people out there who do.

It's hard not to take a short-range view of your own job. What do you do about these guys in specific terms?

BURCH: I think you can feed people. I think you can find alternate ways to go. I believe that if this country stopped nuclear power and suddenly said, "We're not going to have it," that tremendous amounts of energy and funding would go into other resources. We would find ways to do it. Look at—they dumped two billion dollars into building more highways when they wanted to artificially inflate jobs. If they put that two billion into installing solar energy on homes, or insulation, they'd have gotten just as many jobs and would have had something which would continue to pay over a much longer period instead of just adding to the problem with more concrete. But I think people think in very short terms: how much concrete, how much steel? We know, for instance, again from that Ford report, that the fifteen top industries—you know, you've seen those figures, forty-five percent of the energy is used in manufacturing [industries that create only] nine percent of the value added and six percent of the jobs. That's where the energy is going. So the stopping of energy use doesn't mean a stopping of jobs. The [latter is] not a corollary: we can cut down our energy and still have plenty of jobs.

In practical, day-to-day terms, though, your group is out opposing a particular thing, you are not, in fact, out there working for something.

BURCH: Yes, we are. We are pushing conservation and we are pushing alternate energy. We're installing solar units on some of our own homes, and we'll be doing a lot more of that in the year ahead. We are definitely pushing that. And, as I say, we believe in it enough that people have left their jobs and severely cut their life style. We're talking about whatever it takes. We would sell our homes if that's what it takes. We're not committed to preserving a life style. Jobs are one thing, but the safety and the long-term aspects of that are another. You could get a job as a parachute jumper, but if the parachute doesn't work, you're not going to have the job very long.

Twenty-two states have the initiative process for enacting legislation by popular vote, and, in 1975, a Southern California group called the People's Lobby, founded by the late Ed Koupal, a retired salesman, undertook a campaign to put nuclear-safeguards initiatives on the ballot in all twenty-two of those states. Since most of the states with the initiative process are in the West, the effort was called the Western Bloc, and a young man just out of college, John Forster, began travelling throughout the nation to publicize the Western Bloc campaign. On the eve of the vote in California in June, only two other states—Oregon, where a bond-issue to build a nuclear plant had recently been defeated, and Colorado— had qualified a nuclear initiative for the ballot; in most other states there was either no serious effort to collect signatures, or the number fell far short of that required (as in Massachusetts). Limited efforts continued in seven other states, but prospects for passage of the safeguards initiative were dim. Forster continues to express confidence, but he admits that the original motive for the Western Bloc campaign was simply to secure television coverage in California. The theory was that a national campaign would get far better news treatment than a local effort, and this has proven to be the case: the national media have taken the Western Bloc campaign quite seriously. Forster has been successful in persuading groups affiliated with Ralph Nader to give support to the Western Bloc campaign, and he gives the impression that Nader is personally involved in the campaign, which is not true. Forster at times concedes that the effect of the initiatives would be to shut down nuclear powerplants but argues that this would not necessarily be so if adequate insurance were provided. Confusion over the likely effect of initiatives continued in California. Polls showed that many of the initiative's supporters incorrectly believed they were voting for nuclear power. On June 8, 1976, the nuclear safeguards initiative was defeated 65%–35%.
Perhaps the most obvious effect of the People's Lobby campaign has been to summon up a national network of Citizens for Jobs and Energy committees, jointly funded by industry and trade unions; the AFL-CIO executive committee has passed a resolution favoring the construction of nuclear powerplants, and

everywhere nuclear power has become an issue, Democratic candidates have flocked to support labor's position, as Republican candidates have supported industry's position, both in favor of nuclear power.

CHAPTER FORTY-NINE

Reaction

IN 1943, when General Leslie Groves visited the state of Washington to choose a remote spot on which to build the nation's top-secret plant to produce plutonium, the first factory for the transmutation of elements, Mike McCormack was a platoon leader in the infantry. McCormack didn't know anything about the plutonium factory, the Hanford Works; he was busy fighting a war. When he was demobilized at the end of World War II, like many returned soldiers, McCormack, then twenty-four years old, took advantage of the benefits offered by Uncle Sam and went to college. He collected a bachelor's degree in chemistry at Washington State University and then went on to get a master's degree, the key to a good job at the nearby Hanford Works, then in its years of great expansion, the years in which the United States first built its nuclear armory. For the next twenty years, McCormack worked as a technician at Hanford, near Richland, Washington, where the federal government produced much of its plutonium.

In 1956, Mike McCormack was elected to the Washington state legislature and began climbing the ladder of local Democratic party politics; in 1960, he was elected to the state senate, where he served for ten years. His position in the state legislature evidently contrasted in McCormack's mind with a somewhat subordinate role at the Hanford Works. In the course of a long interview, he showed animation at very few points, but he was quite lively when he recounted the following incident:

"I was serving in the state legislature for fourteen years, while I was working at AEC, for AEC contractors [at Hanford], and I can tell you that I caused all sorts of strain in the organization. Because of the fact that I was, on the one hand, in the laboratory, in the rigidly structured organization, and the next moment I was outside, freewheeling in the partisan political world. I was level five in the laboratory at 10:30, and the state senator, chairman of a commission holding a hearing for the manager of the company, a number-one man, at noon, and at 2:00 I was back at level five again. Now, that sort of thing creates extreme strains. I was in a staff meeting [at the laboratory] one day, and the secretary came in and whispered in my ear that I had a telephone call. I excused myself, I went out; I was gone on the telephone call for a long time. I came back and sat down. When the meeting was over, the manager asked me to stay for a minute. He said, 'I wish you'd arrange to have your political telephone calls at other times than at staff meetings,' and I said, 'I'm sorry, sir, that was W.B. Johnson, from Washington, DC, the chairman of the Atomic Energy Commission, I talked to. I'll be glad to tell him I can't talk to him during a staff meeting.' "

McCormack's pleasure in this anecdote is evident, and it is interesting that he tells it with relish to an interviewer whom he evidently regards with suspicion. For Mike McCormack is now a member of the Congress of the United States, and he refers to his past status at Hanford rather grandly as that of a "research scientist"; he sneers at critics of nuclear power who, in his view, do not have acceptable credentials as scientists.

The Congress of the United States caters to the sense of self-importance of its members, and Mike McCormack is now a member of Congress, with an office in the Longworth Building, close to the Capitol. The Longworth Building is an extraordinary example of Washingtonian classical revival: a row of fluted columns is implanted firmly in midair, half way up its façade; the façade itself, ludicrously nonfunctional, is attached to a massive barrackslike building, arranged in a hollow square; inside, a television crew has set up cameras and lights in a corridor, awaiting some still-invisible celebrity; members of Congress are whisked upward on their own reserved elevators, while the mere public crowds into the single elevator reserved for it. The long corridors are heavy with marble; even the public bathrooms are walled in marble. The heavy oak doors of the congressional offices open on crowded anterooms. Representative McCormack's office is like many. His inner sanctum is a windowless room, hushed, softly lit. The massive leather high-backed chair sits behind an imposing oak desk and is flanked by enormous flag standards; behind the chair is heavy velvet drapery of a photogenic blue. In the television studio provided for congressmen, there is a similar dummy office with a mock window showing a view of the Capitol, but McCormack is not yet far enough up the ladder of seniority to rate an office with a window.

The corridors and offices of the Longworth Building are the domain of the middle-aged men who make up the Congress; in its subbasement is a livelier world. Here, in a cafeteria whose walls are chiselled granite, gathers the subterranean army of young women and long-haired young men who drive the typewriters and mimeograph machines of Congress and who carry on much of its real work.

McCormack was first elected to Congress in 1970, from the fourth district of Washington, a lovely rural district that runs along the Columbia River and up to the Canadian border; it includes the Bonneville Dam and the Grand Coulee Dam—and the Hanford Works, which absorb more than two hundred million dollars per year in federal funds. The area has extensive lumbering and farming activity, particularly in the fertile Yakima Valley, and McCormack has campaigned effectively in the rural counties of his district, but the district depends heavily on federal funds for nuclear power. McCormack, himself a worker in the nuclear-power enterprise, represents his fellow workers. He has succeeded to the seat on the Joint Congressional Committee on Atomic Energy held by his Republican predecessor, Catherine May, and he is also chairman of the energy subcommittee of the House Science and Astronautics Committee. McCormack has worked diligently on the house-keeping chores which are required of legislators who aspire to influence (he was chairman of the House's Bicentennial Celebration Planning Committee); and he is slowly working upward in seniority on the Public Works Committee, a locus of real power in the Congress. The AFL-CIO's Committee on Political Education regularly gives McCormack a rating of from ninety to one hundred percent, indicating the frequency of his votes for legislation supported by organized labor.

Mike McCormack is evidently good at what he does; more than twenty years as a legislator and party regular are testimony to political skills of an old-fashioned sort. To an outsider, however, he appears surprisingly ignorant of the substance of his work in government. During an interview, attended by his administrative assistant, John Andelin, McCormack tends to deflect substantive questions by attacking various public figures in intemperate terms. Article I, Section 6 of the Constitution states that members of Congress "shall be privileged from arrest . . . and for any speech or debate in either House, they shall not be questioned in any other place," a salutory provision to protect Congress from the coercion of the executive branch, and which effectively protects members of Congress from suits for libel or slander. Mike McCormack is therefore free to say anything at all about anyone at all, without any fear of legal consequences—which is just as it should be, although a voter unaware of these constitutional protections might easily mistake a loose tongue for unusual courage. McCormack seems not to be heated when he attacks critics of the nuclear-power industry in extraordinarily harsh terms; he tends to shift away from the tape recorder used by an interviewer and to cover a half-smile with a cupped hand. His office distributes two publicity photos: in one, McCormack, staring intently, stabs an accusing finger at the camera—the angry, crusading legislator uncovering malfeasance in high places; the second portrait, of which there are several variations, simply shows a dignified man in his middle fifties, wearing heavy-framed eyeglasses and a half-apologetic smile. The latter, more dignified portrait is now replacing the finger-stabbing investigator; McCormack has served three terms and no longer faces any opposition in primary elections back home. But in talking about the nuclear-power industry and its critics, McCormack still sounds like a campaigner in a very old American tradition. His administrative assistant Andelin and McCormack seem to be on close terms and work easily with each other; Andelin frequently interrupts an interview with interjections of his own, affirming and extending his boss's belligerence.

The belligerence is exacerbated when the conversation turns from the day-to-day work of politics and elections to more substantive matters. McCormack has been on the Joint Committee on Atomic Energy for five years, and an interviewer is interested in his view of the committee's role. The Atomic Energy Commission and its successor, the Nuclear Regulatory Commission, were within the committee's jurisdiction and peculiarly subject to its control. Independent regulatory commissions—like the Federal Power Commission—whose heads serve fixed terms and are not subject to dismissal by the president, are a curious kind of fourth branch of government, over which Congress can exert far more control than over the ordinary executive agencies. In theory, at least, the Joint Committee on Atomic Energy has considerable responsibility and authority to oversee the Nuclear Regulatory Commission, which is an independent regulatory agency.

The Atomic Energy Commission was not solely a part of the executive branch, am I correct about that?

MCCORMACK: I don't know. That's a question of first impression for me, and I'm not just exactly sure how contitutional experts would define the role of the Atomic Energy Commission, as far as its regulatory functions were concerned . . . I don't want to try to handle that question without doing quite a bit of thinking and maybe some research on it. Normally, we think of regulatory agencies as being part of the executive branch. The Federal Power Commission, the Environmental Protection Agency, all these are part of the executive branch.

John Andelin angrily asks the interviewer whether he considers the Federal Power Commission to be part of the executive [it is in fact an independent agency like the NRC]; he asserts that agencies which submit their budgets through the Office of Management and Budget are part of the executive branch, apparently unaware that recent congressional reforms allow independent agencies, and in some cases require them, to submit budgets directly to Congress.

Although McCormack frequently refers to himself as one of only two scientists in the Congress, he is not well informed on some technical matters. The AEC's massive reactor-safety study, headed by Norman Rasmussen, is the source he most often quotes, but he is not aware of that report's estimate of the probability of a fuel meltdown; when the interviewer quotes him the figure and gives Carl Walske's estimate based on the report that there is one chance in three hundred, in any one year, that one of the fifty-five existing reactors would suffer a meltdown, McCormack at first denies the figure (Andelin, apparently quite angry, challenges the interviewer on this point even more strongly than McCormack does) and then offhandedly dismisses the figures, simply saying that the Rasmussen study is wrong.

Do you think the government's public-education role has been adequately carried out?

MCCORMACK: No, obviously not. If it had been, we wouldn't have all this silly antinuclear fanaticism we have around the country. People would have a mature and responsible attitude towards it, and we wouldn't have people running around making money on their antinuclear platforms.

Why is there a debate about nuclear power?

MCCORMACK: There are a lot of legitimate questions surrounding all nuclear energy that are debatable issues, such as whether uranium enrichment [plants] should be publicly or privately owned, what technology should be used, the whole nature of the [plutonium] breeder program. Many related questions. These are all legitimate questions for nuclear debate. But nuclear energy is a subject that is not understood by most members of the public, and this has been taken advantage of by unscrupulous individuals, who have written and used the lecture platform to sensationalize the subject, to frighten people half to death; and, in the absence of truth, a large number of impressionable people have been sucked up in this thing and have given support in numbers or money, or what not, to this sort of sensationalism and fear-mongering. This has caused people to be excited about it, and so this has been the genesis of the debate. Add to that the fact that almost half the Congress, more than a third of the Congress, is new in the last four years. These members have never had the opportunity to participate in any of the more thorough debates of the past, that led to understanding of the program. And much of what they come with is just impressions they gained from somebody trying to brainwash them, usually Ralph Nader types. And also, the fact that you still have a large number of writers for the popular press who find that they can get a story, a poor story, in print more easily if it's sensationalistic, than they can a good story, if it is not.

You say, "Nader types"; do you include Nader?

McCORMACK: Oh, yes, oh, yes.

What are the issues that are being raised without justification?

McCORMACK: Well, there is a spectrum of them we can go back in history and look at: exploding powerplants, deformed babies, infant mortality, too much uranium, not enough uranium, the ease of theft, the presumed ease of theft of material in the manufacture of weapons, what Nader recently referred to as the core-disruptive accident with breeders [this term is actually used by the Energy Research and Development Administration as a euphemism for explosions in breeder reactors]. Let's see, this sort of thing, this whole spectrum of, this is a sort of a new McCarthyism, it's almost a fad with some people, it's an antinuclear McCarthyism, and a lot of the young people who ran out of a cause when Vietnam was over had to find something—and they have, a lot of them have got sucked up in this sort of a faddism and they go around reciting these things without any knowledge at all of the subject.

Is Ralph Nader really criticizing nuclear power for the sake of lecture fees?

McCORMACK: Sure. It's an ego trip for him. I don't believe that Nader believes what he is saying, I really don't, and I frankly think that between lecture fees and ego trip and the attention he gets—I think he gets a kick out of scaring people and sensationalizing on the subject. It's an easy lecture for him to give.

Have you met Mr. Nader?

McCORMACK: I have met with him and I have debated him.

Your feeling is based on your acquaintance with him?

McCORMACK: I had a number of good statements and watched him, video replays of what he has said on the subject of nuclear energy. And he is a charlatan on the subject.

There are scientists who are widely identified as critics of the nuclear program. Do you have a similar opinion of their views?

McCORMACK: It depends on the individual, I would decline to call every individual as being irresponsible. Some of them make very valid criticisms and I take an example such as, is it Ted Taylor? Ted [Theodore] Taylor made criticisms which were constructive, to which the Atomic Energy Commission promptly responded, and Taylor promptly came forward and said essentially that the problems that he had cited had been very substantially mitigated, and in his opinion the nuclear program should proceed. There have been other people. . . .

The debate surrounding nuclear energy quickly runs out of control when it becomes a property of a bunch of uninformed [inaudible]. Then it loses all control, and you have people participating in debate who haven't the slightest idea what they are talking about. As Eric Hoffer said, they're the true fanatics, they have no concern for the truth and they feed on each other. Let me give you a simple example. I spoke at a seminar for the American Enterprise Institute one night, and we were talking about various energy programs, and at recess a couple of the guys contacted me—we were having a short recess period between two halves of the program—all agitated because I supported the breeder, and said, Don't you realize the breeder produces deadly plutonium? I said—the context of the conversation was that water-cooled reactors were all right—well, I said that don't you realize that water-cooled reactors produce plutonium, too, and these guys went all white-faced on me and were just literally stunned. I could have slapped them in the face and produced the same effect, because they knew so little of their subject, they didn't know that a light-water reactor produces plutonium, and they're attacking breeders because they produce plutonium.

Ralph Nader said that plutonium [for] weapons is not the same as plutonium from reactors, plutonium for weapons is not as harmful as plutonium from reactors. *(McCormack seems to be unable to believe this was said seriously.)*

ANDELIN: You can niggle over plutonium 239, 240. . . .

MCCORMACK: But the argument runs in our direction, it runs against Nader.

In fact, Nader was correct on this point and McCormack was wrong. The plutonium produced by civilian reactors is a mixture of several isotopes, because it is left in the reactor, exposed to neutrons, for a longer period than would be the case if more nearly pure plutonium 239 for weapons manufacture were desired. According to a statement issued by the Westinghouse Electric Corporation, Advanced Reactors Division, July 14, 1975, "reactor plutonium, a mixture of isotopes . . . is five times more hazardous than Pu-239," the form of plutonium used in weapons.

You use very strong language in describing the critics of the industry, language that one rarely hears in public discussions of any problem nowadays. Do you think that the critics of nuclear power are so much worse than other people in public life that this is called for, or why is this?

MCCORMACK: Because most of them know better, the ones I am referring to generally know better. And to do this, they are making money, they're making money, it's a parasitical action on society and they're making money and doing a great disservice to society.

Now if the difference were that we're going to have a merry-go-round or not in the city of Boston, I couldn't care less; but when it impacts heavily upon this nation's energy programs, on which your whole society and the whole social structure actually depend, then it becomes a little more serious, a lot more serious.

Now it's been reported in the press that John Gofman and Arthur Tamplin and Donald Geesaman all lost jobs at the Atomic Energy Commission because of their criticisms of the Atomic Energy Commission, and Carl Hocevar resigned from employment; it's not obvious to an outsider that these guys are making a lot of money criticizing nuclear energy. (McCormack launches a long attack on the character and credentials of the scientists named. Andelin claims, incorrectly, that one of the scientists was not employed by the AEC.) *Well, I am not asking why they were fired or anything like that, it is all ancient and very dull history, what I'm asking you about—*

MCCORMACK: You raised the point, though—

Yeah, the point of this is, you're saying these guys are doing it for personal gain, and yet on the record they are losing jobs, for whatever reasons, you know—

MCCORMACK: Losing a job at a laboratory is nothing to those guys.

Well—

MCCORMACK: Do you know that I could make three thousand dollars a night lecturing?

Do you have any evidence that they are making three thousand dollars a night lecturing?

MCCORMACK: I don't see why they couldn't. Nader picked up almost six thousand dollars in one weekend in Washington state.

Does Arthur Tamplin get lecture fees like that?

MCCORMACK: I can guarantee you he can make a lot more lecturing. . . .

ANDELIN: I can make more lecturing than working here.

Well, but do *they, is the point.*

MCCORMACK: They are out on the lecture circuit all the time. They are out all the time, writing articles, among other things.

Do you know for a fact that they make more . . . .

MCCORMACK: Well,—

You're telling me these people are only criticizing the industry because of personal gain of some kind or another, and I'm trying to find out why you think that. [The scientists fired by the AEC had difficulty finding other employment, and did not, in fact, earn lecture fees or writing fees even remotely approaching the figures given by McCormack.]

MCCORMACK: I think that, because of the fact that points they have made have been so thoroughly analyzed by thousands of their peers, all over the country, all over the world, in intense studies, and have been found to be without any substance, and in spite of that they go on going around peddling them.

Let me ask you about other people; for instance, among your colleagues—there are members of Congress who introduce bills frequently to impose a moratorium on nuclear-power construction. Why does that happen? Why are those bills being introduced?

McCORMACK: Well, let me answer the question this way: I don't know how many bills have been introduced so far, but I guarantee that you can get a bill on almost anything; I've introduced bills to make the square dance the national dance, and the apple blossom the national flower, and I guarantee you that you can get a bill that's almost anything. Now those people that—one of the easiest things for a congressman to do [is to introduce a bill] to placate some noisy group or excited group, or something like that: some kids on a college campus, they want a bill. Nobody in Congress is pursuing this subject. [Rep. Morris Udall was, in fact, holding hearings on nuclear power at this time.]

Well . . . certainly Senator Gravel, for instance, has been very consistent in publicly calling for a moratorium.

McCORMACK: That's right, he certainly has, and he's not doing badly, either.

CHAPTER FIFTY

Politics

IN WASHINGTON, where Congress is the local government, cab fares are low on the routes taken by congressmen; the drivers get along by carrying as many passengers at one time as they can. On one sunny afternoon, there are two passengers in the back seat of a cab which stops to pick up a third, who gets into the front seat beside the driver. The new passenger notices, cannot help noticing, a mock-up of a two-dollar bill bearing the portrait of Martin Luther King within a glassed frame hanging from the cab's sun visor. At the time there is some speculation in Washington that the Treasury Department will issue a two-dollar bill for the Bicentennial celebration; the new passenger thinks it is an interesting idea to put Martin Luther King's portrait on such a bill and says so. The cab driver, a middle-aged black man with strong features and receding gray hair, replies that he was the first to have proposed a King portrait on a new two-dollar bill. The mock-up on the sun visor was made up according to guidelines provided by the Treasury Department to avoid the counterfeiting statutes. The driver had written about his proposal. In fact, it seems the driver is a writer in his spare time; here is a copy of an earlier book of his. A question elicits the fact that the new passenger is a reporter, and this information brings forth a flood of further biography.

The cab driver's name is Victor Endres Holt; his business card describes him as a free-lance writer ("Books and Short Stories; Also Creative Subjects and Points of View; Trouble Shooter for D.C. Cab Drivers Grievances"; the address of his "studio" is also given as "Headquarters, Project King $2.00 Bill"). The King Project apparently is occupying much of his time. Holt had written a story on his idea for a black newspaper which had been picked up by the *Washington Star*. On the strength of that publicity, Holt wrote to every member of Congress and received eighteen favorable replies. Senator Birch Bayh of Indiana apparently was among those responding, for Holt says that he now "practically lives" in Senator Bayh's office while lobbying the other members of Congress. Mindful of the need to generate pressure from the grass roots, Holt appears on radio and television whenever he can and carries a thick book of press cuttings with him; he rattles off a list of local radio and TV appearances.

The campaign to get Dr. King's portrait on a two-dollar bill was just getting up some steam, according to Holt, when "the women" started a backfire; "the women," meaning the National Organization for Women, began to lobby for a new two-dollar bill, with a portrait of Susan B. Anthony on its face. Holt frowned and shrugged; he was obviously finding it hard to compete with the larger and more amply funded lobbies entering the field in the women's cause. But Holt was expanding his efforts to meet the challenge; he had incorporated his effort and was selling T-shirts, buttons and framed mock-ups of the proposed new King bill. And he was willing to find a common ground with the women: "I'm willing to compromise; how about a black woman?" But the new passenger has arrived at his destination and must depart.

In 1976, the Treasury Department struck its own compromise and announced plans to issue a two-dollar bill bearing the portrait of Thomas Jefferson.

A more successful special-interest group, among the thousands which make their home in Washington, DC, is the Natural Resources Defense Council, which rents two floors of a soon-to-be-demolished office building in downtown Washington. A visitor finds the usual receptionist at a bank of telephones and is ushered into a simply furnished office to meet with Gustave J. Speth, an NRDC staff attorney in his thirties, a good-looking man who speaks with a soft Southern accent.

Where did you go to law school?

SPETH: Yale.

Did you like it?

SPETH: No *(laughter)*, I didn't like it at all. I thought it was a bore, and I thought that I just couldn't imagine being a real lawyer.

Well, are you a real lawyer now?

SPETH: No. Being a real lawyer, I think, of any form would be—of any of the traditional forms—I don't know how to describe it, but it's an unpleasant kind of a life, I think. You know, whether you're writing wills in a town of fifteen thousand people, or up on Wall Street worrying about the Securities and Exchange Commission and floating another bond, or here in Washington trying to hack the Food and Drug Administration to death. . . .

Did you spend any time in a conventional practice before you got into this?

SPETH: No. We very early—well, late in the second year of law school, and particularly the beginning of third year, I remember reading an article in the [*New York*] *Times* about the environment. This was in 1969. And I also remember seeing something around the same place in the paper about the NAACP Legal Defense Fund, and the fortunate juxtaposition of those stories made me think that maybe there ought to be some kind of environmental legal defense fund. So, I started talking it around with some buddies in law school, and we formed this little group in law school. And the first thing we did was to apply to the Ford Foundation for a grant—because we only had one year before the end of that school year, to decide if we all knew what we would be doing. So we put together a funding application and went to the Ford Foundation. But it turns out that the Ford Foundation was interested in this precise subject. We had no knowledge of that at all, when we started, but they were extremely interested, or had had a similar idea. Then we ran into a snag. Which had to do with—the foundation said, We like the staff, but you fellows don't have a board of trustees. (*Laughter.*) And so we were sent forth in the world to find a board of trustees.

Well, it so happened that at that same time there was another group which had the same idea in New York City, growing out of the Scenic Hudson Preservation Conference [which for ten years battled Con Ed's plans for a pumped-storage reservoir]. They had a board of trustees all set up and a corporation all set up, but only one staff person who they had hired to be the executive director. And no troops. By that time, we had lined up a group of trustees, too, people who said they would be trustee people. So we agreed to merge our trustees with theirs and put us on the staff. And together we all went back to the Ford Foundation. It's too quick to say that we got the grant, because right at that time Wilbur Mills [former chairman of the House Ways and Means Committee that controlled tax legislation] decided that he was going to take over the whole foundation world. I don't know if you remember the Foundation Reform Act of '69, and that whole thing. One provision of the proposed legislation, which he released, said that it would have prohibited the precise kind of activity that we wanted to engage in. So rather than give a grant in the face of that problem, Ford told us all to come back later. So everybody went out and found other jobs for a year. And when we came back, that issue had blown over and we got the grant. And so, that was how NRDC got started.

What did you do for that year?

SPETH: Clerked for [Supreme Court] Justice [Hugo] Black.

Did that intrigue you, or seduce you, toward more conventional work?

SPETH: Clerking? No . . . I've just never been attracted to it [conventional practice] at all!

NRDC is not a law firm?

SPETH: It's a membership, environmental organization which has decided to spend most of its time achieving the environmental objectives with legal activist strategies.

Originally, it was a law firm, when it started, though.

SPETH: That's right, that's right. We were set up very much like a law firm, and we thought about operating like a law firm. It was never as much of a law-firm model—if you're familiar with the Center for Law and Social Policy—as they were. They really never thought about doing anything other than responding to people who sort of came in there with cases, and that sort of thing. We always had a notion of developing our own priorities, and this kind of thing, and always thought that on occasion we'd be representing ourselves as well as representing other people. But, by and large, you're right, it was a lot more like a law firm representing other people in the earlier stages than it is now.

We had decided prior to the time [of] the Sierra Club decision [in which the Supreme Court held that a group may sue on behalf of its own members] that we wanted to operate more like an environmental group than we did like a law firm, and in the wake of that decision we expanded on something which we'd thought about doing a lot before, which was getting members and having a real membership organization. So we developed a membership, and it's up to around twenty thousand now. It's not a big membership, but it's adequate.

What benefits do the members get, a warm feeling?

SPETH: A warm feeling, a newsletter. They get their questions and complaints treated better when they say they're a member. We get things done the right way and get back in touch with them and answer their mail and all of that.

How much of the funding do they provide?

SPETH: I think it's about twenty percent.

And the rest is what, foundation grants?

SPETH: And individuals, big donors. The Ford grant is still the biggest that we can get, though their contribution is down to about twenty-five percent, now.

What's the total budget?

SPETH: For this year [1975] we think it's going to be about $1.7 million.

Have there been any problems with the ethical standard of attorneys that says they should not stir up litigation?

SPETH: I don't think it's a problem, because there have been a couple of instances in which the bar associations have faced the question of whether in a non-fee-paying public-interest context, it's proper to advise clients as to what their legal rights are. And the Supreme Court's also faced it in the [NAACP v.] Button case. And every time that issue's been faced in this kind of context—

where you're talking about a public-interest kind of thing, you're not talking about advertising for financial remuneration—it has been determined that it's perfectly all right to do. So I don't think it's a serious issue. I think everytime it's been faced it's been held to be not only ethical, but constitutionally protected. On the other hand, there have got to be a good number of lawyers out there in the world who still think that it's a cardinal sin and don't like the Supreme Court decision and don't like these new rules.

Do you get any flack from your colleagues in the law profession for stimulating litigation or generally creating trouble?

SPETH: In almost all these cases we've been a party. And if you are a party yourself, this question of stimulating litigation sort of vanishes, since you're clearly pursuing your own interest and not trying to find clients or anything. So, in almost all cases, it just doesn't come up, because we were a part of it. I would say our relationships with the bar are extremely good. We have a long list now, it must be more than a hundred people, who represented us on a *pro bono* [*publico*— without fee] basis in some of the best, biggest firms in the country. We get a lot of *pro bono* help. I would say we've gotten over two hundred thousand dollars worth of *pro bono* assistance from Arnold and Porter here in Washington over the past few years. It's been a big boost—in fact, we couldn't carry anything like the case load that we have now if we weren't receiving *pro bono* help to do it. So it's made a world of difference, and it's made us a lot more effective.

You have been dealing with federal regulatory agencies as a kind of special-interest group—

SPETH: Yesterday ERDA [the Energy Research and Development Administration] had a meeting, they had a series of meetings; and one of them at two o'clock was for business groups, and the one at four o'clock was for special-interest groups. *(Laughter.)* This was what we were told, when we were called up, maybe one of their secretaries just blew it. We were told that we were invited to the one at four. But we went to the one at two. *(Laughs.)* Just to see what was being said. But I interrupted.

That's okay. The federal agencies aren't very well staffed and they don't have very much money and they're kind of small, which is the situation that businesses take advantage of. Are you guys taking advantage of the same situation?

SPETH: ERDA? You mean the AEC? My heart bleeds for the AEC. Those poor understaffed, overworked . . . .

Now, but there isn't an AEC any more.

SPETH: Now there are *two* giants, two monoliths out there.

So, you think ERDA can handle you?

SPETH: I think there is a serious imbalance of resources in Washington, between the corporations that are to be regulated and contracted with and the people that are trying to regulate them and contract with them. I think that in some instances, unfortunate as it is, given our [limited] resources, [we have as many] resources to meet on an even keel with some of the federal regulatory agencies; there are times when we can pull together about as many toxicologists to look at water pollutants as EPA can. On the other hand, that doesn't apply in the context of the nuclear agencies, because they aren't nearly as—stretched nearly as thin as EPA is, for example. I mean, my heart bleeds for EPA, sometimes, the job they're trying to do with the people they have and the limited resources available to them. And we can make an impact there by doing better science than they do when they get out a bad standard or helping them when they're interested in coming out with a good one. But we haven't got anything like the resources that we need to have the best desirable public-interest presence on the nuclear issue. Just right here on our staff, we have basically in Washington here the equivalent of two people. Tom [Cochran] working full time and [Arthur Tamplin and] myself half time on the nuclear question. We also have a little bit of funds to go outside occasionally and get an ouside attorney to work on something. But that's at most another half person. And look at the things that we've got to do. Just this month: we've got to intervene in the Clinch River Breeder licensing and initiate that whole thing; we've got to comment on the tentative decision to stop licensing plutonium [fuel facilities]; we've got to meet with the National Academy of Sciences on hot particles; we've got to testify on the shipment of plutonium by air; we've got to decide whether we want to intervene, and if we do—we have to comment on the Barnwell spent-fuel receiving- and storage-station draft impact statement; and we've got to testify before the Joint [Congressional] Committee on Atomic Energy on the breeder. And that's just the, kind of, highlights and we've got two people and it's just back-breaking work. It never goes away.

Have you had much pro bono *assistance from the lawyers in town in the nuclear area?*

SPETH: Well, I still think it's extremely hard to get, but I can't make the blanket statement that I used to make because we do have a firm here in town that is representing us in a case that we fought a long time ago, which is still kicking around, on the Vermont Yankee plant. . . . A firm is representing us on that *pro bono*, but it's still slim pickings.

So when people see that you have a budget of $1.7 million, that sounds like a lot of money, but, in fact, in terms of what a law firm of your size would be handling in terms of dollar volume on a commercial basis, you'd be talking about a good deal more money, right?

SPETH: Yeah. We could just decide to go profitable one day. In fact, we just decided we're just all going to sell out. (*Laughs.*)

The Coalition for the Environment, in St. Louis, is another special-interest group working through litigation, propaganda and political effort to achieve certain environmental goals. The Coalition office is a store front on a busy street, like the old-fashioned, and now almost extinct, clubs of the urban political machines. But the Coalition is a more modern sort of political entity. Its membership of about nine hundred are affluent, white, middle-class people who live in the suburbs. The group was formed in the 1960s and for several years concerned itself largely with the preservation or creation of park lands, the principal interest of its principal contributors. In the 1970s, the Coalition broadened its interest over the whole spectrum of environmental concerns that burst upon public attention in those years and, under the leadership of a young executive secretary, Ben Senturia, began to have a broad impact on the community. The Coalition's most impressive victories are still in preserving park land; in 1975, the group spearheaded a drive to save a portion of St. Louis's enormous central park, Forest Park, from the encroachments of a sports stadium. The victory over the city's Democratic party leadership, who were committed to the sports-center expansion, was impressive and highly visible and made it clear that the Coalition had become a serious force in local politics. A year before, the Coalition had undertaken to defeat the proposal of Union Electric, the local power company, to build a nuclear powerplant. When another group, Alberta Slavin's Utility Consumers' Council of Missouri, went into the state courts to stop the plant, the Coalition agreed to carry the fight forward in federal proceedings. Ben Senturia, who directs this and other efforts of the Coalition, expresses doubt that it will succeed. In the cluttered store front, at a desk drowning under papers and surrounded by the heaped work tables of an organization that depends on volunteers, Senturia talked about himself and the organization he works for. He is a heavy-set, affable man of about thirty, who spent two years in graduate school, studying biology in a program directed by Barry Commoner at Washington University, in St. Louis.

Do you think of yourself as a scientist?

SENTURIA: No, I think of myself as an environmentalist who is capable of standing between the scientific community and the lay community and can translate either way.

Did you go to graduate school with the intention of becoming a research scientist?

SENTURIA: It was with the intention of staying out of the draft. (*Laughs.*) And also with an interest in science, but no particular intention of becoming a research scientist, although I learned to enjoy the research, but my head was turned, I guess in 1968, by politics, and when that happened, my activities slowly took on the cast of more the community and activist side of scientific issues.

What happened in 1968?

SENTURIA: Well, the people who I was working with in the lab one day asked me if I wanted to go on a march to stop the war in Vietnam. I explained to them that I couldn't do that, because if we stopped the war in Vietnam, why they'd be in Hawaii and then they'd be in the Philippines, and I laid out the entire domino theory. They said, Let us sit you down and talk to you for a minute. And from there, I marched and got involved in the presidential primary and a petition drive, and then I ended up in the McCarthy campaign and went to Chicago, got a little whiff of tear gas, and that was enough to start things rolling. I came back and started being involved in politics and in environmental kinds of things.

How did you get involved in the McCarthy campaign? I mean, what actually happened?

SENTURIA: I enjoyed my involvement in the petition drive, to put on the [Missouri] ballot a presidential primary, and many of the people involved in that also overlapped with the McCarthy campaign. After we finished the petition drive, in the middle of the summer of '68, I just sort of naturally got involved in the McCarthy campaign, since that was where a lot of the people were going from the petition drive.

After the 1968 campaign, what were your political activities like?

SENTURIA: They were fairly intensely in the New Democratic Coalition, the formation of the New Democratic Coalition [a reform group within the Democratic Party that evolved from the work of McCarthy supporters].

And you began to be active in environmental issues?

SENTURIA: Right. Well, partially through my being around Barry Commoner and the people who he related to and my activism in politics, it was sort of a natural that I became interested in environmental issues. I'm not sure I can remember exactly how I first got off into that. Something makes me remember air pollution, air-pollution meetings and getting involved in air pollution, for some reason. I'm not sure why it was that; that was the start. [*Senturia then explains how, after leaving graduate school, he was offered a job with a local environmental group, the Committee for Environmental Information, where he continued his political activity.*] I got very involved in the Schramm campaign. Jack Schramm was running for lieutenant governor and I spent, I would guess, six to eight months very actively involved in Jack's campaign.

That was in 1972?

SENTURIA: Yes.

How did that campaign turn out?

SENTURIA: Well, the primary was great. The general election was a little frustrating. We lost by about four or five thousand votes. I might say that was an incredible opportunity for me to learn an awful lot about how a pretty good political campaign is run, and also some basic skills about fund-raising and organizing. So, that was just an additional contribution to my organizing skills, a good learning experience.

You're working with Jack Schramm again?

SENTURIA: Yeah. I'm working with Jack Schramm at this point, because Jack is one of the few guys around who has a real, real honest sort of reaction to issues and an ability to deal with them in a substantive way, that most candidates don't. Most candidates are good at getting elected. Jack is reasonably good at getting elected and he's also good at dealing with problems.

What's he running for this time [in 1976]?

SENTURIA: Congress. The second district [of Missouri].

He will have one serious opponent in the primary, state senator Robert Young.

SENTURIA: It's presumed at this point, yes.

Young is a union member, a blue-collar worker, and quite likely environmental issues will arise in the campaign.

SENTURIA: I would expect.

How do you think the lines will be drawn between the two primary candidates?

SENTURIA: I'm not sure to what degree the opposing candidate will be willing to sit down and exchange views on the issues. I'm not sure that's where the lines will be drawn. I think that Jack will endeavor to take some strong positions on important issues. I think they may or may not be picked up in the campaign. But there will be an attempt to characterize Jack as sort of a central St. Louis County lawyer, who doesn't understand the problems of people in north St. Louis County [a working-class area], and . . . Jack will attempt to make clear that his history of involvement in politics has been that of worrying about the problems of citizens. Not just worrying about it in a political rhetoric sort of sense. That he's the kind of guy who has sat down and in an arduous way worked [out] the details of providing financial assistance to senior citizens, and that kind of thing. So, I think it will be a tug of war in many ways as to who is characterized in what way.

There are some specific issues on which organized labor and environmentalists have been on opposing sides. Large construction projects in the St. Louis area that have aroused a lot of strong feelings—the nuclear plant, the proposal for a dam on the Meramec River, and a locks-and-dam [project] on the Mississippi. I guess it's reasonable to assume that Young will be in favor of large construction projects.

SENTURIA: I would guess that would be his position.

Peoples' assumption will be that Schramm would side with the environmentalists in opposing at least some of these projects [Schramm in fact opposed the Meramec Dam and the nuclear powerplant during the campaign, and Young favored them] . . . . Do you feel uncomfortable about this kind of class division?

SENTURIA: I might feel uncomfortable if I thought that Bob Young really was a good representative of the blue-collar workers; but, in fact, I really don't and so, no, I don't feel uncomfortable in this situation.

Do you think that if Jack Schramm campaigned on a strong environmentalist platform in the north part of the county, which is viewed as a blue-collar area, that he would do well?

SENTURIA: I can't answer that. It depends on the issues and how the issues are cast. I don't know. If he took a position of "To hell with jobs, we're opposed to the projects," that would be sort of a silly position, I would think. It wouldn't be very well thought through. It wouldn't take into account all the factors, and I would think that he would not take that kind of position.

This polarization, organized labor against environmentalists, is happening elsewhere. Why has there been this conflict appearing repeatedly around the country?

SENTURIA: Well, I suppose the answer is that it depends where the interests, the interest groups, are on given issues. Obviously, the issue of jobs and the issue of adequate income is a higher priority to the people who don't have jobs, the people who don't have adequate income, than is something that can't be pinned down, such as air pollution or preservation of open space. That's not something they see as a particularly vital issue, because they've got a job problem or an income problem right in front of their face and it's very difficult to see the importance of an open-space issue, and it's also very difficult for them to see that there's any alternative way of generating jobs or income at the moment. That is to say, Locks-and-Dam 26 is the issue of the moment. It's the issue that will create jobs and it's difficult to question about, about why it is that that project is the project that's being proposed in the first place. Why aren't there other kinds of projects proposed that also would generate jobs, that would also generate increased income and a higher standard of living, that wouldn't have the negative environmental effects? And I think that the environmentalists have not been very successful in making that point, that the conflict doesn't have to exist. That there are things that one can do, that aren't absurd kinds of projects, that do answer the real human problems that exist.

CHAPTER FIFTY-ONE

Community Organizers

WASHINGTON'S DUPONT CIRCLE, the center of a prosperous zone of town houses and embassies that have been engulfed by crime-ridden slums, is now emerging as a kind of capital of nonprofit enterprise. Drawn by cheap rents for spacious quarters in the old brick residences of the area, and drawn also by each other, a multitude of impecunious representatives of the public interest have settled in the streets that radiate like spokes from the little park in DuPont Circle, where three-piece suits and indescribable costumes mingle.

An old commercial building of four stories, just off the circle, houses the Ripon Society, the Urban Environment Conference, the Movement for Economic Justice, the National Clean Air Coalition, an architect, and what seems to be a law firm. The lobby is guarded only by a workman applying black adhesive to the floor before laying linoleum tile. A visitor picks his way around the adhesive, up a narrow staircase to a doorway, and then up a rather grand circular stair that begins on the second floor. The effect of the broad staircase and the elaborate molding on the doorframes which open at landings is diminished by the unswept linoleum and by the dirt on the walls. At the fourth floor, a six-foot banner stretches across the landing, displaying in bold letters, "The Movement For Economic Justice." Beneath the banner, like a shrine of the Movement, is a copying machine.

There is a small window opening on a tiny cubicle, through which young women once surveyed visitors before permitting them to enter the offices beyond; but the receptionists' cubicle is empty now, and a visitor walks into the headquarters of the Movement quite unchallenged. Beyond the vacant sentry-post is a large room, impossibly cluttered with file cabinets, wall posters, desks, racks of metal shelves holding heaps of printed literature, a sofa, a drafting table, and papers strewn over every horizontal surface. The sound of typing can be heard. A visitor ventures among the file cabinets and soon comes upon the source of the typing, a young woman, Madeleine Adamson, working at a desk almost obscured by paper, posters and slogans.

The Movement for Economic Justice was founded in 1973 by George Wiley, the remarkable man who headed the Welfare Rights Organization until his death in a boating accident that same year; Wiley, who had a doctorate in chemistry, founded the first national interest-group for welfare recipients and built it into a serious political force. The Movement for Economic Justice was the beginning of his attempt to break out of the restraints of special-interest politics in Washington. After his death it was headed by Bert De Leeuw, whose title was "coordinator" in a group that eschewed titles and authoritarian arrangements. De Leeuw allowed himself to be seduced back into the stream of conventional politics in 1976, when he left the Movement to work in Fred Harris's campaign for the Democratic presidential nomination. "My decision does not lessen my commitment to the basic grass roots, multiissue organizing you have been doing and supporting," De Leeuw told the Movement through its newsletter, "but I am convinced that Fred Harris's campaign is a natural partner in our struggle." The new coordinator of the Movement was Madeleine Adamson, who had also worked for George Wiley as a volunteer in the Welfare Rights Organization, and who had been editor of the Movement's newsletter, *Just Economics*, since the group's founding.

When interviewed, Adamson was still editor of the newsletter's professional-looking twelve-page monthly, reporting on the activities of groups with bewildering acronymic names all over the United States. Adamson is a slender young woman of twenty-four with long, shining brown hair which she frequently pulls back with both hands, severely outlining a handsome face with intent eyes. She dresses plainly, carries what she needs in the pockets of her blue jeans, walks to work and seems embarrassed to be called an editor. She and the other half-dozen members of the Movement's staff do not use titles. The group functions on a fluctuating budget of about $150,000 per year, provided in large part by the handful of foundations and church groups who provide the funds for many nascent enterprises of the political Left. In its earliest days, the Movement gave small, two-thousand-dollar grants to local community groups to help them organize but no longer has enough money to continue the practice. Adamson explains that the Movement grew out of the realization that the Welfare Rights Organization and other special-interest groups trying to help the poor would be unable to create the kinds of profound social changes which are needed to eliminate the evils they addressed; that to create social change, a majority of people would have to be mobilized. The Movement for Economic Justice was founded in an attempt to create a broader constituency for the advocates of social change and followed principles that had been developed and put into effective practice by the late Saul Alinsky, an architect of community-based political organizations who had trained and inspired a whole generation of organizers. The Movement for Economic Justice has developed its own means of working out the fundamental idea of uniting the poor, the working class and the more affluent around common economic interests.

ADAMSON: In general, our purpose is to help build local citizen-action organizations around the country. Organized around a variety of economic issues, and electric-power utilities is one of them. Our effort is geared to trying to involve a broad base of people. So the organizations on a local level try to involve poor people and middle-income people in the same organization, by appealing to them through a choice of issues. As I mentioned, utilities has been one that bridged the gap probably better than any other one. I think everyone's been hard hit by the increase in [power] costs. So all of our work is geared toward building local bases of citizens, which ultimately, we hope, will have some national impact.

Do you send organizers into communities?

ADAMSON: We usually describe ourselves as a clearing house for citizen organizations. That means that we provide information and expertise on organizing. We put out materials, how-to-do-it kind of things, as opposed to technical research. The newsletter every month reports on what local groups are doing, so they can see what else is happening, you know, and not feel so isolated, and get ideas from people. We're involved in recruiting and training organizers. We've helped initiate some local projects through helping to find a staff, helping to make an initial organizing plan, helping to raise initial money, and so on. The relationships vary, you know, from place to place, depending on the needs of the local organization. But mostly, we are more or less a center for information and ideas, so that if someone is interested in utilities, or they're interested in local property taxes, or whatever the issue is, they'll come to us and we can point them in directions, tell them what other people have done, what has worked and what hasn't.

How many community groups have you worked with?

ADAMSON: We're not a membership organization so there's no affiliation as such, so I can't say there's eighty chapters or anything like that. Some groups we have worked with more closely than others. It's in the hundreds, but again, these groups vary in their own effectiveness, in size and so on. Some of them are well along, some of them are just getting started.

The Movement for Economic Justice seems to respond to interests expressed by local groups and is not aggressively pressing for programs of its own. Its newsletter carries information about a bewildering variety of consumer issues and groups, but the most frequent themes seem to be local tax reform and utility-rate fights. The Movement has not involved itself in fights over nuclear powerplants;

none of the groups with which it works has taken a position in favor of a nuclear plant, but only two or three that Adamson can think of have interested themselves in the subject. The primary concern, so far as electric-power companies go, is to lower rates for residential consumers. This is the most common community concern that cuts across class lines, and a strategy has emerged among the Movement's affiliates, a campaign for "lifeline" rates.

"Lifeline" is a fixed low sum for a minimum amount of power use. All power companies charge higher unit costs to customers who use less power, and this means that poor people generally pay higher electric-power rates than the wealthy, and that all private homes pay higher rates than businesses or factories. The lifeline rate would reverse this situation for the lowest category of use, by setting a low fixed bill for, for instance, the first two hundred kilowatt-hours of power use each month. The effect would be to lower all residential bills somewhat, with the benefits increasingly important as usage and income decline. The proposal has considerable appeal and has been picked up by groups across the country, in part through the Movement's efforts. Late in 1975, California was the first state to impose lifeline rates on its utilities, and other state legislatures and utility commissions were studying similar proposals.

In California, the successful fight for lifeline rates was led by an organization called the Citizens' Action League (CAL), whose leader is an enormous, bearded, deep-voiced bear of a man named Tim Sampson, a teacher at the University of California, where he conducts classes in community organizing, and a former associate of George Wiley. Sampson's offices are large, bare, white-washed rooms in an office building in San Francisco's commercial district, seemingly deserted except for Sampson when a visitor arrives. Sampson, who seems not to be sure he wants to be interviewed, continues accepting numerous phone calls and stuffs folded circulars into envelopes while he talks. A girl of about twelve wanders into the office occasionally; she seems to be Sampson's daughter (it is summer and school is out), but he cannot spare much attention for her, either. She does not seem at all put out by this.

What is the relationship between your group and the Movement for Economic Justice, if any?

SAMPSON: Bert De Leeuw and I are old friends; and their national information network and support for local organizing efforts like ours is valuable and useful to us. There's no formal relationship.

What sorts of things have they done for your group?

SAMPSON: They provide us with information and contacts with people who are doing stuff like us around the country. They let other people know what we are doing, so that we have a good information exchange, and they keep us apprised of some of the developments in Washington that might be of use to us, and they assist us in contacts with national funding sources. In general it's kind of a cross-fertilization and communication.

What is the Citizens' Action League?

SAMPSON: The Citizens' Action League is a multiissue, multiconstituency mass-based citizen-action organization, that started here in the Bay Area, that eventually we hope will be engaged in building a majority economic-justice agenda organization in California.

When you say it's mass-based, is it a membership organization?

SAMPSON: Yes, it is.

What is its membership now?

SAMPSON: Well, we don't deal in numbers like that. We're in a phase in which we're just beginning to build a base of the organization in terms of membership, but in terms of mass action, for our major actions we have turned out numbers of people ranging from five hundred to a thousand. So it's clear that in this business, we're the kind of outfit that has troops, as contrasted with the kind that doesn't.

Are you based in a particular neighborhood?

SAMPSON: Why don't I just tell you the background of our efforts, so you'll understand what it is. In January of 1974, a group of us who have been active in various kinds of community organization—the peace movement, and the civil rights movement and in my case, in welfare rights; various kinds of community organizing—were living in the Bay Area. Many of [us who] had extensive experience with that kind of organizational work here, got together and talked about the possibilities of a new and broader effort, around some of the economic issues of the seventies and tried to compare our experiences in the sixties, and look towards various possibilities for, you know, continuation in learning from that work. And out of that effort, we decided to try to form a [San Francisco] Bay Area–wide, regional effort that would try to put together some of the constituencies that had been divided—and to try to develop some ways of cutting the issues and forming the organization that kind of melded these constituencies, instead of organizing them separately or allowing them to become divided in some of the classic kinds of divisions that have been a problem in these kinds of efforts.

As the issue around which to begin a campaign, to develop that kind of organization, we selected opposition to a major rate hike—the largest in their history, up to that time, that had just been requested by the Pacific Gas and Electric, which is the second largest private monopoly utility in the country. It's second in size, in net worth, and so forth, only to Con Edison in New York. And we selected that issue because it was one that had a lot of appeal for a lot of people. And because it was very topical in terms of the energy crisis at that time, and inflation; it represented something we believed we could do something about, and because there had been relatively little opposition up to that time in our Public Utilities Commission. We felt we had the leverage with our politicians in the Bay Area . . . .

So we set out—we developed a platform of opposition to this rate hike . . . what we call "lifeline," a basic amount of energy for a low, fair, stable price for everybody. Which was another example of the difference in strategy [from earlier organizing efforts]. We were not after a special break for the poor, or for senior citizens, to be paid for through taxes on working people or middle-class people. We wanted something that would unite [the poor and the middle-class], and so we wanted to change the [utility] rate structure in a way that would benefit the majority of people. So the other half of lifeline is redistribution around the rate structure, what we call "fair share" rates for the large users, business and industry. Those things fit very well, because as we got into and studied utility economics, they were [in accord] with what needed to be done [to reverse the declining rates for larger users of power] and the way that the whole structure of utility economics—raising those kinds of issues was very timely. . . .

In addition, we wanted to oppose the utility stance of essentially passing on cost increases laid on them from the oil companies, and the natural gas companies; and we wanted to expose the basic connections of the banks and insurance companies and large manufacturers, large corporations—what their interests were in the rate structure, and how they controlled the situation. We wanted to open up the Public Utilities Commission, which had been run on a pretty closed basis, with a rubber stamp, and passed along very large rate increases . . . . So we set out with a campaign that we called Electricity and Gas for People: E&GP, which is what happens to the initials when you turn PG&E around, which is the slogan that we adopted—"Turn PG&E around." You might say the constituency's slogan is what our button says, "You and me versus PG&E."

And so we set out. Our tactics were to break these issues into the public arena and to do things which could begin to arouse the public and give them an outlet, to see where they could do something about these things. So we picked things like attacking the closed-door decision-making meetings of the Public Utilities Commission, by the expedient of taking a hundred people into the secret meeting. We got those meetings opened up, the politicians hopped on the bandwagon, put legislation in the hopper. The governor [Ronald Reagan], who vetoed it before, finally found it expedient to sign a bill opening up the Commission, and I think most observers would credit us with really setting that in motion by actually taking people to do it. . . .

We also entered formally into the rate case that we were opposing, and while we didn't sit there day by day as the case went on (and the case is just up for decision now), we intermittently would raise a particular issue. We attacked PG&E directly and asked them to negotiate with us about the rate structure and some of these matters and put a lot of pressure on them. Finally got two vice-presidents of theirs to meet with us, and a large number of our people exposed the fact that they didn't have any constructive rate-structure proposals and weren't willing to entertain ours. We subpoenaed the chairman of the board of PG&E before the Public Utilities Commission in this rate case. Hauled him in there, and we had six of our grass-roots people able to question him, and we exposed to the public what his salary was, and what his position on a variety of things was, and the fact that he sits on General Motors' board when he also plays a substantial role in determining the rates of a customer like GM, with a large plant at Fremont, for example, which is one of the largest energy users in the PG&E service area. They profit from the rate structure and we showed that it's what's commonly known as a conflict of interest; actually that's a wrong notion. It's really kind of a community of interests: it's just not *our* interest. In other words, this is the kind of style we took.

Then, we—not simply being critics or protestors—we developed our own proposal to turn the rate structure around and create a lifeline rate, and we're now in the process of getting it through the legislature. Actually we've had a very complicated process, whereby we've used the legislation, its introduction, getting support from prominent politicians, getting it past the various committees in the assembly, and past the full assembly now. So our bill in the legislature, that bill which first helped the Public Utilities Commission begin to seriously consider lifeline, is now getting support for passage from the same Commission, and it's kind of a combined kind of process. So that's the kind of thing that we've done. [The bill was later passed by the senate and signed by the new governor, Jerry Brown.] That process has, in turn, generated pressure on the Public Utilities Commission.

Meanwhile, just to to fill out the picture, as that campaign got rolling, and as we developed the ability to raise money, and after we developed our own ideas, the organization that we had in our vision when we started that campaign has now begun to take shape. The Citizens' Action League is what we called that. . . . E&GP created the Citizens' Action League, and now the Citizens' Action League is going on to other issues, and to create its own organizational base. We're organizing chapters of the League in San Francisco and San Mateo. And we eventually will have chapters in a number of communities. Both up here and, I think, eventually statewide.

Those chapters now have begun to take on other issues. The largest example is here in San Francisco, we've gotten in a major fight on the property tax. The underassessment of downtown [commercial] property. And the complications and inequities of the property tax. We've mounted a major campaign to "do some short-range things," and we hope eventually to be as successful as we're presently being on lifeline, with coming up with our own kinds of alternative proposals that we also can get through. So that's kind of an example of the structure and the flavor of the campaign.

What long-range goals does the group have?

SAMPSON: Well, I think our basic long-range goals are to unite the majority of people in the state, in a strong, continuing organization which uses the power of citizen action, mass action, people getting together to do stuff; informed by effective and able research. To act in our interest. I think you would say that our vision is of making the big corporations and government responsive and account-able to people. And if there is any word that I think typifies the kind of economic-justice agenda [we support], it probably has to do with redistribution. And restructuring of things like the rate schedule. And, you know, how those kinds of things are put together.

We're very much in the tradition of building a democratic organization, and an organization of large numbers of people, and not one that has got some narrow ideology or some particularist solution to the problem. In that sense I think we're—without, clearly, without the racism—we're in the mainstream of what most people would refer to as a kind of an American populism.

A couple of times you described PG&E as a monopoly. Are you contrasting that with some other, better kind of organization of utility?

SAMPSON: No. In using that, I'm just trying to take care to correctly identify it. I think there's a lot of misinformation in this country that's been disseminated by covering up these terms, you know, that are really accurate [characterizations] of the situation, with fuzzy ones. So that, for example, you arrive at the ridiculous situation where the Public Utilities Commission of the state of California does not regulate public utilities. They only regulate private utilities. And private utilities have become large protected monopolies. Private monopoly utilities are what the PR boys of those utilities have translated into "investor-owned utilities," and other such euphemisms. AT&T took out institutional advertising last week in the *Nation* and the *New Republic* to explain to us why we have to have one telephone system and how glorious that is. Well, I think there's beginning to be a backlash from that kind of thinking. I'm not, wouldn't go so far as to say we ought to have competition between twenty different electric utilities in, you know, in one given city; but there certainly is room for questioning that we have to have the giant concentration that we do in the utility industry, in the precise pattern that we do, and that that's immutably written in the law that's called "natural monopolies." I think that all those things are ideological explanations for convenient ways of making money. And that is an example of, for example, what we find in the declining block-rate structure. In the dim, dark past it was figured out if you sold electricity cheaper to large users—the one who figured it out was Edison and his financial organizer Sam Insull—that you could then get, you could then rationalize the building of large central electric generating facilities. And since you wanted to have those productively occupied all the time, anybody you could get to contract to take large amounts, it was in your interest to reduce the rates. Because you were trying to do that. Well, now, the proper phrase is "in your interest." It was in your interest. It was how you made money. It was not for the public good. The fact that making money in that way electrified America, many centuries later we'll decide whether that was good for American life or bad. Most people are inclined to think that electricity is a help for man, although we're finally getting some dissent on that.

In any case, the industry was for a long time in a declining-cost situation. The more that was made, the cheaper it was to make it. So there was some fit [between costs and rates]. However, there wasn't the ideal fit that they've always talked about, and in part it was an after-the-fact rationalization. That it was cheaper to sell this stuff to the big users. So that rationalization—which has some correspondence to reality, you know—more or less became the conventional wisdom and now the "dumb" utilities people—dumb, in quotes—think that that's actually how costs are in the industry. They think all the rationalizations about how expensive it is to sell to residential consumers, and how cheap it is to sell to big industrial users, they think all those rationalizations which were built in to serve a particular pattern and set of interests are God's only interpretation of what reality is. The smart ones know that it's just rationalizations, and that cost accounting is an art . . . . So we're in a situation where [utility] people are saying that residential consumers have been getting a subsidy all these years from the big users. I mean, well, if you stack up the costs—however you want to stack up the costs—you can prove anything by cost-accounting figures of that nature. I think any true look at it would show that when you're continually expanding the facilities, to keep selling more, and more, and more, that some of your customers don't need all of that stuff. And to charge them for all of that stuff, and then say you're being subsidized, you know, it is a little much. And so we're just trying to cast a little bit of doubt on some of those old-time utility economics.

Now some of this is at the center of the discussion over nuclear power.

SAMPSON: . . . Now, some people, using the initiative process, have gotten the nuclear initiative on the California ballot. We were not involved in that. We don't oppose that [the initiative] and I think the majority of our people have come to distrust the utility industry so substantially, by our experience with them, that anything that they want, nuclear power included, we'd be suspicious of. But we have not taken a stand against nuclear power.

We have taken a stand against building any new facilities before reforming the rate structure and developing other effective conservation measures [that] cut down the need for [new plants]. In fact, it's clear to us that the utilities cannot be trusted to give us information about what we need in the future, since they have a vested interest in increasing their plant. In the case of nuclear, because the nuclear plants are even more expensive, there's tremendous advantage to [the utilities] in terms of the increases in their rate base. The other factors such as cost are much less of concern to them, because they can pass costs along [to the rate payers] all the time. But when their rate base goes up, that's the heart of their profit machine. That's part of the basic equation of rate-making. So we distrust them substantially on conventional, to say nothing of nuclear. And we think that if you change the rate structure around properly and begin to move toward the rate structure that's going to promote conservation—given the kind of response that people have made to the energy crisis and can make if we get straight information and if we can get it reflected in our bills properly—we think that there won't be the need [for nuclear plants] and that, therefore, that part of the equation, that people are concerned about, we strongly support. The part of the equation in which people are attacking safety is outside of our current organizational interests. Of course, many of us are personally persuaded. I'm personally persuaded that building nuclear powerplants is a classic example of how we're going to destroy ourselves. You know, we clearly don't know what to do with all the goddamn waste, and we don't know how to do anything with it—so we're going to make more. And you know, sometime or other, we're going to have a problem with it.

I want to go back to the traditional economics of the utility industry. As you say, since Samuel Insull's time, the monopoly form of the company is justified because it's more efficient to sell big blocks of power and build big plants. What has changed about that? It's still a capital-intensive industry. It still makes sense to use the machines all the time if you can.

SAMPSON: In the first place, a number of things were wrong with it from the beginning. In the past, the *quid pro quo* for privately owned monopolies in utilities was that we'd get to regulate them. But we all know the sharp limits of regulation. That, as a matter of fact, the utility interests wrote the regulatory laws and they wrote the regulatory laws to avoid being taken over. And to perpetuate their franchises. And now . . . they once again forget what the deal is; so that PG&E argues that it has inalienable rights under our Constitution to tell, to put political messages in the rate packet. You know, because you can't interfere with their freedom. But we made a deal with them, way back when: they could have the goddamn monopoly franchise, but, we said, We'll interfere, we'll give you a monopoly franchise, which interferes with our freedom, if you allow us to interfere with yours. And that's the deal. They forgot about that deal.

But corporations are persons. They're entitled to all of these constitutional privileges, aren't they?

SAMPSON: They're entitled to constitutional privileges. But they're not entitled to economic advantage. That's what they're not entitled to. For economic advantage they have to trade something. If they say, "We will refrain from doing X, Y and Z, so give us this economic advantage," and we say, "Okay," then in the same breath they can't say, "Now we've got the economic advantage and we're going to do as we please." That's what the nature of the argument is by these utilities. . . . If it's more efficient for the public to have only one company, then who do we award that to and on what basis? Certainly that's an open question.

Now most people have no knowledge that that's a matter of a deal. They think that PG&E, and PG&E people think, that it's an inalienable right that PG&E has the franchise to sell us our gas and electricity. Well, who says that they should have, you know? Let's get some competitors. How long should we give that to them? Forever? Under what conditions? That's all I'm saying. I'm not saying I even want to take it away from them. I'm not saying that I know for sure what the scale of an economic unit is. But I would say this, I think that at least, the very least, we could reexamine that whole area. It may well be in the public interest that we have only one distribution facility. Or one generating facility. But it might well be that we want to give franchises for billing and distribution to local companies. There's all kinds of ways to slice the deal. I mean New York State subsidizes Con Ed by buying their unbuilt powerplants and owning those. Massachusetts is considering a state power setup. We have a bill in the California legislature. There's all kinds of arrangements. There is not only one arrangement. It was not mandated by God. It's a deal of men. And citizens are going to reexamine those deals.

On his way out, the interviewer is surprised to find Sampson's daughter operating the elevator. She asks, "What floor, please?"

"Are you taking care of the elevator?"

"Yes. My Daddy says we should start helping people when we're young." This is said in an uncommitted way; clearly, she isn't buying it just because Daddy says so. But she has a yellow legal-sized pad on which she is collecting names, and she seems to be having a very good time.

VI.   THE FUTURE

Tom Cochran, physicist, author of a critique of the Liquid Metal
Fast Breeder Reactor (Chapter 52)

Ralph Nader (Chapter 55)

CHAPTER FIFTY-TWO

Plutonium

U ranium prices have been too low and demand too uncertain for private firms to do much exploration for uranium in recent years. The government was forced to provide a number of subsidies and guarantees to keep open the few mines and mills that remained after the boom of the 1950s Cold War expansion. Nor has there been much effort to develop new mining or refining techniques; the known reserves of uranium ore that can be profitably extracted in the US with current technology are therefore quite limited and insufficient to meet the needs of powerplants already planned. On March 5, 1974, Frank Baranowski, then director of materials and production for the Atomic Energy Commission, testified before the Joint Congressional Committee on Atomic Energy that domestic uranium reserves and estimated additional resources of high-grade ore were only 1.5 million tons, but that US demand for uranium would amount to 2.4 million tons over the next twenty-five years (the figures are in tons of U_3O_8). The federal government has relaxed its former ban on uranium imports, and it now seems likely that US utilities will be dependent on foreign supplies of fuel—uranium— in the 1980s if present plans for nuclear expansion continue. Most uranium reserves abroad are in nations friendly to the US—Canada, South Africa, Australia and Zaire—but these nations are already discussing restricting exports to maintain prices and preserve resources for domestic use. There are ample reserves of low-grade uranium ore in the US—uranium is a common element found in granite and many shales—and now that uranium prices have begun to rise rapidly, exploration and development have been renewed at a modest pace, but it will take many years to replace current pick-and-shovel mining techniques with the advanced technology needed to exploit these ores at costs reasonable in dollars and environmental damage.

Another fuel-supply problem is the enrichment process, in which ordinary natural uranium is processed to enhance its content of uranium 235, the isotope which is actually burned in a powerplant. The government's three present enrichment plants, built for the Cold War development of nuclear arms, are not now running at close to their capacity, but future commitments of nuclear plants will soon overwhelm them. As of August, 1975, despite the cancellations and deferrals of the past two years, the government had firm contracts to enrich uranium fuel for powerplants with a total capacity of 307 million kilowatts, both in the US and abroad. (The figure gives some indication of the speed with which nuclear power is expected to expand: the entire electric-power generating capacity of the US was just under 500 million kilowatts at the end of 1975.) Existing enrichment plants could supply uranium fuel for only 290 million kilowatts, however, at their nominal rating. Even pushed to their limits, existing plants would be hard pressed to serve any more than existing commitments. If there are to be any further orders for powerplants, they will have to be served by new and as yet unbuilt enrichment facilities. A new enrichment plant will cost at least $2.5 billion dollars, and several more will be needed in the next few years if the nuclear industry expands on schedule. There were no firm plans, early in 1976, to build any such plants.

There is uncertainty as to which of several presently competing enrichment technologies will eventually prove to be most economical. It would be imprudent to make major investments in this field until federal research has established the costs of the various techniques. There is even more uncertainty as to what the demand for nuclear fuel will ultimately be. Private investment in uranium enrichment has been difficult to obtain, and it will be many years before a private commercial plant is built. In the summer of 1976 Congress began action on legislation to permit and subsidize new uranium enrichment plants, but construction of new plants was still several years in the future. Even if action were taken fairly soon, which is not to be expected, there will probably be a shortage of enriched uranium in the early or middle 1980s, when many of the powerplants now on order are expected to go into operation. There is simply not enough time to organize and build new enrichment plants to meet the planned demand.

If a shortage does materialize, uranium-fuel prices will rise and some power reductions will be required. This is precisely what nuclear power was intended to avoid, of course, and would reduce its supposed advantage of lower operating costs.

During normal operation of a powerplant, a small portion of its fuel is converted to plutonium. The nuclear industry proposal was to use this plutonium to substitute for some of the enriched uranium which will be demanded. Testing in past years established that a part of the uranium in present reactor fuel can be replaced with plutonium without making changes in the powerplant. Since the plutonium was extracted from reactor fuel in the first place, this is called "plutonium recycle."

One should bear in mind that the uranium fuel (only slightly enriched as used in present reactors) cannot be turned into a nuclear explosive without massive enrichment plants costing billions; the fuel is, for all practical purposes, not usable in explosives.

But plutonium *can* be made into explosives, and it *can* be removed from reactor fuel by fairly simple chemical means. (The task of processing perhaps a ton of fuel would be difficult in practical terms, of course.) Adding plutonium to reactor fuel, therefore, makes every power company the custodian of potential nuclear weapons and every shipment of fuel a target for terrorists.

The extraordinary damages which could result from the theft of nuclear-explosive materials hardly need to be emphasized. In 1974, when the Atomic Energy Commission formally proposed plutonium recycling, the draft proposal said quite clearly that present security measures would not be adequate to prevent these damages if plutonium recycling were implemented. That a federal police force to guard nuclear fuel is now considered as one of the ways of dealing with this problem is an indication of its gravity and also of the urgency of the fuel situation which creates the motive for plutonium recycling.

It is solely economic pressures which are creating these unmanageable risks. Plutonium need not be extracted from powerplant fuel for recycling; it can be left in the mixture of intensely radioactive wastes produced during powerplant operation. While the wastes themselves are poisonous and pose problems as to storage and ultimate disposal, it is difficult to imagine their successful theft for terrorist purposes. Critics of the nuclear-power program generally argue that it is only after plutonium has been extracted from fuel wastes that there is a threat of theft.

In an industry developing at a more leisurely pace and without the present emergency, nuclear fuel might be stored and ultimately disposed of without any effort to extract the plutonium and enriched uranium which remain. The economic penalty would be small with proper planning. But with the present shortage of enrichment facilities, no other source of fuel than recycled plutonium is visible in the near future. Over the longer term, plutonium fuel is the means by which utilities hope to keep fuel and operating costs low even though high-grade uranium ore deposits are depleted.

For these reasons, the next stage in nuclear development after plutonium recycling is expected by the industry to be a new form of reactor, known as the "Liquid-Metal Fast-Breeder Reactor" (LMFBR), a powerplant that would be yet another extension of the effort to reduce fuel costs and increase capital investments, the strategy of the power industry of the past eighty years.

The LMFBR is also an extension of the Cold War technology of the nuclear industry; it would rely on plutonium fuel and could make profitable use of the plutonium now being produced in current reactors, not all of which can be recycled. The fuel-reprocessing plants being planned to supply plutonium for recycling will, in a few years, be able to supply plutonium for the LMFBR as well.

This congruence is not coincidental: the "breeder" is the oldest dream of the nuclear-power business. An early plutonium-fuelled reactor, Clementine, used for military research, was the first nuclear reactor to produce electricity—enough to power a light bulb—in 1947. Plutonium-fuelled reactors, of a type similar to that now again being proposed, were expected to be the first to produce commercial electric power. The Enrico Fermi plant, on Lake Erie near Detroit, was an LMFBR the Detroit Edison Company began trying to build in the 1950s. A lawsuit brought by the United Auto Workers and others, guided by Leo Goodman of the AFL-CIO's industrial-union department, sought to halt the construction, but the power companies' plans were upheld by the US Supreme Court. During the years of litigation, an experimental breeder—EBR-I—blew up at a test facility in Idaho. Then a military reactor, the SL-I, exploded, killing three men. Finally, in 1966, as the Fermi reactor was being readied for power production, a piece of metal broke free within its cooling system, and a portion of the reactor's core melted. The reactor was shut down without mishap, fortunately, and the plant was kept out of operation for a year while its owners tried to find out what went wrong. The incident attracted very little public attention until *Environment* magazine, in 1967, published a full account of the accident and its disturbing implications.

The accident at Fermi was a reminder that the LMFBR, unlike the reactors in use today, can undergo a nuclear explosion. Not, certainly, an explosion comparable to that of a weapon, but one severe enough to burst the protective structures and release the billions of curies of radioactive material inside.

Such an accident may not be likely—it is difficult to guess about its probability—but it cannot be ignored. A small nuclear explosion must therefore be added to the other hazards of meltdowns and vapor explosions which LMFBRs share with conventional nuclear powerplants. Richard Webb, another of the scientists whose criticism of the nuclear industry has made it difficult for him to find work, wrote his doctoral dissertation on the subject of explosions in LMFBRs, and described several possible means by which explosions as large as that of thousands of pounds of TNT might occur.

These additional hazards arise from the fundamental principle of the LMFBR. Its purpose is to generate more plutonium than it burns—to "breed" plutonium, and this seemingly magical trick is accomplished by a theoretically simple technique. The initial plutonium (or enriched uranium) fuel is surrounded by ordinary uranium. In the process of splitting and producing heat, the initial fuel generates neutrons, tiny particles which carry on the nuclear reaction; these neutrons bombard the ordinary uranium and convert its larger part, otherwise useless uranium-238, into plutonium. If the reactor is properly designed, more plutonium will be produced than is burned. The practical effect is to make a larger portion of ordinary uranium usable as fuel, effectively enlarging existing high-grade ore resources fiftyfold.

Plutonium, unlike the uranium mixtures used for present powerplants, can sustain a chain reaction without the need for water or any other moderator. The fuel of an LMFBR, therefore, can undergo uncontrolled chain reactions more easily and can in theory suffer a nuclear explosion.

Such an explosion might result, for instance, if the cooling liquid were cut off from a portion of the fuel of an LMFBR, as happened in the Fermi accident. The LMFBR is cooled by liquid sodium—an opaque, highly combustible metal that reacts violently with water, a characteristic which greatly complicates the design of an LMFBR. The cooling liquid, sodium, might be lost through a pipe break, as in a conventional reactor, or through a sodium fire or explosion, or through one of the instabilities found in LMFBR fuel. Relatively small disturbances, like a blockage of a fuel channel, may spread throughout the reactor's fuel, causing widespread melting and collapse of the fuel. Once the process of widespread melting has begun, there is a possibility that energy releases in one part of the fuel will compress other parts, or that a large portion of the fuel will collapse into itself; in either case, a nuclear explosion of small dimensions would result. The further result might then be a catastrophic release of radiation to the atmosphere.

After its accident, the Fermi reactor was shut down and eventually dismantled; there is no breeder reactor now operating in the US. France and Britain, which have no domestic uranium reserves, each have an experimental LMFBR operating, as does the Soviet Union. Despite thirty years of effort, there has been no successful commercial powerplant based on the LMFBR design. The French reactor, built without regard to cost, is the only one so far to operate without difficulty. But the Ford administration, early in 1976, announced its intention of committing ten billion dollars to development of a commercial LMFBR. The first step, the building of a billion-dollar test facility, in Representative Mike McCormack's district in Washington, was already underway; the second, the building of a two-billion dollar prototype plant, at Clinch River, Tennessee, was to begin in 1977. Commercial use of breeder reactors is at least twenty years in the future even if all these billions are expended.

If it is successful, the breeder reactor would be the basis of what Glenn Seaborg, former chairman of the Atomic Energy Commission, called the "plutonium economy of the future." Seaborg, codiscoverer of plutonium, seemed to welcome the prospect.

Plutonium, as Seaborg knew, is nasty stuff to handle: it is extremely poisonous; it can burst into flame when divided into fine particles; and, when inhaled, it is one of the most potent causes of lung cancer known. And, of course, it can be used to build nuclear weapons. An industry founded on this material—material shipped across the country in thousands of separate packages, handled, fabricated, processed, disposed of—would surely represent a considerable extension of the hazards of the present nuclear industry. These increased hazards are generally conceded. The only question is whether they are worth the benefits to be derived from LMFBRs, which in turn are dependent on rapid expansion in the nuclear industry. A slowly expanding industry, or one which was not growing at all, could not use or afford an LMFBR: only the demand for rapid increases in fuel supply over a long period justifies the breeder.

The LMFBR is the nation's highest priority energy program, in terms of dollars spent. The private utilities have grudgingly agreed to contribute two hundred million dollars to its development, and the federal government is spending roughly five hundred million dollars per year, a figure which will continue to rise under current plans. The breeder reactor is nevertheless almost invisible. No efforts have been made to publicize it; it has received little attention in Congress, and the public is almost entirely unaware of the vast sums being expended. Very few details of the program would have been published at all if it were not for a lawsuit by the Scientists' Institute for Public Information in which Gustave Speth, of the Natural Resources Defense Council, representing the Institute, succeeded in persuading a federal court to order the AEC to prepare an impact statement describing the program as a whole. The AEC, not surprisingly, looked upon the LMFBR and found it good, but the AEC's report was rejected by other federal agencies, and by the Scientists' Institute, as inadequate. At the end of 1974, the AEC managed to produce another draft impact statement but could take no final action: Congress had moved to dissolve the AEC, and the breeder program passed, on January 19, 1975, into the hands of the new Energy Research and Development Administration. In January, 1976, ERDA announced that it would accept the AEC's draft as its own final statement, thereby assuring further litigation; the Natural Resources Defense Council brought suit to halt construction of the prototype plant.

The most effective criticism of the LMFBR program has been that made by Thomas Cochran, a young physicist now working for the Natural Resources Defense Council, whose book, *The Liquid Metal Fast Breeder Reactor*, prepared for a Ford Foundation project, is a devastating attack on the program's safety and economic justification. Cochran outlined the explosion hazards and cost overruns which haunt the LMFBR. More recently, Cochran and Arthur Tamplin, a former AEC scientist, have argued that plutonium is even more poisonous than had been believed; that tiny particles of plutonium, if embedded in the lung, are a hundred thousand times more likely to cause a lung cancer than other forms of radiation.

This "hot particle" theory was put forward with very little evidence to support it, and now seems to be incorrect, although the simple experiments that might decide the matter conclusively have not been performed. Unlike their past criticism of nuclear power, the "hot particle" theory advanced by Cochran and Tamplin is not an extension of widely accepted principles; it is a theory about the still-unknown cause of cancer and, as such, is not a theory which has secured support among those engaged in cancer research (who presumably have no bias for or against nuclear powerplants fueled with plutonium). The lack of experimental evidence and the lack of theoretical support have not inhibited Cochran and Tamplin, however, who have called for drastic modifications of safety standards and, by implication, of the LMFBR program. It seems doubtful that the LMFBR could be built at all if the hot-particle theory is correct, and so the

Cochran and Tamplin assertions have been widely viewed as an effort to shut down development of this industry. Congressman Mike McCormack, whose district is, so far, the principal beneficiary of funds spent on the LMFBR, uses the hot-particle theory as his prime example of irresponsible statements made by critics of the nuclear industry. But Cochran's work is otherwise so thorough and effective, and the undisputed hazards of plutonium so impressive, that the brief furor over hot particles seems likely to evaporate with little consequence.

Very few people have any direct experience of plutonium, and it is difficult for even those associated with the present nuclear-power program to gain any clear idea of what conversion to plutonium fuel would mean for the industry and the general public. The only plants now handling plutonium are closely guarded military installations, like the facility at Rocky Flats, Colorado, just northwest of Denver. Rocky Flats was built in 1951 as a key element in the country's developing atomic-weapons production system. It is still a unique facility, and all of our military plutonium passes through it at one time or another. Here, plutonium is cast and machined into the shapes needed for the assembly of nuclear bombs. The conventional explosive detonators are made in Miamisburg, Ohio, and the electrical components are made in Kansas City. The various components of A-bombs and H-bombs are shipped from these scattered locations to Amarillo, Texas, where they are assembled at the Pantex plant; all of this assembly line, scattered across the country for security reasons, is directed by the Albuquerque operations office of the Energy Research and Development Administration, which spends two billion dollars per year on military nuclear development.

Plutonium comes to the Rocky Flats plant from a variety of sources, all of them quite distant. The reactors at Hanford, Washington and Savannah River, South Carolina, produce plutonium; from these sources the plutonium arrives dissolved in nitric acid. Plutonium metal scraps, the left-overs from bomb fabrication or the experiments of the bomb designers, are also returned to Rocky Flats. The plant at Rocky Flats is a production facility, and in some ways it operates as the reverse of the fuel-reprocessing plants. The nitric-acid solutions of plutonium are treated to extract impurities, including Americium, and the plutonium itself is separated, reduced to metallic form, and then cast and machined into the requisite shapes. These operations are carried out by largely unskilled workers who are exposed to very large risks; not because of the unique properties of plutonium, but because of the difficulty of working on an assembly-line basis with any radioactive material. Pat Kelly says you couldn't pay him enough money to work on these production lines, although he works with plutonium in other surroundings.

The Rocky Flats facility, twenty-seven miles from Denver's Stapleton Airport, is operated by Rockwell International's Atomic International Division. Rockwell replaced Dow Chemical, which had operated the plant over a period of years when the facility was wracked by accidents and scandals. In 1969, a fire on the plutonium production line put the nation's weapons-production system out of operation for months, cost about a hundred million dollars and exposed a number of workers to plutonium; plutonium powder, exposed to oxygen, had spontaneously burst into flame.

The plant is on the flat desert plateau near Denver, an enormous table rimmed by the distant Rocky Mountains. The twelve-square-mile facility is surrounded by barbed-wire fencing and patrolled by armed guards, but a recent report commissioned by the governor of Colorado, Richard Lamm, and Representative Timothy Wirth, said that the security measures were inadequate and that the plant could be entered at will. A visitor can, in fact, enter through the main gate, but access to the different building complexes is limited to those wearing the proper identifying badges. Shipments of plutonium to and from the plant, shipments which crisscross the country, have never been hijacked so far as is known, and there has never been an attack on Rocky Flats.

Pat Kelly went to work there in 1956. After high school and the army, he had worked for a while in the coal mines of his native Wyoming, as his father had. Kelly, now a middle-aged man with twenty years seniority and a responsible position on the health-and-safety committee of the Steel Workers' local that represents workers at Rocky Flats, recalls the coal-mining days without rancor. He knew that all the miners would come down with black lung, as his father and uncle had; he knew the underground work was dangerous, although even then underground coal mining was more highly mechanized and less physically hazardous than uranium mining. But Kelly did not stay in the mines. He moved on and eventually found himself at Rocky Flats, where his work is still dangerous, but is clean, dignified and highly skilled. Kelly is now a metallurgist; samples of plutonium and other metals come to him from the production lines and also from the research efforts at Rocky Flats to improve production techniques. Like any manufacturer, Rocky Flats is constantly experimenting with new production techniques to cut costs and increase productivity. Kelly receives samples of the plutonium, grinds and polishes them and then photographs the surfaces to show crystal structure or the presence of impurities, inclusions of metal or other qualities.

What do you do to get admitted to where you work?

KELLY: I have a security badge, which I must show getting into the plant, and then I must show it at another gate, an exclusion-area gate, before I can get into that area.

Before you went to work for the plant, were you given a clearance of some kind?

KELLY: A "Q" clearance.

That is the highest clearance available?

KELLY: Right.

How was that done?

KELLY: Oh, first of all, I had to fill out a PFQ, which is a questionnaire consisting of four to six pages, giving my life history; where I was born, who my parents were, who my brothers and sisters, if any, were, my entire scholastic history, my service history and my work history.

Were there investigations?

KELLY: Oh, certainly. I imagine it was the FBI that conducted most of the investigation, but they did; they checked evidently everything that I had put down on the questionnaire.

Okay, you come to work, and you change clothes. Is that in a common locker room?

KELLY: Yes, there are just—

Are the clothes unusual in any way?

KELLY: Well, really, the unusual part is that you must wear the company-supplied safety shoes before you go through the "double doors," which is an air lock into the back area, or the work area. You must put on booties, and you must be carrying a respirator.

Booties?

KELLY: They are a canvas shoe cover. They also have plastic ones, but for the most part, they are canvas shoe covers, with a rubberized sole.

What is the purpose of them?

KELLY: In case you get in a contaminated area, the only thing, basically, you are going to get contaminated is the booties, and this—to throw those away or to wash them—is much cheaper than a pair of shoes.

That's in case there are traces of plutonium on the floor? What is the reason for the complete change of clothes?

KELLY: Again, personal protection, so that you don't carry any contamination from the area, the work area into the locker area, so that there is no possibility of getting your personal clothes contaminated and, in turn, carrying them or taking them out with you—home.

At the end of the day, then you undress and shower before you get your own clothes?

KELLY: It is mandatory that you shower.

Right. So you have a complete change and a shower?

KELLY: Yeah.

Before you get out of the—

KELLY: Before you leave the area, if you are going in and out of work areas within the building, you self-monitor; or you can have a health-physics monitor monitor you. They have what they call a "combo" which has two different probes on it, one for your feet and one for your body and skin areas, and you are supposed to self-monitor with that. It has a dial marked with a red line; it also has an audio signal. If you are contaminated, it will give an alarm, which is nothing but a series of clicks, you might say.

Is this fixed in a doorway?

KELLY: No, it is an instrument, [a] self-contained instrument that is near the exit. By near, I mean within a matter of feet. There are also these type of instruments in the various lab areas, or the various work areas where you will monitor yourself out before you leave one work area to go to another one.

When you have changed clothes, where do you go next?

KELLY: Well, in my particular case, I go back to what we call the Met Lab, which is a metallurgical lab, which is a laboratory area consisting basically of two rooms. Rather small in our case, but you have quite a bit of equipment in those two rooms, including the dry-box system, where we actually perform the work.

If I walked in, it would look like a laboratory?

KELLY: It wouldn't look like any laboratory you have probably ever seen.

What does it look like?

KELLY: You know what a dry box is, or a glove box?

If you would describe that, it would help.

KELLY: Well, a glove box, in reality, is an enclosed stainless-steel box with numerous glove ports where there are the arm-length rubber gloves. Normally these gloves are what we call lead gloves. They are a rubber that is impregnated with lead and this is a closed system. In other words, the only access, the only way of getting anything into that box, you would use what we call an air lock; to get it out, you take it out through a guide port. [Kelly would work on materials within the glove box by inserting his hands and arms into the leaded gloves that extend into the box.]

What else is in the room with you?

KELLY: Various pieces of equipment are inside the dry box.

I see.

KELLY: In other words, the wheels, the various driving wheels, and the various final polishing wheels are inside the dry box. All of the equipment that we need to perform the job, other than actually looking at the final result, photographing the final result, are enclosed in the dry box.

How big is this dry box?

KELLY: Well, in effect, it's a series of dry boxes. Really, you might call it a line and our line is really a T-shaped line. On the leg of the T there are four, three A boxes and one B box. On the top of the T, there are about six A boxes and one B box.

What actually happens then, what do you do inside the boxes? What are you likely to begin doing?

KELLY: Well, in our particular work, of course, we are dealing in relatively small amounts of material, and we are engaged in determining why there might be a failure. We are engaged in bringing out the grain structure of various metals, including plutonium. So, we section, maybe.

You're getting samples of metal?

KELLY: Yes, of various kinds. Plutonium is just one of several kinds of metals that you handle.

When you say small amounts of material, how much is that?

KELLY: Well, we might be engaged in a specimen that would weigh anywhere from one gram to thirty or forty grams.

How big is that physically?

KELLY: Not very big. You might be dealing with a piece of—that might, for instance, be the diameter of a dime or a nickel, and maybe sometimes thinner, much thinner than a dime, and sometimes maybe two or three times the thickness of a nickel.

These are not themselves functional components of any kind, these are samples taken from larger pieces. Why are you examining them?

KELLY: Again, to determine defects, to determine composition, to determine grain structure, to determine, maybe, if the method used to form that piece was right or wrong. We do some quality-control work as a service-organization aspect. This is somewhat of a sideline, because again, most of our work is involving research and development. We do some quality-control work, we might get a specimen that might not even be plutonium, and we might then determine metallographically if a weld is sound, if proper penetration has been achieved, various things like that.

So the samples you are getting are not always necessarily then from the production line, so to speak.

KELLY: That's right. We are an aid to production. We might determine that they are using the wrong temperature, they are using the wrong time, what impurities are being injected into the system which shouldn't be there.

But, it is not like testing each tire as it comes off the assembly line?

KELLY: Right.

So, you have a dime-shaped piece of thing. What do you do to it in the glove box—in the dry box?

KELLY: Depending upon how they want it done, if they request a certain sectioning, then we section it that way, and then we mount it in an epoxy resin, and then we polish that. You rough grind it on coarse—various grits of coarse paper—and the final mechanical polish is . . . you might use a one-micron, or clear down to a quarter of a micron, diamond-paste polish to it.

The very same equipment is used throughout many industries, where they want to find the same type of thing about any metal.

Depending on what you're looking for, the mechanical polishing might be sufficient. If you want an even finer surface, and you want to bring out certain things, you might electropolish it, which is using common, ordinary electropolishing equipment. You electropolish it and then you might electroetch it, using a chemical solution and an electrical current.

In all of these operations, you're reaching through ports and your arms are encased in rubber gloves?

KELLY: Right.

Basically, all that's between you and the material you are handling, then, is the rubber glove?

KELLY: In that case. First of all, you put on surgeons' gloves before you go into the rubber gloves, and this is to protect your hands in case one of the lead rubber gloves gets a hole in it, or whatever.

And the lead impregnation is sufficient to protect you from any radiation that there might be?

KELLY: Well, not from any radiation. . . Depending upon the type of radiation and how far it will penetrate, and what it will penetrate. For instance, alpha radiation can be shielded by just a piece of paper. Of course, now, gamma radiation cannot, and they speak generally in terms of soft gamma and hard gamma. Soft gamma is not as penetrating as hard gamma. And then you also have the neutron problem. Now, primarily in our area, we really don't have the problem with neutrons. In production areas they do, but then there is additional shielding also.

What sort of monitoring equipment do you have on your person, that lets you know—

KELLY: On your person, first, on your wrist you wear a wrist badge which gives you the exposure to your extremities, your hands and your arms. You also wear a body badge.

These are film badges?

KELLY: No. Film badges are somewhat archaic. These are what is known as the TLD, well, part of these. The wrist badge has two TLD crystals. The body badge has several TLDs, plus other indicators. A TLD is a thermoluminescent dosimeter.

What is the purpose of having several crystals?

KELLY: Well, to determine the types of exposure. In other words, for instance, one of them which is sulphur wafer, I believe, upon exposure—for instance, in the case of a criticality occurrence, this then changes into, I believe it is a sodium isotope, which then emits radiation. So, that can be immediately checked, merely by the monitor checking it with the probe on his instrument, in the case of the criticality situation.

You also have various other *shielded* TLDs. In other words, to determine highly penetrating radiation, like a hard gamma, you might have a TLD which is shielded with aluminum or with copper foil. You have others that are not shielded, which will give you [a measure of] almost any external radiation that you are receiving.

Now I'm sure that these descriptions are going to make these sound like bulky devices, but, in fact, they are not. Right?

KELLY: No. Really, the face of the badge is maybe as wide as a package of cigarettes, maybe a little bit longer and [they] are maybe, about a quarter of an inch, or maybe five-sixteenths thick.

Now if you look at it yourself, you can't tell anything, right?

KELLY: No.

These are collected periodically. How often do you turn them in?

KELLY: It depends on the area in which you work. If you work in a plutonium area, you are on no less than a monthly badge exchange. If you work on certain areas in production with plutonium, you are on a two-week cycle.

How often are your dosimeters collected?

KELLY: Once a month.

Okay. Now, to get back to the dry boxes, ordinarily you're working with your hands in the gloves and the only exposure you are getting then is from the gamma radiation that comes through the gloves.

KELLY: Basically.

Do you know what exposure you are getting, do they tell you when they read the dosimeter?

KELLY: They furnish an exposure read-out monthly, which tells you what your monthly dose is, both wrist and body, the penetrating radiation. It tells you your quarterly dosage, and your accumulated dosage.

Do you remember any of those numbers offhand?

KELLY: On myself? No, because mine has been very low, it's low enough that I'm not concerned with it. Now, I do check it so that I can determine any rise. And mine is extremely low because of my activity. I'm gone a lot.

Because of your union activity?

KELLY: Sure.

You mention the possibility that sometimes there are holes in the gloves. Does that happen often?

KELLY: Compared to what?

How often does it happen?

KELLY: It happens frequently. Because gloves do deteriorate, and depending upon what type of atmosphere is in the particular box, the glove-life might be anywhere from a week to many, many months. . . . The monitors do check the gloves routinely. Plus, when you come out of the gloves, there is a self-monitoring device there which you are supposed to use, too. Before you go from this set of gloves, for instance, into this set of gloves, you are supposed to check your hands and arms.

Do you check every time?

KELLY: I would say that probably everybody doesn't every time. But this is the rule, they are supposed to. And I'll say this, that Rockwell has recently come out with a reemphasis of that, saying that, by God, check yourself everytime you come out of the gloves.

These are the new contractors?

KELLY: Right. But, see, this is also a personal thing. I guess experience is somewhat of a determining factor. For instance, depending on the operation you're doing, you might go in and out of a certain pair of gloves thirty or forty times in an hour, and you might not check yourself, and, depending upon what the operation is, you know some operations are more conducive to getting a puncture wound, for instance, than others. So, this is a judgment factor, I guess.

How often has it happened to you, that you had a hole in the glove and came out of it and discovered that you had some contamination on your hands?

KELLY: Over the years, it would be real difficult to hazard a very accurate guess but it has happened to me several times.

Less than a dozen?

KELLY: Oh, I'd say more than that, in a period of about sixteen or seventeen years of working in plutonium areas, I'm sure it probably happened more often than that.

In those instances do you get or are you getting plutonium on your skin, or is it just all on the gloves?

KELLY: Most of the time it's on the surgeons' gloves, depending, of course, where the glove failure is. If the glove failure is merely through glove fatigue . . .

[the failure] is up higher on the arm, [but] in all probability it is still not going to be on your skin, it is going to be on your coverall.

You're working with a variety of cutting and grinding tools; do you ever just cut right through the gloves and cut yourself?

KELLY: Well, that has happened. I have received one puncture wound. Luckily, it proved to be noncontaminated, but it could have been contaminated.

In that case you would get some plutonium in you?

KELLY: Yes.

And this happens occasionally to people who work at Rocky Flats?

KELLY: Correct.

How hazardous is that?

KELLY: That depends entirely—from everything I as a layman know, and I have listened to a lot of expert people—it depends entirely on how you acquire this. And in what form it is. For instance, you get surface contamination on your skin, your hands, your face, or whatever, this can be very readily and easily removed, sometimes merely by simply washing your hands or your face. If you ingest it, this is another thing. From everything I've heard and read, the worst thing to do is to inhale it.

In the grinding operation, anyway, you generate some powder, generate some fine particles, and I guess they are in the air inside the dry box?

KELLY: That is right.

Do they ever get in the air outside the dry box?

KELLY: The only way it can is if you have a breach of the glove-box system.

Does that happen?

KELLY: Yeah, it has happened.

Have people inhaled plutonium?

KELLY: Oh, yes.

Do you know what the results have been?

KELLY: Well, the results, at least as far as clinical evidence, so far, to the best of my knowledge, there has been no evidence, no clinical evidence of damage.

How often has this happened, to your knowledge, that people have inhaled plutonium?

KELLY: Again, without looking at the explicit records, and this certainly is not a hard number, but I would have to estimate roughly, eighteen to twenty-five people.

Over what period of time?

KELLY: Since the plant started operation in 1953.

This is out of, roughly, what total of people employed?

KELLY: If I remember correctly, the number of people who have passed through that plant is in excess of eight thousand people.

There were some very well publicized incidents at the plant when Dow was operating it; do you think they were at fault? For the fire, for instance, do you think that could have been prevented?

KELLY: Well, this is kind of Monday-morning quarterbacking.

Well?

KELLY: But, yeah, I believe that several things contributed to the fire, and maybe it would have happened, but I believe personally, based on a lot of things, that the extent of the fire at least could have been controlled, or contained, if they maintained the controls that they should have.
     The building where the fire was, was really separated into two buildings. And then the barriers, as you might want to call it, were taken down and it was made into one building. And, of course, the smaller you can contain a fire the easier it is to control; the smaller the area of contamination, the easier it is to control and the easier it is to clean up. I believe that the precautions of installing fire doors in the dry-box system would have prevented, to a large extent, the spread and the damage.

What did the damage wind up to be?

KELLY: Well, that is kind of an argumentative figure somewhere between fifty and one hundred million dollars. It was classified as, not the largest industrial fire, but the most expensive industrial fire [ever to occur in the US].

How many people were exposed in that incident?

KELLY: In reality, I think that there were really two people that received significant amounts, if you want to call it exposure. There was inhalation, and there was contamination, but I believe that one of those was one of the fire-fighter people.

There was some release to the environment, too, in that fire, right?

KELLY: Yeah, they calculate that there was some release, they classified it as minor, and I guess in reality it was, I guess, and very luckily so, in that particular incident. In other words, the building itself was not breached. There was no explosion which blew out walls and stuff like this.

It was seepage from the ventilating system?

KELLY: Right.

As I recall, plutonium was found in the soil surrounding the plant. It didn't, in fact, come from the fire, but there was a big outcry. The plutonium actually came from waste barrels left outdoors.

KELLY: I don't believe that Dow would ever have voluntarily released the information. I believe they finally got into a position, I *know* they finally got into a position, where they had to attempt to hide it, attempt to belittle it, or let it all hang out.

Do you think there was any significant exposure to the public from any of these releases?

KELLY: No. If you take the aggregate, I would have to say "no." Knowing what the readings are on the various sampling sites, both on the plant site and off-plant sites, compared to what you would have to consider a significant value, my answer would have to be "no." I just don't believe that there has been.

Now, finely divided plutonium metal, is that particularly hazardous material?

KELLY: Yes.

Is that likely to catch fire?

KELLY: That is where you will most likely have the problem, because plutonium . . . self-generates heat. It doesn't self-generate heat to where you will really get, as a general rule, spontaneous combustion, but, yes, finely divided plutonium is very pyrophoric.

And this, in fact, is what you work with?

KELLY: Yeah, I *create* finely divided plutonium.

Doesn't that make you nervous?

KELLY: No, I am not afraid of it. I hold it in high respect, but I am not afraid of it.

This is also the form in which it is most hazardous to be exposed to—this is the kind that can be inhaled. Right?

KELLY: Very finely divided. In other words, when plutonium burns, as does any other pyrophoric metal, you will get very fine particles, which can be suspended in air. And, if you breathe that, then you are inhaling particles of plutonium.

Could it happen that the material you are working with would catch fire and you would inhale some of the plutonium outside?

KELLY: It is possible, but not very probable, because you have a lot of control. In other words, you have to realize first of all, the whole building itself is pressure negative to [has an air pressure less than] the outside atmosphere. And, then the work areas, including the hallways, are pressure negative to the outside areas of the building. The rooms are pressure negative to the hallways, for instance, and then the dry boxes are pressure negative to the room atmosphere.

What does that mean in practical terms? If there is a hole in the glove box, what happens?

KELLY: It will not come out, because of the negative pressure. It will go through the filtering system. The air will flow from the room into the glove box. Unless you would have something that would cause the pressurization of a box.

What could do that?

KELLY: Well, you could have pressurization through a ventilation-system failure. You could have pressurization if you were using something in the box which would cause the pressurization, like an air line breaking or something like this; and that is why, for instance, it is taboo, generally speaking, to have an aerosol can, for instance, in a box.

After you're done with the material you're working with in the glove box, what happens to it?

KELLY: We then will do what we call "scrap it out," which means it is returned to the system; it might be burned, controlled burning, reduced to what we call burned-off size, and then will go through the recovery process and ultimately . . . again be reduced to metal.

So you're just, you physically carry it back somewhere. That's how you got it in the first place?

KELLY: It's a very controlled system, and a very controlled method of transportation and a very controlled method of packaging.

How does that work? When you want to charge out some plutonium, do you sign for it?

KELLY: Oh, yeah, very definitely, you sure do. You sign—you know it's not like going to the tool room and signing a little chit for a tool. This is called accountable material, Nuclear Materials Control controls this, they monitor it. We have to inventory, have to account for every gram that we have.

Do you go to a window and ask for your plutonium?

KELLY: Oh, no. Most of the time the material we get is sent to us by someone else, an engineer, or what have you. It is physically transported by an individual, usually in what we call "production control," and then when we get it from him we on the spot sign for that. Then we bring it into our area, we enter it into our log, we enter it into the dry-box log, so we know . . . at any given moment how much material we have in that area.

Could you steal some of the plutonium?

KELLY: Well, I guess anything is possible, but there is a safeguard system to prevent that from occurring. You have to understand, the first safeguard system—I don't know if you want to call it the *quality* of people that are employed, but at least these people have to undergo a very comprehensive security check, so therefore, if they have any background of being a kook or a radical, or whatever terminology you want to use, they are probably not going to get clearance. I guess that's the initial safeguard system. The other safeguard systems are the inventory control, the metal control, how it is handled, and, of course, there are other safeguard systems of detection which, if everybody knew about them they no longer would be a safeguard.

If somehow one individual were employed at the plant—I am not asking you how it could be done, because I wouldn't want to know even if you were willing to tell me—but do you think it is a practicable thing, if a person who were motivated to do so became employed at Rocky Flats, is it a practicable possibility to steal enough plutonium to make a weapon?

KELLY: In one fell swoop?

Say, over a period of months or years.

KELLY: Well, again, I'd have to say almost anything is possible, but it is very, very improbable, for many reasons. Number one, if it is done over a period of time, in all reality, [it] would somewhere show up on an inventory. Also, detection systems would prevent any *magnitude*—and if you are talking about getting enough to make a weapon, you would need a considerable amount, and this would take considerable numbers of trips, over a considerable length of time, and again this would probably be caught.

Could you do it in one fell swoop? Grab it and run?

KELLY: I would say no. (*Laughs.*) You would have to run pretty fast. I think with the detection systems, and the way the physical makeup of the plant is, unless he could run faster than a speeding locomotive and faster than a .45 bullet, I doubt if he could make it.

I get the feeling from what you say, that you feel the problems in handling plutonium are manageable.

KELLY: Yes.

Would you characterize the whole nuclear industry that way?

KELLY: I would—well, I'd have to say, depending upon, number one, do you have a responsible and responsive contractor? I guess that is where the first safeguard comes, you might say. Secondly, you have to have adequately trained and, certainly, well-screened people. You have to have very advanced technological equipment. You have to have very competent technicians.

Now, Rocky Flats maintains a very high level of personnel. Do you think it is possible to maintain this high degree of screening and security and training in the whole power industry, if it were to be based on plutonium?

KELLY: Yes. Because, when you're talking about the whole power industry and the people that operate it, again, only a percentage of the people [are] engaged in handling the fissile material [plutonium]. And I believe that with the proper safeguards, there shouldn't be any problems.

Pat Kelly is probably right that a plutonium industry is feasible, although the costs will clearly be large. It is also quite possible to have a nuclear-power program without ever recycling plutonium or building breeder reactors. The fuel of current powerplants could simply be disposed of intact—the "throwaway fuel cycle" which both Donham Crawford of the Edison Electric Institute and Carl Walske of the Atomic Industrial Forum are willing to consider as a possibility. The Nuclear Regulatory Commission will not decide until 1977, according to its present schedule, whether it will permit the recycling of plutonium. If it decides not to do so, present reactor fuel would go unprocessed and would be disposed of as waste. Such a throwaway fuel cycle would add to the cost of reactor fuel in future years, worsening the difficulties suffered by utilities. If there is no plutonium recycling, further doubt will be cast on the next proposed step in nuclear power development, the plutonium-fueled breeder reactor. The decisive question for the plutonium economy of the future will be whether power companies buy and build breeders, and this will be determined by whether the power companies can continue their traditional mode of doing business, building ever more expensive plants for the sake of ever smaller savings in fuel costs.

CHAPTER FIFTY-THREE

Slowdown

THE ELECTRIC-POWER INDUSTRY has difficulties that threaten to engulf nuclear power and shake the foundations of the nation's economy. The economic forces which once coincided with military and foreign policy to bring nuclear power into being have spent their initial impulse; economic realities have changed, and the new high prices of fuel and capital, which seem to be permanent features on the landscape of the next few years, may mean that the central-station power monopolies are traveling the course of rail passenger service—a necessary service that becomes increasingly unprofitable as competitors absorb the more lucrative industrial business. The utilities' long fight against public ownership may end in defeat—through receivership, as in the case of Penn Central, rather than socialism—unless the power companies take aggressive action to adapt themselves to changing conditions. But there are no signs of such adaptation yet, nor of any recognition that it is soon to be required. Worsening the chances of change, the federal government continues to bring strong pressure to bear to expand the use of nuclear power, for military and foreign-policy aims quite different from those that created the nuclear industry in the early years of the Cold War.

The troubles of the power industry began in the 1960s, when electric-power rates stopped their seventy-year decline and began to rise faster than the prices of other goods and services. The reasons for the reverse are not entirely clear. In part, the incentives toward waste built into the power industry seem finally to have overcome the increases in efficiency in generating plants. Tinkering with the old steam-turbine design had reached its limits, and increases in efficiency through higher steam temperatures and better turbine design could no longer be achieved. Not only did the efficiency of new plants stop increasing, but for many companies, and in some years for the industry as a whole, the efficiency of the newest plants, measured in terms of "heat-rate," began to decline. Interest rates and construction costs, critical cost elements for the power industry, began their slow rise after World War II, and by the 1960s these, too, seemed to be overcoming the efforts of utilities to maintain the growth of their earnings. Accelerating rate increases were needed, and power costs began to rise.

The Arab oil embargo and the oil cartel's dramatic increases in fuel prices precipitated a severe crisis both in the economy and in the power business: by 1974, the utilities seemed to be on the verge of bankruptcy. High fuel prices and high interest had wiped out power-company profits and set off a scramble for unheard-of rate increases. The rate increases were slow in coming, however, and Con Ed of New York missed its first dividend: utility bond and stock prices consequently plummeted, and the nation began to fear for the solvency of its largest industry, measured by capital investment. Total sales of kilowatt-hours stopped their growth and began to decline.

Massive rate increases in 1975 boosted industry's revenues more than thirty percent, however, even though sales of power had increased only very slightly; in almost every state, public-service commissions granted huge increases in rates to power companies and gave permission to pass through fuel price increases to customers automatically. Some states relieved power companies of high interest rates by the simple expedient of forcing customers to lend utilities money without charge. Missouri was among the states which decided to allow power companies to finance the construction of nuclear powerplants by charging customers in advance. Net income of the utilities increased more than ten percent, interest rates moderated somewhat, and utility stock and bond prices recovered, along with the national economy, in early 1976. Industry executives breathed a sigh of relief and let it be known that the crisis was past.

While the immediate emergency was over, the enduring problems of the industry had not been addressed, let alone solved. Demand was levelling off because of higher prices, and the cost of building powerplants was rising; fuel costs and other operating costs were rising even more rapidly, insuring that higher prices would continue to dampen demand. Power companies throughout the country moderated their estimates of future power growth from the traditional seven or eight percent per year to a more modest five or six percent; powerplants were cancelled or deferred in large numbers. In 1975, cancellations of nuclear plants exceeded the number of new orders. In the two years following the oil embargo, about half of the two hundred nuclear powerplants on order were cancelled or delayed for periods ranging up to eight years.

The power companies had been caught in the midst of one of the industry's periodic expansion periods and found that many of the plants already under construction would not be needed for many years. This problem was particularly severe in New York and New England, where utilities had hoped that nuclear power would rescue them from the pressures of air-pollution control and the difficulties of fuel supply. Peter Stern of Northeast Utilities said in 1975 that his system already had fifty percent overcapacity and had another huge nuclear plant, Millstone III, under construction. For several power companies, the remedy was to sell off plants or their interests in plants under construction. Stern said that Millstone III would be built, not for his company, but for "New England," and late in the year the utility did, in fact, begin efforts to raffle off shares of its investment in nuclear power. Millstone III's completion was set three years further in the future—which meant that the company's stockholders and ratepayers would pay three more years of interest charges on several hundred million dollars of unproductive investment. The company announced it would also try to find purchasers for a seven percent interest in Millstone III and would try to sell its twelve percent ownership in two Seabrook nuclear plants in New Hampshire and a thirteen percent share of the Pilgrim plant near Boston. Leland Sillin, Jr., the company's chairman, said that he was "confident" other power companies would buy these shares, but, in fact, most major utilities seemed to be in similar straits. United Illuminating Co. in Connecticut began negotiations for three percent of Northeast's Millstone III but at the same time announced its plans to sell off a ten percent share of the Seabrook plants. The power companies of New England seemed to be engaged in a game of musical chairs in which the local governments were to be left standing; Con Ed of New York, finding no local takers for its incomplete powerplants, sold one nuclear and one oil-burning plant to the New York state government. For several years the utilities had been adding generating capacity, planned long before, at a rate of growth close to ten percent per year; but demand for electricity had been almost level for two years.

The sudden wave of cancellations and deferrals set up reverberations throughout the US economy. General Atomic, which had secured orders for six of its gas-cooled reactors and had hopes for many more orders of this adventurous new design, finding itself reduced to work on a single plant, announced its cancellation of the remaining contracts and its withdrawal from the nuclear-power business.

In announcing the withdrawal of his firm, which had become a joint venture of the Gulf and Royal Dutch/Shell oil companies, a vice-president of General Atomic heartily damned the federal government for launching an industry as complex as nuclear power on a scale and with a speed too great to be managed by private industry.

Other firms were severely troubled. Westinghouse, which had shared eighty percent of the reactor business about equally with General Electric, seemed severely shaken, and there were rumors its board of directors had discussed withdrawing from the nuclear business entirely. Off-Shore Power Systems, a Westinghouse venture with Tenneco to build nuclear plants on artificial atolls off the Atlantic coast, collapsed and was left with a single, tentative order from a New Jersey utility. Westinghouse had other troubles. It had used highly favorable fuel contracts to sell powerplants of conventional design to utilities but had failed to secure enough uranium at low prices to meet its commitments. In 1975 the company was forced to abrogate its contracts with dozens of power companies—including Missouri's Union Electric—which promptly brought law-suits to compel the company to live up to its obligations. Westinghouse had promised uranium at eight dollars per pound or less, but market prices were over forty dollars per pound by the end of 1975. Union Electric was forced to supply its fuel needs at almost fifty dollars per pound. Uranium prices continued their mysterious rise in 1976, following the rise in oil prices, although demand for uranium, like that for oil, was actually declining. Reflecting the cutbacks of plant orders, in August, 1975, the federal government announced that contracts for enriching uranium fuel had declined substantially. No shortage of uranium re-sources caused the price increase: In the summer of 1975, miners were still taking uranium out of the ground by the pick-and-shovel techniques of the nineteenth century; few new mines had been opened and the ore-processing facilities were ancient and wasteful. There was obviously a glut, rather than a shortage, of uranium in 1975, and this would be the case for many years to come, before real costs would justify the use of advanced technology.

US reactor manufacturers relied heavily on export sales to carry them over the crunch of 1973–1976, a period during which export sales, encouraged by the federal government, rose briskly. Westinghouse sold a dozen reactors to Iran, both directly and through Framatome, the French company licensed to produce reactors of Westinghouse design. The French government bought eighteen reac-tors. But in late 1975, Westinghouse was forced to give up its forty-five percent ownership of Framatome: France was determined to take its energy industry out of foreign hands. In exchange, Westinghouse received a much-needed opportuni-ty to buy three million pounds of uranium through the French government, which, in turn, had secure supply sources in the former European colonies of Africa.

General Electric, a far larger firm, seemed less shaken by the crisis, although it, too, moved to secure supplies of natural resources. On December 16, 1975, the federal government announced it would investigate the antitrust aspects of Gen-eral Electric's plan to purchase—for $1.9 billion in stock—the mining firm of Utah International, which holds extensive foreign and domestic supplies of coal, uranium, iron ore and copper, the basic raw materials of the power industry. In 1974, General Electric was the eighth largest industrial corporation in the US with assets of almost $10 billion; its merger with the billion-dollar mining com-pany would create a still stronger firm that would further overshadow troubled Westinghouse (the nineteenth largest industrial corporation in 1974, with assets of just over $4 billion).

Exacerbating the problems created by declining demand and high interest rates were a series of difficulties experienced in building and operating nuclear plants. Construction costs, usually a well-kept secret, are known to have in-creased substantially over early estimates. A team of businessmen and engineers from MIT, in February, 1975, published for the first time some contract prices and actual costs of building nuclear powerplants. The report stated that contract prices for new plants had risen from $130 per kilowatt in 1965 to more than $700 per kilowatt in 1974, which was known, but also that actual costs of construction overran contract prices by two or three hundred percent in some cases. These overruns resulted in massive losses to Westinghouse and GE on some early plant orders that had been sold on a fixed-price basis; in more recent sales the power-company customers were paying for the overruns. The increases in costs were attributed to a variety of factors, including higher construction costs, unrealistic early estimates, higher interest rates, and increasingly severe safety regulations.

Once in operation, nuclear plants were plagued by dozens of minor and major accidents which took them out of service for hours or months. Consolidated Edison, the Chicago utility with a heavy investment in nuclear power, had suffered a series of equipment failures and some sabotage by its employees; the performance of its nuclear powerplants was so bad during 1975 that the actual figures were witheld from official tabulations. At Browns Ferry, Alabama, the world's largest nuclear power station, owned by TVA, was put out of commission for months by a fire in control wiring, set off by a workman searching for air leaks with a candle. Nuclear plants were available to generate power only about seventy percent of the time during 1975, on the average, compared to eighty-two percent for fossil fuel–burning plants, and the accidents and failures were all well publicized.

This concatenation of difficulties gave great glee to critics of nuclear power, who began saying that the industry was about to collapse of its own weight. The usually boosterish magazine *Business Week* carried a cover story, November 17, 1975, headlined "Why Atomic Power Dims Today"; the *New York Times* followed a few days later with a story headed, "Hope for Cheap Power From Atom Is Fading."

Like the celebrated report of Mark Twain's death, the obituaries were premature. The power companies correctly claimed that many of their problems were temporary. The breakdowns would diminish in frequency, and capital costs were high for all forms of power. Partial recovery of the nation's economy in 1975 and early in 1976 brought a renewed surge of demand for electric power; and while growth rates seemed unlikely ever to return to the historic rate of expansion of the industry, growth in demand would continue at a reasonable rate for many years into the future. The hiatus in plant orders which resulted from the change in growth rates would eventually pass, orders would pick up again and the surviving nuclear suppliers would resume business. Utilities and their suppliers, however, began to press for federal assistance during the transitional period; the federal government, according to one scheme propounded by Nelson Rockefeller and espoused by Gerald Ford, might guarantee loans or purchase nuclear plants to keep the flow of orders steady until demand caught up with existing capacity. State utility commissions expressed themselves willing to grant any rate increases the power companies might need, and power prices continued to rise at more than twelve percent per year in 1976. Profits and stock prices returned to close to normal.

Demand began to rise because energy savings brought on by the oil embargo reached their limits in large areas. Households and small businesses have only limited means of conserving energy; once the thermostats for air conditioners and heaters are reset and lighting levels reduced, there are no further savings of importance available to most individuals. The massive readjustment of public habits which followed the embargo therefore showed itself simply as a brief plateau on the rising curve of individual energy use. After two years, the pressures of population growth and economic expansion again began to drive up power use in homes and shops, despite continued conservation efforts.

Over the longer term, it is not clear how long consumers can or will continue paying the rapidly increasing costs of electric power, however. At an inflation rate of more than twelve percent, electricity will quickly become an intolerable burden to the family of ordinary means. Agitation for rate reform and for lifeline rates and the simple refusal of growing numbers of consumers to pay their bills reflect an increasing resistance to the price of electricity. Customers are particularly incensed when they find their bills growing even though they continue to make intensive efforts to conserve energy within the limited means available to them. During Con Ed's crisis of 1974, fully one-third of its customers' accounts were in arrears.

However, there are far more serious threats to the power business: the private monopolies of power generation built by Samuel Insull and his colleagues seventy-five years ago now seem to be collapsing in the face of competition.

The competitors of the power companies are now, as they were in 1900, the power companies' own customers. Large commercial and industrial power customers—the big office buildings, department stores and factories—have been reducing their demand for power at a dramatic rate. In 1975, when the economy in general was recovering and residential and small commercial customers were increasing their demand by about six percent per year, the large industrial customers were *cutting* their demand by six percent below that of 1974; industrial demand for all forms of energy declined seven percent during 1975. Large industrial customers, unlike the home-owner, could and did make substantial and cumulative cuts in power demand. These cuts almost exactly equalled increases in demand from small customers, keeping power sales almost level during 1975.

There are no published reports or surveys accounting for the dramatic decline in power use among large customers, a decline which seems to be accelerating. But some factors which prompted it seem clear. Large customers have always been lured by extremely low rates—prices just higher than the actual operating costs of power companies, about one twentieth of one cent (one-half mill) per kilowatt-hour. But higher fuel costs and general inflation drove up costs over one mill. Because they were already so large a component of the low industrial rates, fuel prices and other operating costs, when they rose, raised industrial rates dramatically; and for the first time, the gap between the prices paid by small and large users began to narrow. Industrial customers responded by cutting their power purchases—conserving energy and generating their own power. Higher fuel prices, paradoxically, make the latter option more attractive, for an on-site industrial powerplant can be used for several purposes. What would otherwise be waste heat from a powerplant can be used to provide space heating and air conditioning through the absorption process; spent steam may be useful directly in some chemical-processing applications, and the heat may be used for others. Calculations published by the Lawrence Berkeley Laboratory in December of 1975 told the story: even a commercial development of one million square feet—a shopping center—could supply its own power, heating and air conditioning at less cost than that of separate heating and purchased electricity. Fuel savings of forty percent more than overbalanced the additional capital costs of building the small power station that would be needed. The "isolated plant" which Samuel Insull and Thomas Edison sought to drive from the market place is returning, pressed forward by the rapid growth of fuel costs that overshadow the modest economies of scale which utilities are able to obtain in their powerplants.

"Total-energy" systems, as the combined heating-powerplants are called, still are rare in shopping centers, office buildings and apartment complexes, but they are increasingly common, despite problems of initial cost and fuel-oil supply problems. While industrial demand for power rose slightly in December, 1975, and January, 1976, and thereafter remained stable, the *number* of customers for power declined by about two percent from the previous year. Increased demand from some customers during a time of economic expansion was balanced by the flight of other customers.

In industry, total-energy systems are more frequently seen, particularly in the chemical industry, the second largest consumer of power, the paper industry and other purchasers of power who require process heat as well. Several chemical companies now offer energy-saving design services; Dow advertises the system it claims allowed it to cut its own energy use by eight percent.

Simple efficiency accounts for much of the energy saving. With higher prices, the cost of insulation, of more efficient motors, of more efficient air conditioners, all begin to be justified. Unlike the residential consumer, the industrial power customer has considerable control over the use of electricity. He can choose or design efficient machinery or process technology; when building a plant, he can specify insulation, wiring, lighting and air conditioning and heating equipment to minimize lifetime costs; he can switch from energy-intensive processes to others; he can build his own powerplant and substitute labor and capital for the higher costs of energy. Energy savings in this sector seem unlimited, depending only on the price of power.

The power industry, therefore, loses some of its industrial customers and other large customers, on which its very existence depends. Central-station monopolies depend heavily on the large customers: factories run round the clock, shopping centers nearly all of the time, but homes and small shops draw power for only a few minutes or a few hours out of the day. Private power monopolies derive their advantage from their ability to combine these diverse demands into a relatively steady flow of power, which can then be provided by the giant central plants of the utilities. But if the large, steady customers withdraw, the advantage of the power monopoly degenerates very rapidly.

This is already occurring. The utility industry's "load factor" has dropped steadily in recent years. The load factor measures the efficiency with which the industry's generators are used; it is given by the power actually generated as a percentage of all the power which the companies could have generated. After rising steadily for many years, the load factor of the industry as a whole reached a level of about sixty-five percent in the late 1950s, a plateau which it maintained for some years. In 1968, the load factor began a decline, however, just as power prices began to rise, a decline which has continued in each year since and which accelerated after the oil embargo. The industry's load factor was down to 61.3 percent in 1974, the last year for which figures are available, and the continued steep decline in industrial sales in 1975 guaranteed a further drop in this key indicator. In the early months of 1976, industrial sales were static while power consumption by small customers rose, presaging further divergence between these groups.

In Donham Crawford's words, "load factor is the name of the game" for power companies. If a billion-dollar nuclear plant can only be kept working slightly more than half the time, its effective cost is nearly doubled. Because of the high cost of nuclear construction, utilities are shifting as much of the load to nuclear plants as they can, leaving coal and oil facilities idle; yet the load factor for nuclear plants, their capacity factor, is only about fifty-eight percent.

A continuing decline in load factor would spell disaster for the power industry. The private monopolies would be forced to rely more and more heavily on residential and small commercial consumers, who already bear the brunt of the financial burden. Power sales would increasingly shift from industry to small customers—a trend that has already begun—while industrial power rates rise even more rapidly, accelerating the flight of industrial customers. To stem the tide of rising fuel prices the utilities have committed themselves to nuclear power and very large coal-burning plants. Generating capacity continues to grow more rapidly than demand, but the continuing flight of industrial customers creates growing gaps between the peaks and valleys of energy use.

Nuclear powerplants, once planned as a means of reducing operating costs and so retaining industrial customers, have greatly increased the hazards the utilities face. By raising construction costs and the time required to build a plant, the utilities have greatly raised the stakes in their gamble that industrial customers will continue to increase their demands for power.

It now appears that gamble has been lost, but so far the power companies have refused to do more than modestly hedge their bets. Plants have been deferred, and in a few cases the massive base-load coal and nuclear plants, designed to supply blocks of continuous power, have been replaced in utility-construction planning by peaking plants—turbines designed to operate only a few hours per day. Unfortunately, peaking plants burn oil, not uranium, and so their use runs directly counter to federal policies.

The Federal Power Commission reports that utilities have not accepted the apparent trend of decline in industrial demand, and that they still plan to expand their generating capacity at about 6.2 percent per year, just slightly lower than they have in the past. There is some evidence, in plant cancellations and deferrals, however, that these claimed plans are already being cut back in concession to economic reality.

To forestall worsening economic difficulties and impossibly large rate increases, the power companies would have to change course in radical fashion. When industrial customers depart or cut their purchases, the only possible way of retaining the industry's load factor and efficiency is to improve the load factor of residential and small-business customers. This can be done by selling more power during off-peak hours (spring or fall nights), or by cutting present peak demands, during summer afternoons.

Electric space heating and water heating were intended to improve load factors in this way. Despite their gross inefficiency in energy use, electric heating devices make good economic sense for the power business and can be supplied with power at industrial rates. Unfortunately, construction of housing and commercial space proceeds so slowly, and even the lowest rates available from utilities are now so high, that there is very little hope that space heating or water heaters will expand fast enough to make an important difference.

A more immediate and effective technique would be to reduce the peaks, rather than to fill in the valleys, of power demand. The peaks of demand in almost all parts of the country are created by the use of air conditioners. The thirst for power of these machines is such that average power sales in summer weeks are more than twenty percent above those in winter; for any given power company, the lowest point of demand on a winter night may be only one-fourth of the highest demand on an August afternoon.

Air conditioners could easily be made more efficient; machines on the market now vary over a threefold range of efficiency. Switching all new sales from the average to the most efficient air conditioners now available would wipe out the need for all the nuclear powerplants expected to go into operation over the next five years. As Barry Commoner pointed out in testimony before the Missouri

Public Service Commission—which thought he was joking—it would be cheaper for the local power company, Union Electric, to buy efficient air conditioners for its customers, as power companies once supplied light bulbs to encourage consumption, than it would be to build the additional generating plants needed to supply power for the inefficient air conditioners now in use. Power companies could encourage the purchase of efficient air conditioners through cash rebates, deductions on utility bills, or just by publicizing the over-all savings from purchase of an efficient air conditioner.

The power companies might try to revive that aboriginal source of their justification, electrified transit systems. Electric transit systems within and between cities have been allowed to decay but could be revived; electric buses, too, are now close to technical feasibility and could be charged up at night, balancing domestic daytime loads. For the more distant future, electric automobiles would be more efficient in their use of fuel than present gas-burners and would not pollute city air directly.

A combination of peak-shaving and load-building in off hours could restore the economic viability of power companies by increasing their load-factors and efficiency if undertaken quickly and vigorously. Power companies could avoid building new, expensive powerplants for several years and might even make a net contribution to environmental quality. There is no indication yet that the power companies plan to depart from their traditional devotion to the growth of generating plants, nor that the government is prepared to assist with the complex planning that would be required.

What is needed is a hiatus of several years in orders for plants, while demand peaks are reduced and the valleys of demand rise slowly; but this would strike hard at the companies which make powerplants, which might not easily survive a five-year moratorium on growth.

Federal policy in this field ideally would be to help the utilities and manufacturers through the difficult years ahead, while vigorously pursuing the new technology which eventually must replace the private power monopoly's behemoth plants. Central to future needs is an effective means of energy storage. On a large scale, this would accelerate the trends toward solar power and fundamentally more efficient combined-purpose plants for industry and commerce. If available on a small scale, energy storage would relieve the small customer of his disadvantages and allow combined heating-cooling-lighting-power fuel or solar plants in every home and office building. These would be enormously more efficient in their use of scarce resources than the present central-station plants and would employ far more labor. Cheap and efficient energy storage would eliminate almost all of the disadvantage in capital costs a small plant would otherwise suffer.

There is no sign that the federal government, in any of its branches, perceives the need for either the short-term support of the utility industry or the long-term development of more efficient replacements for it. The executive, which traditionally originates such long-range plans, has not so far produced any over-all energy policy. Such policy as we have is dictated by the Department of State, rather than by an agency with domestic concerns. The US is committed to a course of lessening its dependence on foreign oil supplies by conserving energy and by increasing the rate at which the power industry builds coal-burning and nuclear powerplants. This over-all policy is dictated by complex global considerations. Thomas Enders, then assistant to Secretary of State Henry Kissinger for economic and business affairs, described the rationale for US policy in the journal *Foreign Affairs* in July, 1975. We were concerned to reduce our demand for Mideastern oil not, it seems, for any domestic economic purpose—the United States was among those least affected by higher oil prices, and our competitive position internationally was enhanced by the price rise. To support our allies in Europe and Japan, and most particularly to prevent the emergence of an Arab bloc capable of assuming leadership of the Third World, it was a matter of first importance to break the power of the OPEC cartel. This could be done if the US and the other industrial nations freed themselves of dependence on oil imports; by contracting demand for oil, the price could eventually be driven down, and the unity of the cartel of producing countries could be broken in the early 1980s.

In order to accomplish this end, Enders notes, the US and other industrial nations must foster construction of nuclear and coal-burning plants by guaranteeing construction loans and by removing environmental restraints on power production; conservation must be encouraged; and alternative sources of oil—from coal and offshore reserves—must be developed.

Foreign policy, still in the shadow of confrontation with the Soviet Union, once again dictates the development of nuclear power, which again becomes the afterthought of a program far removed from electric-power engineering. Present policies whipsaw the US power industry between incompatible requirements: fuel costs must go up, and energy conservation must be encouraged, which makes it impossible to retain industrial power customers. But utilities must nevertheless build nuclear and coal-burning behemoths, which can be justified only if large blocks of power are sold to industrial customers.

The power industry is responding to the government's demands with its customary patriotism by building more nuclear and coal-burning plants. The eventual outcome of this policy may be the bankruptcy of the industry.

CHAPTER FIFTY-FOUR

Proliferation

ALMOST ANY NATION can become a power capable of mass destruction by several means. Biological warfare—the deliberate infection of enemies with disease—has been practiced sporadically for centuries. Chemical warfare, which is more recent, is both more effective and more reliable. The nerve gases developed by German scientists before World War II are inexpensive and easily mass-produced. Poison-gas warfare was most recently employed in Yemen, and the Egyptian and Syrian forces probably have missiles armed with poison gas; Israel claimed to have captured some of these in the Sinai in 1967. The United States maintains huge stockpiles of poison gas, as presumably do its potential adversaries.

Chemical and biological agents are easy to prepare, using information available in published literature; nerve gases are chemically similar to pesticides, and some commercially available pesticides are themselves quite toxic enough to serve directly as weapons. A terrorist could, in theory, hold an entire city hostage with such agents, and such threats have been a staple of fiction for many years. In reality, however, the easy availability of chemical and biological warfare agents has not resulted in terrorist activity, and these weapons have not been employed very often even in warfare between states that possess the ability to devastate each other.

Now that nuclear weapons are becoming almost as easy to obtain as earlier means of mass destruction, it is reassuring, in a limited way, to recall that weapons of mass destruction actually seem not to be used very often.

Perhaps there are internal limits that keep even the most determined murderer from killing millions of civilians; the history of the past forty years, however, does not give much support to this idea. A more discouraging appraisal would be that weapons of mass destruction simply are of no practical use except to a major industrial country engaged in all-out warfare. The mass destruction of civilian populations, if it ever has any military function, certainly does not serve the limited aims of small countries engaged in combat, nor does it seem well calculated to secure political support for a revolutionary movement. The terrible certainty of retaliation, of genocidal warfare which no small country or group could hope to survive, certainly must operate as a deterrent. How else are we to explain the almost complete abstention from the use of weapons of mass destruction in the worldwide violence of the past thirty years?

The spread of nuclear weapons does not seem to make any fundamental change in this situation, but atom bombs do represent a new and unpredictable factor. It is probably not much harder to build a single nuclear explosive—once the needed plutonium or uranium has been acquired—than it would be to make a nerve-gas weapon. Weapons-grade uranium is not even risky to handle, and a primitive nuclear weapon could be made with a shotgun and a section of pipe. Plutonium is hazardous and more difficult to build into a weapon, but Theodore Taylor, an old associate of Carl Walske's in the atom-bomb design labs, has recently made it known that a small team, perhaps only one ingenious and well-educated person, could build a plutonium bomb, using only the information in published literature. The implosion-type bomb is usually described, but, in fact, this is not the sort of bomb which the US now builds, and a truly ingenious person might hit upon the simpler designs that are possible.

Handmade, primitive nuclear explosives can be built by individuals or small groups if they can put their hands on enriched uranium or plutonium. It is therefore a growing problem that these materials are shipped to and fro across the surface of the globe. Enriched uranium and plutonium are shipped routinely by commercial carrier within the United States and between the United States and its allies. Much of this activity is military, but the international growth of the nuclear-power industry exacerbates the problem.

Still, it is not at all clear why a criminal group would go to the trouble and risk of stealing plutonium, when so many other means of dealing death in massive quantities are available with less risk. Enriched uranium is perhaps a more likely target since it is non-toxic and is so weakly radioactive as to be far more difficult to detect or monitor. Uranium can be used in the simplest gun-type weapon, described in every history and encyclopedia that deals with the development of nuclear weapons. Fortunately, enriched uranium is not used in most powerplants; in the US, only the gas-cooled reactors of the now-defunct General Atomic would have used highly enriched uranium.

Unfortunately, the owners of plutonium and uranium must guard themselves against efforts to steal these materials, even if those efforts do not seem very likely to materialize. The risks are too great to omit taking any feasible protective measure. And so the industries which rely on plutonium or enriched uranium must necessarily come more to resemble the Rocky Flats plant. Pat Kelly is proud of his security clearance and the protective measures which keep out radicals and other undesirables; but how far do such measures have to extend when the power industry produces thousands of shipments of weapons materials every year, and plutonium and uranium sufficient to arm a major power is stored in hundreds of locations throughout the country? Carl Walske thinks we need a new federal police agency; the Virginia Electric Power Company recently asked the state legislature for police powers which would allow it to arrest people and search their homes (the authority was refused). And in Texas and California—and almost certainly elsewhere—local police forces have obligingly kept an eye on people at the request of power-company officials.

The risks of the theft of plutonium are genuine, if remote; the countermeasures will be immediate, real and threatening; ultimately, police measures may call forth just the response they were intended to prevent. But there is nothing unique in the nuclear industry in this regard.

The spread of nuclear armaments among nations is quite another matter. There is a great distance between the handmade weapons that a terrorist might construct, or which the United States and other nations built as a first step in developing nuclear arms, and nuclear weapons of military significance. Military strength requires a massive industry capable of producing plutonium or uranium in large quantities and of mass-producing the weapons and the aircraft or missiles to deliver them. It took the United States almost ten years and billions of dollars to construct such a capability from scratch. The designs are not in encyclopedias, the production techniques are closely guarded secrets, and the needed investment still runs into the billions of dollars. Very few nations are capable of mounting such an effort. The US, USSR, Great Britain, France and China have significant military weapons, although China is still struggling to build an arsenal of a few hundred, after ten years of effort. The two Germanies and Japan could certainly make the required effort but have been kept from doing so, until now, by the combined threats and promises of the United States and the Soviet Union. India's tiny nuclear explosion is certainly no more than a symbol of its pretensions. Israel may have a handful of bombs or the plutonium to make them. The Union of South Africa has just put into operation a plant for producing enriched uranium, presumably for powerplants, which could be used as the basis of a military program. Brazil has signed an agreement with West Germany, which will assist the former country to build all of the facilities needed to produce plutonium from nuclear powerplant fuel. Argentina has secured an agreement from India to assist it in nuclear development. The United States, Britain, France, Canada and the Soviet Union are vigorously competing with each other to sell nuclear powerplants everywhere in the world. Japan has agreed to help South Korea build nuclear powerplants. Italy, Spain and Belgium are building US-designed plants, while Poland, Hungary and Rumania build Russian reactors of similar design. Very few of these countries now have the industrial base to undertake a serious nuclear-weapons program, but in the years to come, as industrialization gathers momentum and the necessary scientific and engineering skills are diffused, many nations will find themselves capable of becoming nuclear powers. Some seem clearly to be aiming at that goal. Iran has made the construction of a dozen nuclear powerplants an early step in its planned industrialization; the US has agreed to supply nuclear technology to Egypt and Israel, and apparently to the oil-rich nations with whom we hope to be allied.

In a decade or so, many of these nations will develop the industrial base on which a nuclear-weapons program can be constructed; whether they will choose to enter the nuclear club or whether their presence in that club will make any profound difference are unanswerable questions.

Despite all of the attention focused on plutonium, this unpleasant metal will probably play a diminishing role in the proliferation of nuclear weapons. The United States still focuses its efforts on the control of plutonium and strives for international agreements regulating its use, as we must. The federal government and the private nuclear-power industry are working toward international agreements that would prevent individual countries from controlling the extraction of plutonium from reactor fuel; secret meetings in London in 1975 apparently produced an agreement in principle among the nuclear exporting nations to extend the modest international controls on plutonium extraction, now exercised by the International Atomic Energy Agency, to all the nations exporting reactors. These controls are largely self-enforcing, however.

The growing international apparatus of controls on plutonium is coming increasingly to resemble a Maginot line, already being flanked by advances in weapons research. The Union of South Africa, apparently with assistance from West Germany, has constructed a plant for producing enriched uranium that makes use of a new and less expensive process—the nature of the process is a secret, but the scientific journals have frequently reported that it is an adaptation of the nozzle-diffusion technique investigated by West German researchers. A cheap means of producing enriched uranium would dramatically lower the barriers to entering the nuclear industrial and military club. Gas centrifuges, being built in Europe and the US, are also steps in this direction and may replace the current universal reliance on expensive and cumbersome gaseous-diffusion plants. These new techniques will still be expensive, but they will reduce the difficulty of producing weapons materials and of making the weapons themselves.

A more dramatic change may be created by laser technology, which is just beginning to make itself felt. Lasers can be used to separate the isotopes of uranium and in theory could be used in a simple and inexpensive technique for enriching weapons-grade uranium. Lasers may also be used as a component of weapons themselves.

In 1968, the Atomic Energy Commission quixotically tried to clamp secrecy regulations over all laser research in the United States. Its proposed regulations were soon abandoned, but the effort did draw attention to the weapons capabilities of lasers. One former AEC official suggests privately that the proposed regulations were not seriously intended but were simply an effort of the laser-weapon researchers to publicize their secret program and so obtain more funds.

In an H-bomb, uranium or plutonium is used as a trigger; the detonation of this trigger creates pressures and temperatures comparable to those in the sun itself, sufficient to fuse hydrogen isotopes; this fusion step releases the larger explosion. Surprisingly small quantities of energy can set off the fusion explosion, if they can somehow be compressed quickly into the small volume of hydrogen isotopes needed. The huge explosion of an A-bomb is used, somewhat wastefully, to drive a small fraction of its energy into the fusion materials.

Lasers can deliver energy much more efficiently. A battery of lasers focused on a small pellet of material, firing simultaneously in tiny fractions of a second, could compress and heat the pellet enough to set off a self-sustaining fusion reaction. The pellet could be made of lithium deuteride, an inexpensive, non-radioactive material which is easy and safe to handle.

This technique is being actively explored in the weapons laboratories of the US and the Soviet Union. A portion of the research has been made public on the theory that it will ultimately lead to controlled-fusion powerplants; but the original motive for the research and its funding are once again military.

The bottleneck at present is the laser component; lasers, while admirably efficient compared to nuclear explosives, can carry only limited energies, but all the military services are actively working to develop a new design for high-powered lasers. These, the services hope, will be usable as antiaircraft weapons, perhaps as antimissile weapons. The Air Force is vigorously pursuing laser development, hoping to build laser armaments into bombers and perhaps fighters. The Army has deployed a single experimental tank, built upon the frame of an armored personnel carrier, whose principal weapon is a high-powered laser. The most vigorous research is being conducted in the nuclear-weapons labs, however, and is for the most part highly secret. If high-powered lasers of simple design are developed—and there seems to be no reason why they should not—simple, inexpensive nuclear weapons will become a possibility, and for the first time any nation will be able to acquire nuclear weapons cheaply and without handling radioactive materials, a fact which will make it difficult or impossible to monitor the international traffic in nuclear weapons.

It is difficult to see what can be done about any of this.

CHAPTER FIFTY-FIVE

A Conversation

IF RALPH NADER'S enterprises were a holding company organized as the large utility holding companies are, the Center for Study of Responsive Law would be a management-services subsidiary. Even more than the utility management companies, it is retiring and makes no effort to publicize its existence. A reporter seeking an interview with Ralph Nader is answered on stationery which bears only a post-office-box return address; a telephone number is typed in, and the street address is given verbally over the phone.

The address turns out to be that of a shabby seven-story building at the tail end of a row of brand-new skyscrapers in Washington's prosperous commercial center. This building is faced with concrete, scored to look like the granite of public buildings; over the doorway, a scrollwork and flag of cast concrete, mimicking carved stone, gives the building the look of an abandoned branch post office. Rusted bolt holes in the ornamental arch under the bogus scrollwork show that brass lettering once surmounted the doorway; the concrete still bears the ghostly impression of the legend which once surmounted Ralph Nader's doorway: "The John Marshall."

A rickety self-service elevator carries a visitor from the stained-marble lobby to a hallway badly in need of paint, where a hand-lettered sign, fastened with tape between two doorways, indicates the offices of the Center. On the left is the reception room, guarded by a pleasant dark-haired woman who appears to be in her forties; a tiny office behind her, crammed with toppling heaps of paper, reports and unanswered correspondence, is Nader's own sanctum. Nader is running an hour behind his schedule and will not meet his visitor yet. The visitor makes himself as comfortable as can be in the room to the right, apparently a combination of library and mail room, where reference works—federal statutes and regulations, *Who's Who, Standard and Poor's*, congressional directories (including the directory produced by a Nader subsidiary), dozens of congressional hearing reports—and periodicals are displayed. The journals are multinational and multilingual. From Spain, the cover of a glossy magazine proclaims in Spanish "Bread: Giant Fraud"; the London *Economist*; *AMPO*, an English-language publication from Japan; the *Ladies Home Journal*; the *Congressional Record; Consumer Reports; Counter-Spy; People & Power* ("For The Abolition of Nuclear Fission"). A young woman with long blond hair manages incoming calls on one line while carrying on a running conversation with a boyfriend on a second line.

Security measures are more direct here than in most business offices. The file cabinets are secured with steel bars kept in place by heavy padlocks. The offices are extraordinarily quiet. Few visitors apparently are able to find the address or the telephone number of these headquarters.

Despite its ramshackle and deserted look, the Center is indeed a nexus of far-flung activities. Prospective employees of many Nader subsidiaries pass through these offices; each fills out an "Employment Form" which asks and codes such information as "Employer (name of [Nader] group)," personal statistics, date of leaving (for summer interns), rate of pay, job description, health-insurance information, federal income-tax information, and spaces for signatures:

     Your signature _____

     Your Supervisor's Signature_____

     Ralph Nader _____

There is an information sheet which accompanies the form, explaining payroll policy: "Timing: Each check represents payment from the first day of each month to the last day of each month. Checks are disbursed on the 15th day of each month. . . . Taxes: Two of the following four employment tax forms must accompany the Employment Form."

Nader donates his own considerable lecture and writing fees to the over-all holding company, drawing only a modest salary for himself. His home address is not in a prosperous area of Washington, and his clothes are inexpensive; he does not own a car; if, as his critics charge, he is profiting personally in some way from his activities, it is difficult to say what he can be spending the money on. Public Citizen, a fund-raising subsidiary, is another major source of funds for the enterprise. It conducts extraordinarily effective direct-mail solicitations, which produce about half a million dollars per year (beyond the costs of fund raising and administration) for the support of the various operating subsidiaries. A brochure lists a few of the major operating units: the Health Research Group, the Litigation Group, the Congress Watch, the Tax Reform Group, the Capitol Hill News Service, the Citizen Action Group (which coordinates public-interest research groups, staffed and funded by students, in eighteen states); there are also the Center for Auto Safety; the summer research projects (Nader's Raiders) and loosely affiliated parallel structures for reform of corporations and for "whistle-blowers," employees of government and private industry trying to effect change in their employers; and Critical Mass, which sponsors an annual convention of opponents of nuclear power.

One of the outputs of this varied enterprise is a steady stream of books; the publication list runs to four single-spaced pages; most of the books are produced under a royalty-advance arrangement with Grossman Publishers.

At the center of this network stands Ralph Nader, a polite, friendly man who seems relaxed and happy. His personal warmth is a surprise to people who see him primarily through the eye of television. On this occasion, he is red-eyed and tired; it is six o'clock, and after an hour's interview he will move on to greet some new young volunteers and a group of what one would assume were, if Nader ran a business enterprise, visiting Japanese industrialists.

How did you first become aware of nuclear power as an issue?

NADER: Well, apart from generally reading about it, I spent a summer at Oak Ridge [an Atomic Energy Commission laboratory and production complex in Tennessee] in 1964—dealing with a conference on science and public policy—and talked to a good many people there at that time. Simply general questions—and my impressions were that they [the AEC officials] were very self-assured, and they did not concede that there were any serious problems. They never alluded to the maximum potential catastrophe. There was no mention of how this technology might lead to nuclear proliferation abroad, although at the same time they were greatly optimistic about nuclear plants and desalination, making green many arid areas of the world that border on oceans. There was no concern about the problem of sabotage, the problem of transportation.

There *was* a focus on the waste problem. They were not overly concerned about the plant itself, the nuclear plant itself, its ability to be properly managed. And, of course, that was the time when there were probably only, oh, three or four nuclear plants, maybe six, and they weren't that aware of the problems that the utilities would bring to these plants or the problems that the vendors would. They just were a very self-assured technical laboratory that had solved a lot of problems. They figured that they would continue to solve all the problems that were required to be solved.

After that period of repeated assurances, I began to hear how they were disturbed, around the country, about a physicist called [Ernest] Sternglass. Who, they thought, was completely unprofessional and exaggerated in what he was saying about the potential genetic deaths, or the real genetic deaths, from low-level radiation emissions. That, of course, was a challenge to the AEC which led them to commission [John] Gofman and [Arthur] Tamplin and it was really their reports [on radiation] that first alerted many of us to the reality of the problem. Then the emergency core-cooling system—and the thermal-pollution [discharges of hot water] issues, around the early, very early seventies—that's what really broke it. As far as I was concerned, when I looked over the details of the emergency-core-cooling-system failures and the lack of empirical testing, and connected that up with what the AEC's own witnesses, technical witnesses, were saying at the hearings, then I knew from experience with other government agencies that there was a coverup. The only question was how extensive and deep was it, in the areas extending to the whole nuclear fuel cycle.

The rush of disclosures in the last three years has been almost geometric, and it's quite clear that they put a technology on stream before they solved the basic problems pursuant to the safety of that technology.

I might say, by the way, that two other factors were two books . . . that came out [in 1969] which raised these problems very early, and people like Leo Goodman, what he was constantly saying. . . .

Now that you've had a chance to learn more and the debate has been going on publicly for a long time, what is it in particular about the nuclear-power program that moves you to oppose it?

NADER: Well, the major hazards to society are the radioactive-waste problem; the problem of a meltdown—whether by accident, sabotage or earthquake; the problem of theft of plutonium, of weapons-grade materials, in transit or in site; the problem of nuclear proliferation through the export of technology abroad; and the accident problems abroad where the infrastructure, technical infrastructure, is so much less than it is in our country, which is having trouble with nuclear power. The constant near misses, quality-control failures, shutdowns, and spills, radioactive spills, which have occurred; the increasing diseconomy of the technology, if you take into account current and rising capital costs—the fivefold increase in the cost of a nuclear plant in the last eight years, at least—and if you take into account the federal subsidy, one of my objections is that this is a technology that is not going to be able to go on stream without massive federal subsidies, and I don't mean just Price-Anderson [federal insurance] types, but direct cash or credit or purchase-and-lease-back type [of] subsidies.

If nuclear power is allowed to continue, it will immeasurably slow down alternative sources of energy like solar—its development—and it will put our economy in a highly perilous reliance on nuclear power during which period there may be a major catastrophe. And then we'll have both an economic and a radioactive crisis, or an energy and a radioactive crisis, splitting the society in two very hostile camps: one camp saying we've had one major catastrophe eradicating a metropolitan area, we must shut them down, and the other saying we can't shut them down because we have to keep the economy going because we've been hooked on nuclear power in terms of the economy's reliance on it.

What about the problems of exporting nuclear power that you mentioned earlier?

NADER: I think the proponents and developers of nuclear power must now take responsibility for the consequences of the export of this technology—which they all, by and large supported—in terms of the nuclear proliferation around the world. It's quite clear that twenty years ago this problem could have been foreseen. It was technically known. They could have projected that these plants were going to be exported abroad. They could have projected that some nations with balance-of-payments problems might [inaudible] for export, like Western Europe, and there'd be a race as to who can export the most and it will be out of control. And so, if there is a nuclear proliferation, as there is every indication there will be, from the sale of these powerplants, that is another burden that must be imposed on the conscience of the government and industrial and professional people who are in the nuclear-power industry.

Of course, there's another problem that isn't being given much attention. That's the very human problem—quite apart from diversion of weapons-grade materials—in the less developed countries. The human problem of accidents. They don't have the transportation securities and the civil securities, and the technical infrastructure that we have, and it's going to be a horrible laceration on our conscience if someday we wake up and we see that one of these nuclear powerplants had a major meltdown between Rio and Sao Paulo, or near Alexandria, or in Pakistan or Formosa. Because we can see that . . . the society there is highly exposed to the myth that if it's safe for Americans, it's safe for Formosans or Brazilians; and if it's being built twenty-six miles north of Manhattan, then why not build it near [some other] metropolitan area of eleven million people, for whom there is really no escape, no possible evacuation, no alternative food supplies, and no sophisticated cleanup treatment or diagnostic treatment for radiation diseases.

There's a report now that Brazil and Germany have signed an agreement that may lead to military technology in Brazil. Is this a concern of yours as well?

NADER: Yes, it's a concern, because my information from Brazil is that it was a second-level decision in the government. That the generals did not take a personal interest in this. That they really relied on second-level ministers and a few, very few technical advisors as an act of faith. And this is what we're seeing all over the world, that is, there's no alternative information about nuclear power in these countries. Good heavens, there was no alternative information in *this* country until about eight years ago. There is no dissent against this. There simply is a very narrow technical line . . . ratified by the highest level of government, basically on the premise that this is something that America is going into big. This is part of the indicia of modern industrialization, and this is necessary for possible defense or nuclear capability. I think that, in some of these countries, terrorists are going to be the only people who give common sense, inadvertently, to the rulers of the country. That is, if Brazil ever says "no" to nuclear power, it will be largely out of concern over the vulnerability of its nuclear-power industry to terrorist penetration.

Isn't it a matter of survival for the United States to free itself from dependence on Middle Eastern oil, and isn't the only way of doing that quickly to build a lot of nuclear powerplants?

NADER: Well, there's a lot of answers to that. One is that even within that narrow framework of reference, it's really a matter of survival for the Middle East oil companies to rely on our consumption of their oil. The dispersal of oil discoveries all over the world is going to make it very difficult for two or three countries to close the door successfully in any future embargo, and any embargo is temporary anyway, and can easily be handled by an emergency oil reserve in our country. This has been considered in Congress recently. [Legislation passed late in 1975 provided for the creation of an emergency reserve of oil in the US.]

We do not need nuclear power. We can do very well over the next forty years, until solar and other forms of energy come on very strong, by more efficient use of what we have. There are enough studies now starting with the [Arjun] Makhijani study in '71, indicating that by the year 2000 we could be using sixty-eight percent—per capita—of the energy we are now using. In other words, a zero energy-growth scenario, and have a more efficient economy, a less inflationary economy, a less polluted economy, a more competitive economy, and a less capital-draining economy. Now, if we want to open up other areas like bioconversion, like burning our wastes—which might produce seven or eight percent of our electricity—more efficient powerplants, etc. Powerplants now lose so much heat that you'd think that their primary purpose was to heat the heavens. These are just a few of the frontiers of conservation. Conservation is a massive technological frontier, in addition to the technological solutions we already have on hand, to build cars to get thirty miles per gallon. And so on.

I also think that we have a far greater amount of fossil fuels in this country than we're owning up to. For instance, whenever you hear of estimates of what our oil reserves are, they never include secondary and tertiary sources—which are about 250 billion barrels. At the present rate of consumption that would take us easily to the year 2000. We don't even take into account the tar sands, which are another three, four hundred billion barrels in Alberta alone, not to mention new technologies which might be able to get oil out of shale without water and landscape desecration thirty or forty years from now. For example, there are 225 trillion cubic feet, estimated, of methane [natural gas] in coal beds. The Bureau of Mines is now getting [natural gas] out of two pilot mines, before the actual exploitation of the mines, so that you make the mines safer. Right now it's being blown away, the methane; 225 trillion cubic feet is ten years supply of natural gas at present rates [of consumption].

Another point: I think that comment can be completely turned around. I don't think that we can survive *with* nuclear power. I don't think that a society can endure the disaster of one major meltdown, and the consequential conflict in the society over whether to shut down nuclear power when the economy is heavily reliant on it or to keep it going and risk further meltdown catastrophes. I think it's one thing for a society to tolerate fifty thousand fatalities a year, in different places, at different times, throughout the year from automobile crashes. I really don't think that our country can tolerate the trauma of a couple of hundred thousand people dying all at once in one place, and many more dying over a period of time from cancer, leukemia, mutations and what have you.

On what do you base these fatality estimates?

NADER: Based on the figures produced by the Union of Concerned Scientists, and the American Physical Society panel report [reviewing the draft Rasmussen study of nuclear safety]. They took, for example, the Rasmussen figures of three hundred fatalities in one scenario accident and raised it to ten thousand, and that was just one. So . . . you're talking about factors of twenty-five or more; and also the fact that the lower estimates are based on incredibly efficient evacuation, which is not to be relied upon; and finally, that the estimates, the low estimates that the government has been giving out, completely ignore the resource contamination, land and water, and the effect on five or six generations . . . into the future. They also do not take into account the sociological consequences of a technological accident. For example, it's been known in studies of disasters around the country, of a more conventional sort, that ancillary accidents occur—suicides occur; that has to be taken into account when you're dealing with a radioactive cloud over metropolitan New York.

Are there other issues that concern you?

NADER: Yeah. I think another argument on nuclear power is, Do we want to rely on a technology, each one of whose parts is so heavily dependent on all other parts? If a coal-fired plant explodes it's not really going to affect other coal-fired plants around the country. If a nuclear plant has a major meltdown it's going to have two consequences: a tremendous revulsion against other nuclear plants, after the public realizes that it can happen, and, second, the regulatory sequence. In January or February [1975], I guess it was, the Nuclear Regulatory Commission shut down twenty-four nuclear plants for two weeks or more, for inspections because they discovered cracks in one plant in Illinois, in the pipes of the Dresden plant in Illinois. So, you have that kind of interdependence. There are other kinds of interdependence. Dr. Ralph Lapp once said he advised the state of Illinois that, if a maniac called the governor's office and said, "I've got a detonating device under a certain part of one of your forty-two nuclear plants"—and this is 1988, let's say—"and unless you give me what I want, it's going to go off," the governor will have to shut down all the plants: (a), to facilitate the search; (b), to minimize the risk. There's your economic crisis. This is the kind of interdependence and vulnerability that operates.

Furthermore, disruption in one part of the society can link up with the nuclear vulnerability much more than disruption can link up with a coal or oil or other fuel capability. It's just that terrorists think in terms of atoms, nuclear bombs, plutonium. They know that the society cannot second-guess them. If they make a call and they say they have three pounds of plutonium that they're going to throw from the roof of a building in the middle of a city, there's no second-guessing there.

It's really technological suicide for a country to produce one thousand, or more, national security problems—which is what every nuclear plant and every transportation vehicle carrying radioactive materials would become. The impact on civil liberties is already apparent, . . . the Virginia Electric Company [unsuccessfully] asked, earlier this year, of the Virginia legislature, to give it [the utility] private police authority to arrest anybody in the state, to have access to confidential citizen records. This is just a glimmer of the kind of garrison mentality that is going to be required. . . .

What do you propose we do about nuclear power?

NADER: Stop it. Period. First, it's too dangerous to continue—for all the reasons I mentioned. Second, it's going to be a massive absorber of capital which we cannot afford in this country if we're going to have capital for other needs. Third, we can get along without it simply by a strong conservation policy. Because we waste so much, as you know, more than twice [the energy] per capita of Sweden. Finally, because we're going to turn our country into a minigarrison state, with curtailments on civil liberties; police everywhere guarding and checking, dossiers, files, security classifications—all to insure that nuclear-power technology is insulated from sabotage, terrorism or other intrusions. And, perhaps a more mundane objection, is that I don't think these plants are working well enough. They shut down too much. The capacity of Commonwealth Edison [Chicago] is twenty-eight percent in the first quarter of '75 for its seven nuclear plants, and the bigger ones seem to have more trouble.

When you say "stop them," do you mean shut down existing plants?

NADER: Yes. Stop construction, shut down the existing plants. . . . If there are countervailing emergencies from the shutdown of a few of the nuclear plants—like hospitals would go without electricity—then we should derate [reduce the power output] them in the process of phasing them out. But, there's a great deal of excess electric generating capacity in this country, for the few days during the year when air conditioners reach a heavy peak. That's something that can be dealt with.

If it turned out to be impossible to give up our reliance on nuclear power completely—for foreign-policy reasons or for some other reason—what changes would you suggest in the program to make it more acceptable?

NADER: I don't think that way. Because, you see, I don't think there's a tolerant level. I don't think there's a middle ground to that magnitude of peril. As a product of our advocacy, we're obviously getting the government to be a little more careful and to inspect a little more assiduously and to [impose] fines a little more effectively, and I'm sure there's a greater alertness to these problems. The safeguards problem has gotten more attention. But, as long as nuclear power maintains its [inevitable] radioactive intensity then there is no alternative than to stop it.

Conservation is one alternative. Solar energy is another. More efficient use of fossil fuels is a third. A much more refined technology, energy-efficient technology, of course, is part of conservation. It isn't just shutting off more lights or driving less. It's more efficient air-conditioning systems, building standards, automobile standards and the like. It's interesting to note that the American Institute of Architects study recently said that with practical, energy-efficient new and old buildings [retrofitting the old], by 1990 we would save more energy, just from that source along, than what nuclear power is scheduled to give us.

You have been working with the Union of Concerned Scientists, Henry Kendall and Dan Ford? Are they the people who you work most closely with in educating yourself on technical issues?

NADER: Yes, but there are others now coming out. We're getting letters, for example, from former nuclear-industry engineers, radiation physicists. . . . I think we'll see in the next two or three years a tremendous diffusion of involvement.

You are a layman, with no scientific background, and you are asking for a really radical measure. You're asking for a shutdown of what has become a major industry.

NADER: Mm hm.

Can you think of any incident since the Luddites [who wrecked the machinery in textile mills] of the early 1800s, in which people have really tried to stop a technology from being developed?

NADER: Yes. The oil industry has been the Luddite for the solar industry. (*Laughter.*) That's one; AT&T has been the Luddite for satellites for many years. The network television broadcasters have been the Luddites for cable TV for many years, and people were Luddites for the SST [supersonic transport]. You know, Luddites come in many forms. GM is the greatest Luddite against mass transit, an efficient internal-combustion, or other-combustion, engine. That's what big industry does. That's their career. They are far more threatened by a new technology of abundance, against their technology of scarcity, than they are by price cutting or competition or any other sort.

Since you mentioned that, how central is solar power to all this? Do you think it's very important to develop it quickly with a lot of money?

NADER: Absolutely central. In fact, it is intricately related now to nuclear power, in the sense that if nuclear stopped, solar would move faster into use, and if nuclear isn't stopped, solar will always play a back seat—for the next two generations. Our society operates very well under stress. The only reason we're even talking about solar now is . . . because of what happened [the oil embargo] in the fall of '73. It [solar energy] used to be considered a Buck Rogers joke. And suddenly, just from the slightest R&D and the slightest interest, enormous frontiers have been opened and ideas and realities and practicalities. You can imagine what would happen when it becomes a number-one priority. The best thermonuclear reactor is the sun. It's well shielded. (*Laughs.*)

How do you get that much confidence in your judgment on what are, to a large extent, technical issues?

NADER: Well, for several [reasons]—the legitimacy comes from the fact that power rests in the people, ultimately, in any society. The power to decide which way a society is going to go politically, economically, technologically, rests with every citizen in this country, and that includes us. That's the premise on which our democracy is based, no matter how abstruse, elite, technical or what not. . . . The people who are going to be exposed to the institution or technology will make the final judgment.
Now, how can that judgment be made in an informed way? By a careful analysis, first, of what the advocates of nuclear power concede. They concede that there is a prodigious amount of radioactive material inside a nuclear plant. They concede that this has to be secured—for all purposes, permanently—from the human environment. They concede that they have not solved the radioactive-waste disposal problem. They concede that the nuclear plants are going to be sending radioactive materials to and fro throughout the country, [materials] which will be exposed to sabotage and terrorism. They concede that these nuclear plants should be built remote from metropolitan areas, if at all possible. Some would advocate building them underground to make them safer. And some do not trust the utilities to manage them safely.

Now, you take all of these concessions and you say, well, there are two things that we can conclude. One, that there are verified problems that have not been resolved, pursuant to the dangers of nuclear plants, even by the admission of their advocates. Consequently, a citizen is entitled to ask why they are putting these machines and these technologies on stream before they solve the problems. The answer is one which a citizen can judge by virtue of being a citizen, because they can either say, well, a society can take the risk—well, they're no more capable of evaluating risk, evaluating the acceptability of risk than anyone else. That's an entirely human judgment, that has nothing to do with whether you're a farmer or an engineer or a mathematician.

Second, they can tell us that they don't think it's ever going to happen—"it" meaning sabotage, accident, major catastrophe. Now, they have absolutely no probability theory which can apply to that judgment, [no theory] that's reliable at all. . . . So, what they're really doing is they're telling you they *hope* this is not going to happen. And it doesn't, again, take a professional degree to be able to say that. That's basically a hope. Anybody can hope. *We* hope it will not happen by not having nuclear powerplants. We think . . . that human history shows that there are going to be failures. Every technology has had a failure in one or more of its parts. Planes, cars, ships, trucks, plants, and astronaut capsules, and so forth.

It's quite clear, then, that when you strip their case down to its essential elements, that whether from an empirical point of view, a probability point of view, or with simple logical reasoning, they cannot sustain their case. Because in the final analysis they cannot say that we should be exposed to such peril, because the survival of the society is at stake. That's their final argument, if they could make it, and they can't make it because we have alternative energy sources, which we've discussed.

Of course, there are other questions which people are entitled to ask, regardless of their degree: that is, if it's so safe, why isn't it insurable—fully—to the maximum potential catastrophe?. Why does it have to be subsidized?

In short, people who are exposed to the risk of the technology, people who have to subsidize the technology, and people who have to be concerned about the technology's effect on their children and grandchildren, are the people who must decide in the final analysis. Einstein said it a long time ago, as you know, when he said these are issues which should be decided at the village square.

We also, I think, have a basis to be very skeptical about experts that are indentured, who are indentured to specific subeconomies or specific mind sets. The experts led us into Vietnam. That's why a book was written called *The Best And The Brightest*. The experts built the [overpriced] C5A. The experts ran the [now-bankrupt] Penn Central. So, it isn't just experts. We have to ask ourselves, Do they have higher allegiances that prevent their expertise from being applied in an objective and humane manner?

There is a kind of related question that I think very often goes along with that one which, bluntly put, is, Who elected you? You are in a position in which you can make your views felt very strongly.

NADER: Mm hm.

And that, in fact, even if you weren't, you went to court and asked the court simply to halt the nuclear-power program. Now, how do you know that, in fact, that this is what people would choose to do, if they were all educated to the degree you are on this issue?

NADER: Well, in answer to your question who elected me, my answer is, the same institution that elected all of us. It's called the US Constitution. A citizen does not have to be elected. A citizen *elects*, elects representatives, elects policy preferences, elects things to support and things not to support. And not all of us can be involved in all of society's problems, but those of us who stake an area out, and work on it, are entitled to have our views heard in the decisional forums made up of the delegated institutions in our society, whether the courts, legislatures or administrative agencies. And if the people do not like a court decision that we win, they can overrule it by their legislation or by initiative. But the option has always to be open for the individual citizen to prevail, in a lawful context.

The forum you often choose is the court, rather than the more massive public-education program or community organizing, or legislation, and that has a kind of uncomfortable sound. That you're asking a very small number of people, perhaps only one person, to make really profound policy judgments and essentially political judgments about what should happen.

NADER: There are three responses to that question. One, is there's a time immediacy for some situations, which the courts, with their injunction power, can respond to. Second, we are trying to broaden the constituency for decision making in nuclear power far beyond the desire of government and business who want to keep it in an administrative agency or behind closed doors. Consequently we, for example, are supporting the initiative drives in the Western States on nuclear power, so the people can be informed and decide under what conditions they want nuclear power, and that's as broad a constituency as you can reach. Thirdly, we are trying to encourage community-action groups, student public-interest groups, and other groups around the country, to take this up as an educational and advocacy program. And finally, we're working on Congress to try and get them to declare a moratorium.

So, you can see we're working on pretty much every constituency affected by nuclear power except the unborn, which we can't do much about at the present time except to try to uphold our trust for future generations.

Has this become a mass movement?

NADER: Yes. I think it has. There have been—there are about a hundred and fifty identifiable groups around the country. Many of them, to be sure, very small, but they're very intense. Like the North Anna group in Virginia, which makes up for small numbers by its knowledgeability and determination and creativity. And I think within another three years there will be five, six hundred. I think there will be more groups staking out this issue, more environmental groups, tax-payer groups, local community groups who don't find nuclear power their main concern but who will participate in supporting the [mass movement]. I think also with the expansion of construction, there will be localized opposition to every plant, so that the more they are built the more opposition they're going to breed—which is basically the only meaningful sense to the term "breeder reactor." (*Laughs.*)

Does the movement have any political content?

NADER: Yes, increasingly. The national petition drive, which has about two hundred thousand names, is broken down by congressional district and has already [had a] discernably registered impact on a number of members of Congress. They either become neutral where they were once automatically supportive of the Joint Atomic Energy Committee's position, or on some occasions they change—and there are names that can be given to you as to the ones who changed. Because when someone walks into a congressman's office and says, "I have four thousand names and addresses of people in your district who have signed for a moratorium," the member [of Congress], especially if he only won by fifty-four percent or so, is going to have to sit up and take notice. Now, there has been considerable antinuclear activity in Iowa in the last two years, and four members of the congressional delegation from Iowa have signed on for a moratorium.

What has the movement accomplished so far?

NADER: Well, there have been some plants stopped. David Pesonen in California stopped the Bodega Bay plant. I think there have been successes in requiring more environmental-impact statements to be filed. There have been enforcement activities as a result: . . . the North Anna Coalition has generated fines.

To what extent are the antinuclear groups aware of, or coordinating with, community groups fighting utilities over rate issues and economic issues?

NADER: The nexus there is going to come through a proposal that we have advanced called *RUCAG*. It's a proposal for a Residental Utility Consumer Action Group, which will establish a statewide core of professional investigators, litigators, advocates, lobbyists and publicists, representing the residential utility consumers who contribute to this statewide group. The novelty of the proposal rests in a consumer check-off system, which piggy-backs your monthly electric, telephone and gas bill, and encourages you, or invites you, to make a voluntary contribution, adding onto your utility bill. Whereupon, after every two weeks, the utility company will amass the contributions and send them over to the statewide consumer-action group, run by a council of directors elected regionally throughout the state by the contributors. One contributor—one vote. Now, that will give a full-time representation—linking it up to a large constituency of contributors who can be communicated with through newsletter and meetings—to deal not only with electric utility prices and service, but also the problems of nuclear power, siting, taxation, subsidy, pollution, you name it. And that's where I think the two will come. Now, there'll also be another nexus. And that is once we get over this numerical shenanigan that people are getting lower increases, [a lower] rate of increase because nuclear power is coming on board. Once the full disclosure of the cost of nuclear power economically is brought home to people, we'll, of course, see a transfer of some of that attention onto the technology itself rather than to the rate-making process.

But some utilities have, in fact, given rate reductions recently because of the claimed savings from using nuclear fuel.

NADER: Well, partially that is sophistry. Partially it camouflages the taxpayer subsidy, which doesn't show up on the bill, and partially it reflects plants that were built eight or ten years ago, when the capital costs were much lower. There are other utilities that are not disclosing the problems of *higher* rates due to nuclear powerplant costs, deferrals, shutdowns and all the rest of it. For example, VEPCO [Virginia Electric Power Company] was in trouble because of its problems with the construction of the North Anna complex and has admitted that some of its rates have gone up because of that problem. The [utilities'] drive to make a nuclear-shutdown adjustment clause, which succeeded in Connecticut a few weeks ago, is going to, in effect, pass the costs on, automatically, of the shutdowns. The "construction work in progress" transfer is also another example that's going to really [educate the public], because when you get [the cost of building] a billion-dollar nuclear plant transferred [to customers], that's quite different than a three- or four-hundred-million-dollar coal-fired plant. And, finally, you've got the enormous costs of repairing these plants once they're shut down. They're much more expensive to repair—to put back into operation—than fossil-fuel plants. And then you've got the costs of safeguards . . . there's no end to the spiral once it starts.

Are you aware of the Movement for Economic Justice?

NADER: Yes.

Do you have any comment about the relationship of what they're doing and the "lifeline rates" to the overall thrust of RUCAG and these other efforts?

NADER: Well, they're getting quite interested in RUCAG. And, I think they should realize that no matter what single issue they prevail on—lifeline [for example]—there's just going to have to be a continual year-round backbone to keep those gains from being eroded and to achieve further ones: that is, they are apples and oranges. One is a substantive policy, the other's an instrument to achieve it, and RUCAG is the instrument.

The RUCAG proposal is oriented toward working at the state level, rather than setting up a different structure of regulation.

NADER: Mm hm.

Was that a conscious decision?

NADER: No. It was a recognition that the bills are sent out on the state level, so they have to be checked off on the state level. There's nothing to stop the state RUCAGs from getting together and saying, We're going to push for a completely different form of utility—structure, ownership and regulation. They can transform them into mutuals or cooperatives, or have federal regulation.

In some of the publications that have come out of your enterprises, there is a discussion of federal chartering of corporations. Do you think that has a bearing on these issues? Should utilities have federal charters, should the reactor manufacturers have federal charters?

NADER: Yes. Oh yes. If they're in the top thousand corporations—that's the rough cut-off in the proposal—and most of them are.

Why would that make it easier to deal with them?

NADER: Because we could rewrite the constitution of the corporation, which was written in the nineteenth century and has not changed much. We could completely rewrite the relationship between the legal fiction called the corporation, and the state who gives birth to it by chartering it. That means we could develop tougher disclosure requirements. We can develop more freedom for employees to disagree on policy issues and still be given due process [of law] rather than [be] arbitrarily fired. We could insure that there is enough deconcentration of the industry. A charter could insure that sanction systems are effective and pierce the corporate veil [the fiction of corporate personality]. In other words, there's a whole list of changes that need to be made which can be made wholesale through a chartering function.

Wouldn't you run into the constitutional protections that have shielded corporations to some extent from these intrusions?

NADER: No, no, they would still have their constitutional protections—barring any amendments—except that some of these protections are not equivalent to individuals. The corporation does not have a right of privacy, for example, and it's not out of the realm of imagination to think that someday there'll be a constitutional amendment distinguishing between a corporation and an individual, and, in effect, keeping many of the constitutional protections which were originally meant for individuals from being applied to corporations. Particularly in the area of equal protection of the law; I don't think corporations should have equal protection of the law accorded to individuals. There may be taxes that should be just imposed on corporations, but not on individuals, that may now be challenged on an equal-protection basis, which really shouldn't be.

Has there been any public effort, or public discussion, of such a constitutional amendment?

NADER: There's been some discussion. Professor Arthur Miller at George Washington University [School of Law] has talked about that; he looks at it from another point of view. He says that the Bill of Rights was designed to protect people from government arbitrariness, and now we need an extension of the Bill of Rights to protect employees from corporate arbitrariness, because the corporation wasn't on the scene in 1789. So that if a nuclear engineer speaks up on a nuclear powerplant defect and is immediately fired, that becomes an issue of free speech. Which it is not, now.

How much effort are you personally devoting to these kinds of generic solutions, like corporate chartering or a constitutional amendment? How important are they in your scale of priorities?

NADER: That's going to be one of our top ten projects, maybe our top five projects, over the next three or four years.

The chartering project?

NADER: Yes.

How about the constitutional amendment?

NADER: That follows the chartering. A tremendous educational process will be required.

CHAPTER FIFTY-SIX

A Final Word

WE ARE BUILDING nuclear powerplants before finding out if they are capable of destroying whole cities. There are worse things in our modern world: Nuclear weapons are far more menacing. Shipments of liquified natural gas and other industrial chemicals may be as hazardous as nuclear powerplants. That we make worse mistakes is scant comfort, however.

We made this particular mistake because the technology of nuclear power was developed during the Cold War for military purposes and adopted by power companies to forestall socialism. Power companies have continued to build nuclear powerplants for a variety of reasons unconnected with the intrinsic merits of reactors: nuclear power's high investment, and low fuel and labor costs, fit the pattern of private power monopoly created by the law and technology of the late nineteenth century.

The people who try to stop nuclear power came rather late to the discussion and are generally unaware of the forces which brought it into being. Those who oppose nuclear power generally do so because of the danger of radiation releases through accident or criminal design. They are therefore attacking an aspect of the industry which its owners and managers have not considered seriously for many years and now think of as a settled matter, resolved in the early years of military development.

Power companies are not required by any law or institution to take the views of their critics into account. The managers of these companies, trained in an industry which has been fighting public ownership for a century, are arrogant in their insistence that they be free from any outside interference. Newly awakened public concerns about radiation are treated either contemptuously or as the product of a conspiracy.

The fight over nuclear power therefore becomes, willy-nilly, a fight over control of the electric-power business. The larger issue quickly overwhelms the narrower one. Ralph Nader and his compatriots are fighting for democratic reforms of industry and not just simply to stop nuclear power.

Organizations of residential utility consumers are the means through which this democratization is most often sought. There is spontaneous interest throughout the country in various forms of consumer organizations which would unite the majority of citizens in some institution capable of influencing the management of power companies. Nader's effort to create consumer-action groups is an institutional reform; the numerous consumer-action programs and citizen-action leagues appearing across the country are building wide support and a political base from which specific policies can be evolved; and all of these efforts are being carried on outside the usual structure of government. Like the democratic reforms in government in the last century, democratic reforms of industry are producing new institutions and political techniques.

The more recent converts to the antinuclear cause are sometimes moved by a religious fervor that will not allow them to wait for the slow institution of reforms, however. Efforts to stop nuclear power therefore have become divisive. Industrial workers, who are forced to accept personal hazards as the price of employment, are being asked to sacrifice their jobs for the sake of speculative risks and abstract calculations. Simply stopping nuclear power, as a negative program, forces organized labor into an alliance with management and places labor in opposition to a large segment of the white-collar middle class which is concerned for its health and the long-term viability of our economy. This polarization, actively fostered and exploited by the nuclear-power industry, has appeared in other political debates and threatens to wreck any chance of securing majority support for needed reforms.

Furthermore, the nuclear industry is a ward of the federal government, and efforts to stop nuclear power immediately are being pressed in state and local channels. To oppose federal programs with state laws or policies is both hopeless and retrogressive. The only hope for rational energy policies and for democratic control of industry is at the national level.

Perhaps the citizen- and consumer-action leagues will someday emerge as a new national force capable of fighting for these important reforms. Right now, however, the fight over nuclear power is dominated by special-interest politics, which is a politics of divisiveness, and in which no one seeks to represent the majority. In the arena of special interests, the opponents of nuclear power are greatly overmatched by its defenders. The need for coherent national programs and for the control and direction of industrial technology is obscured and defeated.

There will be a nuclear-power program for some years yet, therefore, and we must consider its implications. Nuclear power is an industry that seems to require rigid stratification. The distance between the uranium miner and the occupants of Manhattan offices is more than geographical; stratification of sex, age, race, education and wealth seems to be built into the very machinery of the industry. The anonymous and uncounted radiation workers suffer appalling risks, the least of which are associated with radiation, but their representatives in trade unions and the Congress are barely aware of these risks and are not concerned to reduce them. Reactors threaten the public's health and safety, but the threat is remote and difficult to perceive: it is less immediate than other hazards many of us face every day and pales to insignificance against the overwhelming hazards of war. Finally, the enlargement of the industry will mean increasing militarism and regimentation of society.

In all of these respects the nuclear industry may resemble others. But this is not a book about American enterprise, it is a book about nuclear power.

What can or should be done to change the nuclear-power industry?

It is not enough—indeed, it is dangerous—simply to say "stop." But, there is a positive program widely advocated by the opponents of nuclear power which makes excellent sense. Energy conservation and the rapid development of solar power, power from the earth, from the wind and a variety of other sources is widely accepted as preferable to our present arrangement, but cost and feasibility questions must be faced. Without efficient and cheap energy storage, solar and wind power will be severely limited in their application to electricity generation. With effective energy storage, there can be a rapid transformation of the power industry and an end of the central-station monopoly which is presently so costly.

An ideal program would couple research in needed new technology with a practicable energy-conservation program. Competition must be encouraged once again in the power business; the competition between the power companies and their customers is already renewing itself, as large consumers of power and heat increase their efficiency and combine their needs for energy in isolated plants of their own that are far more efficient than the central-station plants of the monopoly utilities. Small homes and business can be helped to do the same, with new technology and new methods of financing.

Perhaps the utilities can be persuaded to participate in the changes which are needed—the alternative for them is a slow slide into bankruptcy. Power companies might reasonably finance the efficient equipment, the insulation, the new construction, even the isolated plants which will compete with them for power sales. If these are economically desirable activities, they may also be profitable. The initial capital costs will be high, and the profits long in accruing, but that is precisely the sort of arrangement utilities are able to deal with. The utilities would be selling energy conservation, rather than power, but the fundamental techniques might not be very much different.

How likely is any of this to occur? The answer depends in large part on our creating new democratic means of planning what our industry does. What those means might be is a question to which some of the people in this book have begun to turn their attention. Perhaps, in time, they will even find a way of controlling the military technology that threatens to destroy us.

APPENDIX A:

A Note On Sources

THE PRINCIPAL source materials for this book are the interviews presented in it, tape recorded during the summer of 1975. The conversations with John Conway and David Pesonen were not recorded, but are presented as they appear in my notes made during and immediately after the interviews. The testimony and cross-examination before the Public Service Commission of Missouri is taken from the official transcript of the hearings in the fall of 1974.

In choosing to present the controversy over nuclear power through the words of the people engaged, I have tried to present material which cannot be conveyed in statistical or derivative forms. On the other hand, the selection of spokesmen and the editing of their words necessarily introduces some bias. To minimize the effects of this bias, I have chosen for interviews those officials in industry and government who speak officially for their agencies or industries, or who set policy. In the case of the opponents of nuclear power I somewhat arbitrarily chose figures who are well known to the public and who head organizations that have been active in opposing nuclear power; the controversy over a nuclear powerplant and electric-power industry reform in Missouri was chosen to stand for dozens of such battles elsewhere in the US for no better reason than that it was the closest to me and seemed reasonably representative.

The interviews generally ran from one to two hours and clearly could not be presented *in toto*. Where verbatim excerpts are given, editing has been kept to a minimum, although some editing was needed to clarify statements which were ambiguous when written, but quite clear when spoken because of the added meaning of gestures and tone of voice. In some cases, phrases used for punctuation ("you know") have been replaced by punctuation marks, and digressions and repetitions have been deleted, as were defamatory statements. For the most part, the subjects' words are here precisely as they were spoken, and in no case has the order of a person's remarks been changed.

The interviewer's questions have been included, in almost all cases. This is a departure from the anonymity reporters ordinarily try to preserve, but in the present case it seemed to me that the elimination of the questions would have given a misleading impression of the responses. The subjects have not been permitted to review the transcripts of the interviews or the edited versions which appear here. Simple prompting questions ("And then what happened?") and statements by the interviewer which explained the motive of questions and which are now incorporated in the narrative portion of the text were also deleted, but otherwise a complete context is given for every statement.

Since the principal sources for the book are the interviews presented in it, and the materials referred to in the text, I have not burdened the book with an apparatus of footnotes. Where sources have not been given, the material is drawn from articles in *Environment* magazine or from the author's earlier book, *The Careless Atom* (Houghton Mifflin, 1969).

The reader interested in pursuing a point further, however, may find the following references useful, as I did. I have not tried to give a comprehensive list of sources. On technical points the reader should consult *Nuclear Science Abstracts*, available in most university libraries, which has an excellent subject index.

Chapter 1

Many of the documents dating from the wartime Manhattan Project have been declassified and have added to the flood of published information on the period. An extremely interesting account, among the many biased versions, is that written by General Leslie R. Groves, the only person who definitely knew the Manahattan Project as a whole and was aware of its limitations as a military enterprise: *Now It Can Be Told* (New York: Harper, 1962). The only attempt at an unbiased and complete record, written with access to all of the documents, is the first volume of the Atomic Energy Commission's official history: *The New World, 1939/1946*, by Richard G. Hewlett and Oscar E. Anderson, Jr. (University Park, Penn.: Pennsylvania State University Press, 1962). On the decision to use the bomb: *Atomic Diplomacy*, by Gar Alperovitz, and *Meeting At Potsdam* by Charles L. Mee (New York: Evans, 1975).

Chapter 2

Barry Commoner's current views on atomic energy and other related matters can be found in *The Poverty of Power* (New York: Knopf, 1976).

Chapter 3

The Smyth Report: *Atomic Energy for Military Purposes: The Official Report on the Development of the Atomic Bomb Under the Auspices of the United States Government 1940–1945*, by Henry DeWolf Smyth (Princeton: Princeton University Press, 1945).

The pitiful state of our atomic armaments in the 1940s is described for the first time in the second volume of the AEC's official history, *Atomic Shield, 1947/1952*, by Richard G. Hewlett and Francis Duncan (University Park, Penn.: Pennsylvania State Press, 1969), which gives the text of the memorandum to Truman. The meeting with Truman, and his surprise, is also described by Lilienthal in his diary of those years, *The Journals of David Lilienthal: The Atomic Energy Years 1945–50*, vol. II (New York: Harper, 1964).

Chapter 5

For the early years of the military nuclear program, see *Report on the Atom: What You Should Know about the Atomic Energy Program of the United States*, by Gordon Dean (New York: Knopf, 1953 and 1959), from which the quotes from Dean are taken; and *Men and Decisions*, by Lewis L. Strauss (New York: Doubleday, 1962), for the Strauss and McMahon statements; and for a more detailed and unbiased presentation, *Atomic Shield*, op. cit. Herbert York recently reported (*Scientific American*, August, 1975) that the Russian test in 1954 was not a "super" but merely an operational version of the augmented A-bombs we had tested.

Chapter 6

The Dean quotes from *What You Should Know*, op. cit. For Strauss's efforts, see his *Men and Decisions*, op. cit. Edward Spencer's speech is given in *Power Reactor Conference: Proceedings of a Meeting Held November 26,27,28, 1956 at Fabrimetal, Brussels* (US Atomic Energy Commission, 1957). For early power-company assessments of the dim commercial prospects of nuclear power, see *Nuclear Power*

Reactors: Reports to the Atomic Energy Commission on Nuclear Power (two volumes); (US Atomic Energy Commission, 1954). For the relevant legislation, see Appendix C.

Chapters 9, 12
The Dartmouth Case is *Dartmouth College v. Woodward*, 17 US (4 Wheat) 518 (1819); the background and attorneys' arguments are taken from the Lawyers' Edition report (4 L. Ed. 629) and from a fascinating history of the case, *The Dartmouth College Cases and the Supreme Court of the United States* by John M. Shirley (Chicago: G.I. Jones, 1895; reissued by Da Capo Press, 1971).

Chapter 14
For a brief biography of John Marshall, see *The Justices of the Supreme Court 1789–1969, Their Lives and Major Opinions* (four volumes), Leon Friedman and Fred L. Israel, eds. (New York: Chelsea, 1969), or *John Marshall*, by William Draper Lewis, (Philadelphia: Winston, 1907). For the contemporary views of the decision, see materials in *The Supreme Court in United States History* (two volumes), by Charles Warren (Boston: Little, Brown, 1937). An excellent review of the legal position of the corporation from colonial times: *The Position of Foreign Corporations in American Constitutional Law*, by G.C. Henderson (Cambridge: Harvard University Press, 1918).
The best history of the robber barons is still the book that put the term into currency: *The Robber Barons*, by Matthew Josephson (New York: Harcourt, 1934). On management of the railroads, past and present: *The Wreck of the Penn Central* by J.R. Doughen and P. Binzen (Boston: Little, Brown, 1971) which includes some interesting historical material.

Chapter 18
On Waite, his colleagues' dim view of him, and the contemporary reaction to his decisions, see *The Commerce Clause Under Marshall, Taney and Waite*, by Felix Frankfurter (Chapel Hill: University of North Carolina Press, 1937); Warren, *op. cit.*; and the biography by Benjamin Rush Cowen, a contemporary, in *Great American Lawyers* (seven volumes), edited by William Draper Lewis (Phildelphia: Winston, 1907). The Grain Elevator Case is *Munn v. Illinois*, 94 US 113 (1877). For the issues in the dispute over electoral votes see *The Proceedings of the Electoral Commission and of the Two Houses of Congress in Joint Meeting Relative to the Count of Electoral Votes Cast December 6, 1876 for the Presidential Term Commencing March 4, 1877* (Washington DC: US Government Printing Office, 1877), and *Reunion and Reaction: The Compromise of 1877 and the End of Reconstruction*, by C. Van Woodward (Waltham, Mass.: Little, Brown, 1951).

Chapter 20
Justice Field's opinions asserting the personality of corporations are from The Railroad Tax Cases: *County of San Mateo v. Southern Pacific Railroad Co.*, 13 Fed. 722 (1882) and *County of Santa Clara v. Southern Pacific Railroad Co.*, 18 Fed. 385 (1883), 118 US 394 (1886), affirmed on other grounds. Waite's dictum asserting that corporations are persons within the meaning of the Fourteenth Amendment is in the report of the latter decision. In *Minneapolis and St. Louis Railway Co. v. Beckwith*, 129 US 585 (1889), Justice Field for the majority said the *Santa Clara* case "held" that corporations are persons, but *Beckwith* was the first case in which the Court actually so held.

For the dispute over Conkling's role and the motives for putting the corporation within the Fourteenth Amendment's guarantee, see *Everyman's Constitution*, by Jay Howard Graham (Madison, Wisc.: State Historical Society of Wisconsin, 1968).

Chapter 24

There is no good over-all history of the electric-power industry. For early technical developments, an unbiased source is *The Rise of the Electrical Industry During The Nineteenth Century*, by Malcolm MacLaren (Princeton: Princeton University Press, 1943); see also *The Electric Power Business*, by Edwin Vennard (New York: McGraw-Hill, 1962, 1970). Industry-sponsored histories are worthless (Samuel Insull has been expunged from General Electric's, for instance). Some of the early organizational developments are given in *The Potentates*, by Ben Seligman (New York: Dial, 1971); *The Robber Barons*, *op. cit.*; and other general histories. The Appleton plant is described in *Let There Be Light*, by Forest MacDonald (Madison, Wisc.: American History Research Center, 1957).

Chapter 26

The best source of information about the early years of the central-station power business are two volumes of the privately printed collection of Insull's speeches: *Central-Station Service: Its Commercial Development and Economic Significance As Set Forth in the Public Addresses of Samuel Insull, 1897–1914*, edited by William E. Keilly (Chicago: 1915); *Public Utilities In Modern Life: Selected Speeches 1914–1923*, edited by William E. Keilly (Chicago: 1924). A pro-Insull biography by a historian who understands the early technological developments is *Insull*, by Forest MacDonald (Chicago: University of Chicago Press, 1962).

Chapter 28, 29

Insull's statements are taken from his collected speeches, volume I, *op. cit.* For the turn-of-the-century merger movement and its present residue, see, for instance, *Economic Concentration: Structure, Behavior and Public Policy*, by John M. Blair (New York: Harcourt, 1972); and Josephson's *The Robber Barons* and Seligman's *The Potentates*, *op. cit.* On the interlocking structure of industry created at that time, see *The Theory of Business Enterprise*, by Thorstein Veblen, (New York: Scribners, 1904).

Chapter 30

Insull's statements and activities are from the first volume of his speeches and from MacDonald's biography. The statements of other industry figures are taken from the *Bulletin of the National Electric Light Association* for the dates in question, and from the annual-meeting issues of that bulletin. Statistics on electric-power sales are taken from the Edison Electric Institute's weekly, monthly and quarterly bulletins.

Chapter 31

There has been an enormous outpouring of literature on new energy sources, especially solar energy. An optimistic view is taken by Commoner in *The Poverty of Power*, *op. cit.* The material in this chapter is based principally on articles which have appeared in *Environment* magazine in recent years.

Chapter 33

The *Proceedings of the Council between General W.T. Sherman and Samuel Tappan, for the United States, and the Chiefs and Head Men of the Navajo Tribe, on May 28, 1868,* an extraordinarily moving document with a brief history of the area, is available from KC Publications, Las Vegas, Nevada (the Barboncito quotation is taken from this publication), under the title, *The Navajo Treaty—1868.*

Chapter 34

The Rasmussen report: *Reactor Safety Study: An Assessment of Accident Risks in US Commercial Nuclear Power Plants,* WASH-1400 (Washington, DC: Nuclear Regulatory Commission, October, 1975). A draft of the report had been issued in September, 1974, and all of the figures and statements used throughout this book refer to the draft, rather than to the final report. The draft was mentioned by virtually everyone that I interviewed, and much of the discussion of reactor safety was conducted in terms of the draft report. Rather than pepper the interviews with corrections, I have simply retained the numbers used by the draft. The final report made a number of minor changes. The estimate of the probability of a meltdown, given in the draft as one in seventeen thousand per year per reactor (about one in three hundred per year for fifty-five reactors) was decreased to one chance in twenty thousand per year per reactor, or roughly one chance in three hundred and fifty for the industry per year. Maximum damage estimates were increased, however, to thirty-three hundred fatalities and forty-three thousand early illnesses, and fourteen billion dollars in property damage.

Other useful sources on reactor safety are *Nuclear Power,* by Walt Patterson (Viking, 1976) and *The Nuclear Fuel Cycle* by Henry Kendall and Dan Ford (mimeographed; Union of Concerned Scientists, Cambridge, undated).

The material on steam explosions and Stirling Colgate's hypothesis is taken from Kevin Shea's important article, "An Explosive Reactor Possibility," in *Environment* (January–February, 1976).

I have also drawn on my own previous work in *Environment* and *The Careless Atom* (Boston: Houghton Mifflin, 1969), and the technical studies cited therein.

Chapter 36

The Gofman and Tamplin estimates were published in *Environment,* April, 1970 ("Radiation: The Invisible Casualties"). The best over-all review of this subject is *Radioactive Contamination,* by Virginia Brodine (New York: Harcourt, 1975). The National Academy of Sciences report referred to in the text is *The Effects on Populations of Exposure to Low Levels of Ionizing Radiation* (Washington, DC: National Academy of Sciences, November, 1972), which collects citations to the literature up to that time.

Chapter 37

For more information on the technology of hot-waste storage, see *Nuclear Power,* by Patterson, *op. cit.,* and "Hot Wastes From Nuclear Power," by George Berg (*Environment,* May, 1973); difficulties in waste storage have been noted in numerous articles in *Environment* magazine since 1970.

Chapter 38

The account of procedures at the West Valley plant is taken from Robert Gillette's excellent article in *Science* ("News and Comment," October 11, 1974).

For economic and technical data see "Expensive Enrichment," by Marvin Resnikoff (*Environment*, July–August, 1975).

Chapter 43

Insull speeches are taken from his collected speeches, *op. cit.*, volume I. Swidler is quoted with approval in the standard text on utility law, *Principles of Public Utility Regulation, Theory and Application* (two volumes), by A.J.G. Priest (Charlottesville, Va.: Michie, 1969), which gives a brief review of early legislation and decisions.

Chapter 45

I am indebted to Steven Emmings's unpublished paper, prepared for the Scientists' Institute for Public Information task force on energy, for locating many of the materials quoted in connection with the early history of propaganda campaigns in the power industry. Official statements can be found in the *Bulletin of the National Electric Light Association* for the years indicated. The Nebraska public power system is described in *Public Power in Nebraska*, by Robert E. Firth (Lincoln: University of Nebraska Press, 1962).

Chapter 46

The quoted passage is from M.L. Ramsay's *Pyramids of Power: The Story of Roosevelt, Insull and the Utility Wars* (Indianapolis: Bobbs-Merrill, 1937).

An excellent review of the fight between Roosevelt and the utility industry in the 1930s is Philip J. Funigiello's *Toward A National Power Policy: The New Deal and The Electric Utility Industry, 1933–1941* (Pittsburg: University of Pittsburg Press, 1973); on the fight over TVA, see Thomas K. McCraw's *TVA and The Public Power Fight, 1933–39* (Philadelphia: Lippincott, 1971); and on the rural electric cooperatives, see Marquis Childs's *The Farmer Takes a Hand: The Electric Power Revolution in Rural America* (New York: Doubleday, 1952).

Chapter 52

Thomas B. Cochran's book, *The Liquid Metal Fast Breeder Reactor: An Environmental and Economic Critique* (Baltimore: Resources for the Future, 1974), is the best available review of the breeder program. The Energy Research and Development Administration, in January, 1976, released a multivolume *Final Environmental Impact Statement* which includes references to all the relevant literatures and incorporates the text of many of the criticisms made of the program. The "hot particle" theory is clearly presented by Tamplin and Cochran and effectively rebutted by Robin Mole, in the May 29 issue of the British journal *New Scientist*.

Chapter 53

Statistics on the electric-power industry are taken from the Edison Electric Institute's weekly bulletin, *Electric Output*, its monthly, quarterly and annual statistical summaries, and the most recent (1974) statistical yearbook and *Year End Summary* (figures for load factors are taken from the latter); all of these publications are available from the Edison Electric Insitute, 90 Park Ave., New York, NY 10016. Statistics for plant orders are taken from the press releases of the Atomic Industrial Forum (see Appendix C).

Chapter 54

A detailed presentation of the problem of plutonium theft and nuclear proliferation is *Nuclear Theft: Risks and Safeguards* by Mason Willrich and Theodore B.
Taylor (Cambridge, Mass.: Ballinger, 1974). On laser developments, see the
author's articles in *Environment* (December, 1968 and December, 1969); more
recent military laser technology is reported at length in a series in *Aviation Week
and Space Technology* during August and September, 1975.

Reactors in the United States and Abroad

A S OF JUNE 30, 1975, the electric-power generating capacity of the United States was 492,300 megawatts (MWe: a megawatt is one thousand kilowatts). There were fifty-five nuclear powerplants operating at that time, which had a capacity of 37,165 MWe, or about eight percent of the US total. During 1975, eleven new reactors were ordered, and thirteen were cancelled, for a slight net decline in total orders; but outside the US orders and plans for nuclear powerplants increased by thirty-three percent. The following tabulations, collected from information published by the Nuclear Regulatory Commission and the Atomic Industrial Forum, is current for June 30, 1975.

OUTSIDE THE UNITED STATES

Total

	1975		1974	
	Net MWe	Reactors	Net MWe	Reactors
Operable	29,175	102	24,293	96
Under construction	59,767	85	50,097	77
On order	54,462	70	56,112	73
Planned	150,874	169	90,073	102
Totals	294,278	426	220,575	348

By Country

	1975 MWe %***		1980 MWe %***		Forecast* 1990 MWe %		2000 MWe %	
Argentina	319	(5.1)	919	(814)	6719	(24)	—	
Austria	—		730	(5.6)	—		—	
Belgium	1375	(15)	3175	(25)	13,000	(60)	26,000	(60)
Brazil	—		630	(2.2)	10,200	(14.6)	13,000	(60)
Bulgaria	440	(—)	1760	(20)	5760	(50)	13,000	(60)
Canada	2513	(—)	6000	(10)	35,000	(30)	100,000	(40)
China (Taiwan)	636	(9.6)	3242	(31.7)	—		35,000	(—)
Czechoslovakia	112	(—)	1800	(7.5)	10,000	(25)	11,400	(55)
Denmark	—		—		4900	(40)		
Egypt	—		—		4200	(—)	8000	(45)
Finland	—		2160	(20)				
France	2861	(—)	7400	(25)				
East Germany	520	(—)	—					
West Germany	3500	(7)	20,000	(24)	120,000	(55)	200,000	(71)
Hong Kong	—		—				—	
Hungary	—		1240	(3)	8620	(9.9)	—	
India	580	(—)	1200	(—)			2000	(20)
Iran	—		—		600	(10)	15,300	(67)
Ireland	—		5500	(12)	3900	(40)	125,000	(85)
Israel	1500	(3.5)	—		47,000	(70)	—	(50)
Italy	3709	(—)	1795	(20)	—		—	(50)
Japan	—		—		—		—	
Korea								
Luxembourg								

	1975 MWe %***	1980 MWe %***	Forecast* 1990 MWe %	2000 MWe %
Mexico	—	1308 (7)	14,800 (75)	—
Netherlands	535 (—)	3500 (20)**	—	—
Pakistan	125 (7)	725 (—)	5800 (—)	15,000 (60)
Philippines	—	—	3200 (51.7)	7800 (56.3)
Portugal	—	—	3300 (37)	8000 (47)
Rumania	—	—	—	20,000 (—)
South Africa	—	—	—	10,000 (13)
Spain	953 (—)	—	3000 (8)	—
Sweden	1260 (—)	8260 (—)	23,600 (—)	—
Switzerland	1006 (17)	3700 (43)	8000 (62)	—
Thailand	—	—	1200 (22)	2400 (20)
UK	7000 (10)	11,000 (15)	43,000 (30)	—
USSR	4062 (—)	31,200 (—)	—	150,000 (60)
Yugoslavia	—	—	4800 (80)	—

*Number of net megawatts planned to be produced by nuclear power, with the percentage of total electric generating capacity in parentheses.

**Figures for 1985.

***Percent of total national power-generating capacity.

About seventy percent of the reactors already chosen are American-designed light-water reactors.

INSIDE THE UNITED STATES

Total

Reactors	Net MWe
55 Reactors Operable	37,165 MWe
63 Reactors Under Construction	64,266 MWe
99 Reactors On Order	111,899 MWe
17 Commitments for Letters of Intent/Options	18,847 MWe
234 Reactor Commitments	232,177 MWe

By 1976, three of the reactors under construction had gone into operation.

The following list of reactors is current as of June 30, 1975, and gives reactors already shut down, those in operation, or those planned. Shut down reactors are principally the first small experimental reactors, none of which performed satisfactorily. No military reactors are included in the listing. Several small military reactors have produced small amounts of power; the "N" Reactor, at the Hanford Works in Richland, Washington, which is primarily for plutonium production, has a power capacity of 850 MWe, and has produced substantial amounts of power for sale by the Washington Public Power Supply System since 1963.

SHUT DOWN OR DISMANTLED

	Location	Type	Capacity (MWe)	Start up	Shut down
Hallam Nuclear Power Facility, Sheldon Station (AEC and Consumers Public Power District)	Hallam, Nebr.	Sodium graphite	75.0	1962	1964
Carolinas–Virginia Tube Reactor (Carolinas–Virginia Nuclear Power Associates, Inc.)	Parr, SC	Pressure tube, heavy water	17.0	1963	1967
Piqua Nuclear Power Facility (AEC and City of Piqua)	Piqua, Ohio	Organic cooled and moderated	11.4	1963	1966
Boiling Nuclear Superheater Power Station (AEC and Puerto Rico Water Resources Authority)	Punta Higuera, PR	Boiling water integral nuclear superheat	16.5	1964	1968
Pathfinder Atomic Plant (Northern States Power Co.)	Sioux Falls, S. Dak.	Boiling water nuclear superheat	58.5	1964	1967
Elk River Reactor (AEC and Rural Cooperative Power Association)	Elk River, Minn.	Boiling water	22.0	1962	1968
Enrico Fermi Atomic Power Plant, Unit 1 (Power Reactor Development Co.)	Lagoona Beach, Mich.	Sodium cooled, fast	60.9	1963	1973
Peach Bottom Atomic Power Station, Unit 1 (Philadelphia Electric Co.)	Peach Bottom, Pa.	High temperature gas cooled	40.0	1966	1974

The following list includes only commercial generating units having a letter of intent/option, on order, under construction, or operable. Utilities having a twenty percent or greater interest in any particular unit are listed in parentheses after the operating utility. Status is indicated by: L—letter of intent/option; O—on order; C—under construction. **Reactors in operation or licensed by the NRC to start up as of June 30, 1975, are indicated in bold face type.** The reactor types listed are: Pressurized Water Reactor—PWR; Boiling Water Reactor—BWR; High Temperature Gas-cooled Reactor— HTGR; Liquid Metal Fast Breeder Reactor—LMFBR. The reactor manufacturers are: Allis-Chalmers—AC; Babcock & Wilcox—B&W; Combustion Engineering—CE; General Atomic—GA; General Electric—GE; Offshore Power Systems (Westinghouse-Tenneco)—OPS; and Westinghouse—W. Asterisk indicates that the plant has been deferred and the date of commercial operation has not been announced.

Operating or Planned

State and Utility	Plant	Location	Net MWe	Type/Mfr.	Operable
ALABAMA					
Alabama Power Co.	Alan Barton 1 (O)	Verbena	1170	BWR/GE	1984
Alabama Power Co.	Alan Barton 2 (O)	Verbena	1170	BWR/GE	1985
Alabama Power Co.	Alan Barton 3 (O)	Verbena	1170	BWR/GE	1986
Alabama Power Co.	Alan Barton 4 (O)	Verbena	1170	BWR/GE	1987
Alabama Power Co.	Joseph M. Farley 1 (C)	Houston County	860	PWR/W	1976
Alabama Power Co.	Joseph M. Farley 2 (C)	Houston County	860	PWR/W	1977
Tennessee Valley Authority	Bellefonte 1 (C)	Scottsboro	1213	PWR/B&W	1979
Tennessee Valley Authority	Bellefonte 2 (C)	Scottsboro	1213	PWR/B&W	1980
Tennessee Valley Authority	**Browns Ferry 1**	**Decatur**	**1067**	**BWR/GE**	**1973**
Tennessee Valley Authority	**Browns Ferry 2**	**Decatur**	**1067**	**BWR/GE**	**1974**
Tennessee Valley Authority	Browns Ferry 3 (C)	Decatur	1067	BWR/GE	1975
ARIZONA					
Arizona Nuclear Power Project [Arizona Public Service Co. (operating utility), Salt River Project]	Palo Verde 1 (O)	Wintersburg	1270	PWR/CE	1982
Arizona Nuclear Power Project [Arizona Public Service Co. (operating utility), Salt River Project]	Palo Verde 2 (O)	Wintersburg	1270	PWR/CE	1984
Arizona Nuclear Power Project [Arizona Public Service Co. (operating utility), Salt River Project]	Palo Verde 3 (O)	Wintersburg	1270	PWR/CE	1986

340

APPENDIX B

State and Utility	Plant	Location	Net MWe	Type/Mfr.	Operable
ARKANSAS					
Arkansas Power & Light Co.	**Arkansas Nuclear One—1**	**London**	**850**	**PWR/B&W**	**1974**
Arkansas Power & Light Co.	Arkansas Nuclear One—2 (C)	London	941	PWR/CE	1978
CALIFORNIA					
Pacific Gas and Electric Co.	Diablo Canyon 1 (C)	Avila Beach	1131	PWR/W	1976
Pacific Gas and Electric Co.	Diablo Canyon 2 (C)	Avila Beach	1156	PWR/W	1977
Pacific Gas and Electric Co.	**Humboldt Bay**	**Humboldt Bay**	**63**	**BWR/GE**	**1963**
Pacific Gas and Electric Co.	unit 1 (O)	—	1168	BWR/GE	—
Pacific Gas and Electric Co.	unit 2 (O)	—	1168	BWR/GE	—
Sacramento Municipal Utility District	**Rancho Seco 1**	**Clay Station**	**913**	**PWR/B&W**	**1974**
Southern California Edison Co. (San Diego Gas and Electric Co.)	**San Onofre 1**	**San Clemente**	**430**	**PWR/W**	**1967**
Southern California Edison Co. (San Diego Gas and Electric Co.)	San Onofre 2 (C)	San Clemente	1140	PWR/CE	1979
Southern California Edison Co. (San Diego Gas and Electric Co.)	San Onofre 3 (C)	San Clemente	1140	PWR/CE	1980
COLORADO					
Public Service Company of Colorado	**Fort St. Vrain**	**Platteville**	**330**	**HTGR/GA**	**1973**
CONNECTICUT					
Connecticut Yankee Atomic Power Co.	**Connecticut Yankee**	**Haddam Neck**	**575**	**PWR/W**	**1967**
Northeast Nuclear Energy Co.	**Millstone 1**	**Waterford**	**652**	**BWR/GE**	**1970**
Northeast Nuclear Energy Co.	Millstone 2 (C)	Waterford	830	PWR/CE	1975
Northeast Nuclear Energy Co.	Millstone 3 (C)	Waterford	1150	PWR/W	1979

State and Utility	Plant	Location	Net MWe	Type/Mfr.	Operable
DELAWARE					
Delmarva Power & Light Co.	Summit 1 (O)	Summit	770	HTGR/GA	1980
Delmarva Power & Light Co.	Summit 2 (O)	Summit	770	HTGR/GA	1983
FLORIDA					
Florida Power Corp.	Crystal River 3 (C)	Red Level	850	PWR/B&W	*
Florida Power Corp.	Orlando 1 (L)	Orlando	1300	PWR/CE	*
Florida Power Corp.	Orlando 2 (L)	Orlando	1300	PWR/CE	*
Florida Power & Light Co.	St. Lucie 1 (C)	St. Lucie County	802	PWR/CE	1975
Florida Power & Light Co.	St. Lucie 2 (O)	St. Lucie County	802	PWR/CE	1980
Florida Power & Light Co.	South Dade 1 (O)	South Dade	1150	PWR/W	mid-80s
Florida Power & Light Co.	South Dade 2 (O)	South Dade	1150	PWR/W	mid-80s
Florida Power & Light Co.	**Turkey Point 3**	**Turkey Point**	**666**	**PWR/W**	**1972**
Florida Power & Light Co.	**Turkey Point 4**	**Turkey Point**	**666**	**PWR/W**	**1973**
GEORGIA					
Georgia Power Co. (Oglethorpe Electric Membership Corp.)	**Edwin I. Hatch 1**	**Baxley**	**786**	**BWR/GE**	**1974**
Georgia Power Co. (Oglethorpe Electric Membership Corp.)	Edwin I. Hatch 2 (C)	Baxley	795	BWR/GE	1978
Georgia Power Co.	Alvin W. Vogtle 1 (C)	Waynesboro	1150	PWR/W	*
Georgia Power Co.	Alvin W. Vogtle 2 (C)	Waynesboro	1150	PWR/W	*
ILLINOIS					
Commonwealth Edison Co.	Braidwood 1 (O)	Braidwood	1100	PWR/W	1981
Commonwealth Edison Co.	Braidwood 2 (O)	Braidwood	1100	PWR/W	1982

State and Utility	Plant	Location	Net MWe	Type/Mfr.	Operable
ILLINOIS, *cont.*					
Commonwealth Edison Co.	Byron 1 (O)	Byron	1100	PWR/W	1980
Commonwealth Edison Co.	Byron 2 (O)	Byron	1100	PWR/W	1982
Commonwealth Edison Co.	Dresden 1	Morris	200	BWR/GE	1959
Commonwealth Edison Co.	Dresden 2	Morris	809	BWR/GE	1970
Commonwealth Edison Co.	Dresden 3	Morris	809	BWR/GE	1971
Commonwealth Edison Co.	LaSalle 1 (C)	Seneca	1100	BWR/GE	1979
Commonwealth Edison Co.	LaSalle 2 (C)	Seneca	1100	BWR/GE	1979
Commonwealth Edison Co.	Zion 1	Zion	1100	PWR/W	1973
Commonwealth Edison Co.	Zion 2	Zion	1100	PWR/W	1973
Commonwealth Edison Co.	Quad Cities 1	Cordova	809	BWR/GE	1972
(Iowa-Illinois Gas and Electric Co.)					
Commonwealth Edison Co.	Quad Cities 2	Cordova	809	BWR/GE	1972
(Iowa-Illinois Gas and Electric Co.)					
Illinois Power Co.	Clinton 1 (O)	Clinton	950	BWR/GE	1981
Illinois Power Co.	Clinton 2 (O)	Clinton	950	BWR/GE	1984
INDIANA					
Northern Indiana Public Service Co.	Bailly Nuclear 1 (C)	Dunes Acres	657	BWR/GE	1980
Public Service Indiana	Marble Hill 1 (O)	Madison	1150	PWR/W	1983
Public Service Indiana	Marble Hill 2 (O)	Madison	1150	PWR/W	1984
IOWA					
Iowa Electric Light and Power Co.	Duane Arnold	Cedar Rapids	550	BWR/GE	1974
(Central Iowa Power Cooperative)					
Iowa Electric Light and Power Co.	Central Iowa 1 (L)	Central Iowa	1220	BWR/GE	—
(Iowa Power and Light Co.)					

State and Utility	Plant	Location	Net MWe	Type/Mfr.	Operable
KANSAS					
Kansas Gas and Electric Co. (Kansas City Power & Light Co.)	Wolf Creek (O)	Burlington	1100	PRW/W	1982
LOUISIANA					
Gulf States Utilities Co.	River Bend 1 (O)	St. Francisville	940	BWR/GE	1980
Gulf States Utilities Co.	River Bend 2 (O)	St. Francisville	940	BWR/GE	1982
Louisiana Power & Light Co.	Waterford 3 (C)	Taft	1165	PWR/CE	1979
MAINE					
Central Maine Power Co.	Sears Island 1 (O)	Searsport	1200	PWR/W	1983
Maine Yankee Atomic Power Co.	**Maine Yankee**	**Wiscasset**	**885**	**PWR/CE**	**1972**
MARYLAND					
Baltimore Gas and Electric Co.	**Calvert Cliffs 1**	**Lusby**	**845**	**PWR/CE**	**1974**
Baltimore Gas and Electric Co.	Calvert Cliffs 2 (C)	Lusby	845	PWR/CE	1977
Potomac Electric Power Co.	Douglas Point 1 (O)	Charles County	1178	BWR/GE	1985
Potomac Electric Power Co.	Douglas Point 2 (O)	Charles County	1178	BWR/GE	1987
MASSACHUSETTS					
Boston Edison Co.	**Pilgrim 1**	**Plymouth**	**655**	**BWR/GE**	**1972**
Boston Edison Co.	Pilgrim 2 (O)	Plymouth	1180	PWR/CE	1980
Boston Edison Co.	Pilgrim 3 (O)	Plymouth	1150	PWR/CE	*
Northeast Nuclear Energy Co.	Montague 1 (O)	Montague	1150	BWR/GE	1988
Northeast Nuclear Energy Co.	Montague 2 (O)	Montague	1150	BWR/GE	1988
Yankee Atomic Electric Co.	**Yankee**	**Rowe**	**175**	**PWR/W**	**1960**

State and Utility	Plant	Location	Net MWe	Type/Mfr.	Operable
MICHIGAN					
Consumers Power Co.	**Big Rock Point**	**Big Rock Point**	**72**	**BWR/GE**	**1962**
Consumers Power Co.	Midland 1 (C)	Midland	527	PWR/B&W	1981
Consumers Power Co.	Midland 2 (C)	Midland	855	PWR/B&W	1982
Consumers Power Co.	**Palisades**	**South Haven**	**800**	**PWR/CE**	**1972** *
Detroit Edison Co.	Enrico Fermi 2 (C)	Lagoona Beach	1150	BWR/GE	*
Detroit Edison Co.	Greenwood 2 (O)	St. Clair County	1208	PWR/B&W	*
Detroit Edison Co.	Greenwood 3 (O)	St. Clair County	1208	PWR/B&W	*
Indiana & Michigan Electric Co.	**Donald C. Cook 1**	**Bridgman**	**1060**	**PWR/W**	**1974** *
Indiana & Michigan Electric Co.	Donald C. Cook 2 (C)	Bridgman	1060	PWR/W	
MINNESOTA					
Northern States Power Co.	**Monticello**	**Monticello**	**545**	**BWR/GE**	**1971**
Northern States Power Co.	**Prairie Island 1**	**Red Wing**	**520**	**PWR/W**	**1973**
Northern States Power Co.	**Prairie Island 2**	**Red Wing**	**520**	**PWR/W**	**1974**
MISSISSIPPI					
Mississippi Power & Light Co.	Grand Gulf 1 (C)	Port Gibson	1300	BWR/GE	1984
Mississippi Power & Light Co.	Grand Gulf 2 (C)	Port Gibson	1300	BWR/GE	1984
MISSOURI					
Union Electric Co.	Callaway 1 (O)	Callaway County	1150	PWR/W	1981
Union Electric Co.	Callaway 2 (O)	Callaway County	1150	PWR/W	1983
NEBRASKA					
Nebraska Public Power District	**Cooper**	**Brownville**	**800**	**BWR/GE**	**1974**
Omaha Public Power District	**Fort Calhoun 1**	**Fort Calhoun**	**475**	**PWR/CE**	**1973**

State and Utility	Plant	Location	Net MWe	Type/Mfr.	Operable
Omaha Public Power District (Nebraska Public Power District)	Fort Calhoun 2 (O)	Fort Calhoun	1150	PWR/W	1983
NEW HAMPSHIRE					
Public Service Co. of New Hampshire (United Illuminating Co.)	Seabrook 1 (O)	Seabrook	1150	PWR/W	1979
Public Service Co. of New Hampshire (United Illuminating Co.)	Seabrook 2 (O)	Seabrook	1150	PWR/W	1981
NEW JERSEY					
Jersey Central Power & Light Co.	Forked River 1 (C)	Lacey Township	1120	PWR/CE	1982
Jersey Central Power & Light Co.	**Oyster Creek**	**Toms River**	**640**	**BWR/GE**	**1969**
Public Service Electric and Gas Co.	Atlantic 1 (O)	Little Egg Inlet (offshore)	1150	PWR/OPS	1985
Public Service Electric and Gas Co.	Atlantic 2 (O)	Little Egg Inlet (offshore)	1150	PWR/OPS	1987
Public Service Electric and Gas Co.	unit 1 (O)	—(offshore)	1150	PWR/OPS	1990
Public Service Electric and Gas Co.	unit 2 (O)	—(offshore)	1150	PWR/OPS	1992
Public Service Electric and Gas Co.	Hope Creek 1 (C)	Salem County	1067	BWR/GE	1981
Public Service Electric and Gas Co.	Hope Creek 2 (C)	Salem County	1067	BWR/GE	1983
Public Service Electric and Gas Co. (Philadelphia Electric Co.)	Salem 1 (C)	Salem	1090	PWR/W	1976
Public Service Electric and Gas Co. (Philadelphia Electric Co.)	Salem 2 (C)	Salem	1115	PWR/W	1978
NEW YORK					
Consolidated Edison Co. of N.Y., Inc.	**Indian Point 1**	**Buchanan**	**265**	**PWR/B&W**	**1962**
Consolidated Edison Co. of N.Y., Inc.	**Indian Point 2**	**Buchanan**	**1033**	**PWR/W**	**1973**

State and Utility	Plant	Location	Net MWe	Type/Mfr.	Operable
NEW YORK, *cont.*					
Consolidated Edison Co. of N.Y., Inc.	Indian Point 3 (C)	Buchanan	1033	PWR/W	1975
Long Island Lighting Co.	Jamesport 1 (O)	Jamesport	1150	PWR/W	1981
Long Island Lighting Co.	Jamesport 2 (O)	Jamesport	1150	PWR/W	1983
Long Island Lighting Co.	Shoreham (C)	Shoreham	820	BWR/GE	1978
New York State Electric & Gas Corp.	Somerset 1 (L)	Somerset	1220	BWR/GE	1984
New York State Electric & Gas Corp.	Somerset 2 (L)	Somerset	1220	BWR/GE	1986
Niagara Mohawk Power Corp.	**Nine Mile Point 1**	**Oswego**	**610**	**BWR/GE**	**1969**
Niagara Mohawk Power Corp.	Nine Mile Point 2 (C)	Oswego	1100	BWR/GE	1982
Niagara Mohawk Power Corp. (Power Authority of the State of New York)**	**James A. FitzPatrick**	**Scriba**	**821**	**BWR/GE**	**1974**
Power Authority of the State of New York	Greene County (O)	Cementon	1200	PWR/B&W	1982
Rochester Gas and Electric Corp.	**Robert E. Ginna**	**Rochester**	**490**	**PWR/W**	**1969**
Rochester Gas and Electric Corp. (Orange & Rockland Utilities, Niagara Mohawk Power Corp.)	Sterling (O)	Sterling	1100	PWR/W	1982–3
NORTH CAROLINA					
Carolina Power & Light Co.	Brunswick 1 (C)	Southport	821	BWR/GE	1977
Carolina Power & Light Co.	**Brunswick 2**	**Southport**	**821**	**BWR/GE**	**1974**
Carolina Power & Light Co.	Shearon Harris 1 (O)	Bonsal	915	PWR/W	1984
Carolina Power & Light Co.	Shearon Harris 2 (O)	Bonsal	915	PWR/W	1986
Carolina Power & Light Co.	Shearon Harris 3 (O)	Bonsal	915	PWR/W	1988
Carolina Power & Light Co.	Shearon Harris 4 (O)	Bonsal	915	PWR/W	1990
Carolina Power & Light Co.	South River 1 (O)	South River	1150	PWR/B&W	*
Carolina Power & Light Co.	South River 2 (O)	South River	1150	PWR/B&W	*

**100% ownership by Power Authority of the State of N.Y.

State and Utility	Plant	Location	Net MWe	Type/Mfr.	Operable
Carolina Power & Light Co.	South River 3 (O)	South River	1150	PWR/B&W	*
Duke Power Co.	William McGuire 1 (C)	Cowans Ford Dam	1180	PWR/W	1978
Duke Power Co.	William McGuire 2 (C)	Cowans Ford Dam	1180	PWR/W	1979
Duke Power Co.	Thomas L. Perkins 1 (O)	Davie County	1280	PWR/CE	1983
Duke Power Co.	Thomas L. Perkins 2 (O)	Davie County	1280	PWR/CE	1985
Duke Power Co.	Thomas L. Perkins 3 (O)	Davie County	1280	PWR/CE	1987
OHIO					
Cincinnati Gas & Electric Co. (Columbus and Southern Ohio Electric Co., Dayton Power and Light Co.)	William H. Zimmer 1 (C)	Moscow	840	BWR/GE	1979
Cincinnati Gas & Electric Co. (Columbus and Southern Ohio Electric Co., Dayton Power and Light Co.)	William H. Zimmer 2 (O)	Moscow	1170	BWR/GE	1984
Cleveland Electric Illuminating Co.	Perry 1 (O)	North Perry	1100	BWR/GE	1980
Cleveland Electric Illuminating Co.	Perry 2 (O)	North Perry	1100	BWR/GE	1982
Toledo Edison Co. (Cleveland Electric Illuminating Co.)	Davis-Besse 1 (C)	Oak Harbor	906	PWR/B&W	1976
Toledo Edison Co. (Cleveland Electric Illuminating Co., Ohio Edison Co.)	Davis-Besse 2 (L)	Oak Harbor	906	PWR/B&W	1983
Toledo Edison Co. (Cleveland Electric Illuminating Co., Ohio Edison Co.)	Davis-Besse 3 (L)	Oak Harbor	906	PWR/B&W	1985
OKLAHOMA					
Public Service Co. of Oklahoma	Black Fox 1 (O)	Inola	1150	BWR/GE	1983
Public Service Co. of Oklahoma	Black Fox 2 (O)	Inola	1150	BWR/GE	1985

State and Utility	Plant	Location	Net MWe	Type/Mfr.	Operable
OREGON					
Portland General Electric Co. (Eugene Water & Electric Board)	Trojan (C)	Rainier	1130	PWR/W	1975
Portland General Electric Co. (Pacific Power & Light Co.)	Pebble Springs 1 (O)	Pebble Springs	1260	PWR/B&W	1980
Portland General Electric Co. (Pacific Power & Light Co.)	Pebble Springs 2 (O)	Pebble Springs	1260	PWR/B&W	1983
PENNSYLVANIA					
Duquesne Light Co.	**Shippingport 1**	**Shippingport**	**90**	**PWR/W**	**1957**
Duquesne Light Co. (Ohio Edison Co.)	Beaver Valley 1 (C)	Shippingport	852	PWR/W	1975
Duquesne Light Co. (Cleveland Electric Illuminating Co., Ohio Edison Co.)	Beaver Valley 2 (C)	Shippingport	852	PWR/W	1981
Metropolitan Edison Co. (Jersey Central Power & Light Co., Pennsylvania Electric Co.)	**Three Mile Island 1**	**Goldsborough**	**830**	**PWR/B&W**	**1974**
Metropolitan Edison Co. (Jersey Central Power & Light Co., Pennsylvania Electric Co.)	Three Mile Island 2 (C)	Goldsborough	880	PWR/B&W	1978
Metropolitan Edison Co. (Jersey Central Power & Light Co.)	Portland 5 (L)	Portland	1220	PWR/CE	1994
Pennsylvania Power & Light Co.	Susquehanna 1 (C)	Berwick	1050	BWR/GE	1980
Pennsylvania Power & Light Co.	Susquehanna 2 (C)	Berwick	1050	BWR/GE	1982
Philadelphia Electric Co.	Fulton 1 (O)	Peach Bottom	1160	HTGR/GA	1984
Philadelphia Electric Co.	Fulton 2 (O)	Peach Bottom	1160	HTGR/GA	1986
Philadelphia Electric Co.	Limerick 1 (C)	Limerick Township	1065	BWR/GE	1981

State and Utility	Plant	Location	Net MWe	Type/Mfr.	Operable
Philadelphia Electric Co.	Limerick 2 (C)	Limerick Township	1065	BWR/GE	1983
Philadelphia Electric Co. (Public Service Electric and Gas Co.)	Peach Bottom 2	Peach Bottom Township	1065	BWR/GE	1973
Philadelphia Electric Co. (Public Service Electric and Gas Co.)	Peach Bottom 3	Peach Bottom Township	1065	BWR/GE	1974
RHODE ISLAND					
New England Electric System	Charlestown 1 (O)	Charlestown	1200	PWR/W	1982
New England Electric System	Charlestown 2 (O)	Charlestown	1200	PWR/W	1984
SOUTH CAROLINA					
Carolina Power & Light Co.	H.B. Robinson 2	Hartsville	700	PWR/W	1970
Duke Power Co.	Catawba 1 (O)	York County	1153	PWR/W	1981
Duke Power Co.	Catawba 2 (O)	York County	1153	PWR/W	1982
Duke Power Co.	Oconee 1	Lake Keowee	871	PWR/B&W	1973
Duke Power Co.	Oconee 2	Lake Keowee	871	PWR/B&W	1973
Duke Power Co.	Oconee 3	Lake Keowee	871	PWR/B&W	1974
Duke Power Co.	Cherokee 1 (O)	Cherokee County	1280	PWR/CE	1984
Duke Power Co.	Cherokee 2 (O)	Cherokee County	1280	PWR/CE	1986
Duke Power Co.	Cherokee 3 (O)	Cherokee County	1280	PWR/CE	1988
South Carolina Electric & Gas Co. (South Carolina Public Service Authority)	Virgil C. Summer 1 (C)	Parr	915	PWR/W	1979
South Carolina Electric & Gas Co.	Virgil C. Summer 2 (L)	Parr	915	PWR/W	1984
TENNESSEE					
Tennessee Valley Authority	Hartsville 1 (O)	Dixon Springs	1233	BWR/GE	1981
Tennessee Valley Authority	Hartsville 2 (O)	Dixon Springs	1233	BWR/GE	1982

State and Utility	Plant	Location	Net MWe	Type/Mfr.	Operable
TENNESSEE, *cont.*					
Tennessee Valley Authority	Hartsville 3 (O)	Dixon Springs	1233	BWR/GE	1981
Tennessee Valley Authority	Hartsville 4 (O)	Dixon Springs	1233	BWR/GE	1982
Tennessee Valley Authority	Sequoyah 1 (C)	Daisy	1148	PWR/W	1977
Tennessee Valley Authority	Sequoyah 2 (C)	Daisy	1148	PWR/W	1977
Tennessee Valley Authority	Watts Bar Dam 1 (C)	Spring City	1177	PWR/W	1978
Tennessee Valley Authority	Watts Bar Dam 2 (C)	Spring City	1177	PWR/W	1979
Tennessee Valley Authority (Commonwealth Edison Co., ERDA)	Clinch River Breeder Reactor Plant (O)	Oak Ridge	360	LMFBR/W	early '80s
TEXAS					
Gulf States Utilities Co.	Blue Hills 1 (O)	Jasper	950	PWR/CE	1983
Gulf States Utilities Co.	Blue Hills 2 (O)	Jasper	950	PWR/CE	1985
Houston Lighting & Power Co.	Allen's Creek 1 (O)	Wallis	1150	BWR/GE	1980
Houston Lighting & Power Co.	Allen's Creek 2 (O)	Wallis	1150	BWR/GE	1982
South Texas Project [Houston Lighting & Power Co. (project manager), Central Power and Light Co., City Public Service Board of San Antonio]	South Texas Project 1 (O)	Matagorda County	1250	PWR/W	1980
South Texas Project [Houston Lighting & Power Co. (project manager), Central Power and Light Co., City Public Service Board of San Antonio]	South Texas Project 2 (O)	Matagorda County	1250	PWR/W	1982
Texas Utilities Co.	Comanche Peak 1 (C)	Somervell County	1150	PWR/W	1980
Texas Utilities Co.	Comanche Peak 2 (C)	Somervell County	1150	PWR/W	1982
VERMONT					
Vermont Yankee Nuclear Power Corp.	**Vermont Yankee**	**Vernon**	**540**	**BWR/GE**	**1972**

State and Utility	Plant	Location	Net MWe	Type/Mfr.	Operable
VIRGINIA					
Virginia Electric and Power Co.	North Anna 1 (C)	Mineral	934	PWR/W	1976
Virginia Electric and Power Co.	North Anna 2 (C)	Mineral	934	PWR/W	1977
Virginia Electric and Power Co.	North Anna 3 (C)	Mineral	938	PWR/B&W	1980
Virginia Electric and Power Co.	North Anna 4 (C)	Mineral	938	PWR/B&W	1981
Virginia Electric and Power Co.	**Surry 1**	**Gravel Neck**	**820**	**PWR/W**	**1972**
Virginia Electric and Power Co.	**Surry 2**	**Gravel Neck**	**820**	**PWR/W**	**1973**
Virginia Electric and Power Co.	Surry 3 (C)	Gravel Neck	882	PWR/B&W	1983
Virginia Electric and Power Co.	Surry 4 (C)	Gravel Neck	882	PWR/B&W	1984
WASHINGTON					
Puget Sound Power and Light Co.	Skagit 1 (O)	Sedro Woolley	1269	BWR/GE	1982
Puget Sound Power and Light Co.	Skagit 2 (O)	Sedro Woolley	1269	BWR/GE	1985
ERDA (Power distributed by Washington Public Power Supply System)	**Hanford—N**	**Richland**	**800**	**Graphite (AEC)**	**1966**
Washington Public Power Supply System	WPPSS 1 (O)	Richland	1267	PWR/B&W	1980
Washington Public Power Supply System	WPPSS 2 (C)	Richland	1103	BWR/GE	1977
Washington Public Power Supply System	WPPSS 3 (O)	Satsop	1240	PWR/CE	1981
Washington Public Power Supply System	WPPSS 4 (O)	Richland	1267	PWR/B&W	1982
Washington Public Power Supply System	WPPSS 5 (O)	Satsop	1240	PWR/CE	1983

State and Utility	Plant	Location	Net MWe	Type/Mfr.	Operable
WISCONSIN					
Dairyland Power Cooperative	**LaCrosse**	**Genoa**	**50**	**BWR/AC**	**1967**
Northern States Power Co.	Tyrone 1 (O)	Durand	1150	PWR/W	1985
Northern States Power Co.	Tyrone 2 (O)	Durand	1150	PWR/W	*
Wisconsin Electric Power Co.	Koshkonong 1 (O)	Lake Koshkonong	900	PWR/W	1983
Wisconsin Electric Power Co.	Koshkonong 2 (O)	Lake Koshkonong	900	PWR/W	1984
Wisconsin Electric Power Co.	unit 1 (L)	—	900	PWR/W	1980s
Wisconsin Electric Power Co.	unit 2 (L)	—	900	PWR/W	1980s
Wisconsin Electric Power Co.	unit 3 (L)	—	900	PWR/W	1980s
Wisconsin Electric Power Co.	unit 4 (L)	—	900	PWR/W	1980s
Wisconsin Michigan Power Co. (Wisconsin Electric Power Co.)	**Point Beach 1**	**Two Creeks**	**497**	**PWR/W**	**1970**
Wisconsin Michigan Power Co. (Wisconsin Electric Power Co.)	**Point Beach 2**	**Two Creeks**	**497**	**PWR/W**	**1972**
Wisconsin Public Service Corp. (Wisconsin Power and Light Co.)	**Kewaunee**	**Carlton Township**	**535**	**PWR/W**	**1973**
PUERTO RICO					
Puerto Rico Water Resources Authority	Isolte (O)	Arecibo	583	PWR/W	1981
OTHER PLANTS NOT SITED					
Tennessee Valley Authority	unit 1 (O)	—	1233	BWR/GE	1982
Tennessee Valley Authority	unit 2 (O)	—	1233	BWR/GE	1983
Tennessee Valley Authority	unit 3 (O)	—	1300	PWR/CE	1982
Tennessee Valley Authority	unit 4 (O)	—	1300	PWR/CE	1983

State and Utility	Plant	Location	Net MWe	Type/Mfr.	Operable
Tennessee Valley Authority	unit 5 (L)	—	1220	BWR/GE	1980s
Tennessee Valley Authority	unit 6 (L)	—	1220	BWR/GE	1980s
Tennessee Valley Authority	unit 7 (L)	—	1300	PWR/CE	1980s
Tennessee Valley Authority	unit 8 (L)	—	1300	PWR/CE	1980s

A Chronology of Federal Nuclear-Energy Programs

AUGUST 13, 1942, to December 31, 1946: Manhattan Engineer District. The Army's secret program to develop atom bombs. Headed By Lt. General Leslie R. Groves. Produced a total of five atom bombs, two of which were exploded in Japan, at a total cost of $2.2 billion.

1946: The Atomic Energy Act of 1946 (PL 585, 79th Congress, First Session, 60 Stat. 755–75) provided for government control of production, ownership and use of fissionable material, and for control of information concerning atomic energy. The bill created a five-person civilian commission to carry out all federal nuclear activities, and a joint congressional committee, with legislative and oversight authority, to supervise the commission. The Atomic Energy Act established strong criminal penalties—five years imprisonment and ten thousand dollars fine—for violations of information controls.

January 1, 1947: The civilian Atomic Energy Commission takes over the military atomic program.

1954: The Atomic Energy Act of 1954 (PL 83–703, 83d Congress, First Session, 68 Stat. 919). A complete revision of the 1946 act to permit private ownership and development of nuclear industry. The five-person civilian Atomic Energy Commission and the Joint Congressional Committee on Atomic Energy were continued, as were the strict security provisions.

1957: Congress enacts the Price-Anderson amendment (PL 85–256, 84th Congress, Second Session, 71 Stat. 576) to add section 170 (42 USC § 2210) to the Atomic Energy Act of 1954. The Price-Anderson amendment limits the liability of holders of nuclear licenses to $560 million and provides federal indemnity for $500 million in any one nuclear accident. This permitted private insurors to limit their exposure to $60 million (since increased to $100 million per reactor) and guaranteed utilities they would not be liable for damages in excess of the insurance available from government and private sources.

1974: The Energy Reorganization Act of 1974 (PL 93–438, 94th Congress, First Session, 42 USC 5801–5891) abolished the Atomic Energy Commission and replaced it with two new agencies. The Nuclear Regulatory Commission, which was to continue as a five-person independent agency, was given authority to regulate the civilian nuclear-power industry. A new Energy Research and Development Administration, headed by a single administrator serving at the president's pleasure, was created within the executive branch to carry on research and development activity in military and civilian nuclear-energy programs. ERDA took over operation of the weapons-production facilities and the program to develop a fast-breeder reactor, which had been AEC's largest activities. The Joint

Committee on Atomic Energy, with authority over NRC and ERDA, is retained. Subsequent legislation assigned to ERDA responsibility for research in other fields relating to energy: coal research was transferred from the Department of Interior, solar research from the National Science Foundation, and so forth, with oversight authority divided among various committees of Congress.

In 1974 the Congress also created a Federal Energy Council, a cabinet-level committee of federal officials with energy responsibilities, the function of which was to coordinate energy policy, but which had no operational authority. Congress also created a Federal Energy Agency to regulate prices of oil and gasoline and to carry out other federal energy policies. Frank Zarb, the first head of FEA, made it into the chief energy agency; he also served as director of the Federal Energy Council.

January 20, 1975: The new Nuclear Regulatory Commission and Energy Research and Development Administration come into existence, and the Atomic Energy Commission is dissolved.

Index